Sex and Death

Sex and

Science and Its Conceptual Foundations

A series edited by David L. Hull

Death

An Introduction to Philosophy of Biology

Kim Sterelny and Paul E. Griffiths

The University of Chicago Press

Chicago and London

Kim Sterelny is currently Reader in Philosophy at the Victoria University of Wellington. He is the author of *Language and Reality* and *The Representational Theory of Mind: An Introduction.*

Paul E. Griffiths is currently head of the History and Philosophy of Science unit of the University of Sydney. He is the author of *What Emotions Really Are: The Problem of Psychological Categories.*

The University of Chicago Press, Chicago 60637
The University of Chicago Press, Ltd., London
© 1999 by The University of Chicago
All rights reserved. Published 1999
08 07 06 05 04 03 02 01 00   2 3 4 5
ISBN: 0-226-77303-5 (cloth)
ISBN: 0-226-77304-3 (paper)

Figure 6.3 is from Karen Arms and Pamela S. Camp, *Biology,* 3d ed. © 1987 by Saunders College Publishing. Reproduced by permission of the publisher. Figure 9.4 is from W. Hennig, *Phylogenetic Systematics,* © 1979 by the Board of Trustees of the University of Illinois. Reproduced by permission of the University of Illinois Press.

Library of Congress Cataloging-in-Publication Data

Sterelny, Kim.
   Sex and death : an introduction to philosophy of biology / Kim
Sterelny and Paul E. Griffiths.
      p.   cm. — (Science and its conceptual foundations)
    Includes bibliographical references (p.   ) and index.
     ISBN 0-226-77303-5 (cloth : alk. paper). — ISBN 0-226-77304-3
(pbk. : alk. paper)
    1. Biology—Philosophy.   I. Griffiths, Paul E.   II. Title.
III. Series.
QH331.S82   1999
570′.1—dc21   98-47555
                                CIP

to Melanie and Kate

Contents

Preface

This book began long ago and far away, in Chicago in 1993 when one of us (Sterelny) tried out the basic idea on David Hull and Susan Abrams, both of whom were supportive. On Sterelny's return to the Antipodes, he continued to think about the project, and decided that a collaborative project would be more fun to do and would result in a better book. So he talked the idea over with Griffiths and with David Braddon Mitchell, an Australian philosopher of science with interests in both philosophy of biology and Australian botany. Thus the basic body plan of the book was laid down; a body plan that, in contrast to some others, has not remained impervious to developmental and other perturbations. After a couple of years of talking, we seriously got down to writing in 1995. Since 1996 this book has probably been the main project of the two survivors, Braddon Mitchell having been submerged by other plans. He did, however, have a major input into chapter 12.

There are, of course, many different ways to write an introduction to philosophy of biology. One option would be to use biological examples to stalk general issues in philosophy of science — the nature of theory and theory change, causation, explanation, and prediction. There is much to be said for such a book, for philosophy of science, in our view, has been too dominated by exemplars from theoretical physics. That matters: for example, the historical explanations central to, say, geology and evolutionary biology seem importantly different from those of physics. Still, that is definitely not the book we have written. This book is very much focused on the conceptual and theoretical problems generated by the agenda of biology, rather than pursuing a philosophy of science agenda through biological examples.

We have also chosen not to approach philosophy of biology by tracking the conceptual and theoretical development of evolutionary ideas, as David J. Depew and Bruce H. Weber have done in their *Darwinism Evolving*. There is an occasional nod to the history of the disciplines concerned, but the organizational spinal cord of our book is the conception of evolutionary biology that was developed in the classic works of Mayr, Dobzhansky, Simpson, and Stebbins in the 1940s. That conception, the "modern synthesis," dominated evolutionary thinking at least into the late 1960s. The current problems of evolutionary theory have been largely, though not wholly, the result of pressures to rethink that conception. We have chosen to call this core conception "the received view" rather than the "synthesis view" because we represent it in a rather schematic and ideal form. The real synthesis was never wholly uniform, of course, and for the most part the variation within it has not been our concern.

We have called our book *Sex and Death: An Introduction to Philosophy of Biology*. First, the subtitle: The reader may have noticed that while it speaks of biology, in the preface we have written of evolutionary biology. Indeed, in the text we have focused on evolutionary biology. That focus is not exclusive: chapter 5 explores connections between evolutionary and developmental biology, and chapters 6 and 7 push this exploration further. Those chapters take up the relationship between the role of genes in evolutionary theory and the molecular biology of the gene. Moreover, chapter 11 is devoted to exploring the interplay between evolutionary biology and ecology. But it is true that we have discussed other areas of biology mainly as they relate to evolution. So evolutionary theory and evolutionary theorists loom large over this work. (So we too can say that if we have seen too little, it is because giants have been standing on our shoulders.) This emphasis is partly, we think, a reflection of the genuinely conceptually challenging nature of evolutionary theory. As we show (we hope) in the text, evolutionary theory really poses a striking compound of conceptual and empirical problems. But it is partly a historical accident, too. We have no doubt that there are similar problems in ecology, developmental biology, and molecular biology (at least), and we hope to have done at least a little to extend the reach of philosophy of biology into those areas.

Second, the title: We chose the title because it was fun. And philosophy of biology *is* fun. The living world is splendid and bizarre—far more bizarre than we, at least, could have imagined—and the conceptual problems posed in understanding it are wonderfully intriguing (and important, as we argue in chapter 1). We hope this book shows that. We nurture the illusion that it will both manifest our relish in the subject and perhaps infect others with the same disease.

The structure of the book is, we hope, evident from the analytic table of contents. Part 1 sets out the scope of the project. Parts 2–4 work through the core debates, as we see them, in evolutionary theory and associated branches of biology. In part 2 genes are at center stage: we discuss both the idea that evolutionary history is really, fundamentally the history of gene lineages and the relationship between the evolutionary and molecular understanding of genes. In part 3 the focus changes to organisms, groups, and species. An important connecting thread is the question of whether groups and species play a role in evolution importantly like that played by organisms. Natural selection is central to part 4, for that part is on evolutionary explanation, and the key controversy about evolutionary explanation is the role of selection. So in a sense parts 2–4 are the heart of the book. Part 5 takes up human evolution, and more particularly, the sociobiological debates and their relatives. Apart from the intrinsic interest of this subject, many of the issues about the nature of evolution and natural selection are nicely exemplified through their application to humans. Part 6 winds up the show. We here attempt to put the central debates about the nature of evolutionary processes and patterns in a broader context by asking whether the characteristic patterns and processes of life on earth are likely to be features of any living world.

We have tried to write a book for three audiences. We wanted a book that would be accessible both to biology students with little or no philosophy and philosophy students with little or no biology. So we have used as little technical jargon as possible. When we have used specialist terminology from either discipline we have explained it in the text (usually immediately after the term's introduction) and, often, included it in the glossary. We have also made fairly liberal use of boxes in the text to discuss and explain more technical material. We have, however, tried to write the text so that no box is essential to following the flow of the argument. So readers should be able to skip the boxes if they like without losing the thread of the ideas. Many of the issues discussed in the book interconnect, one with the other. We have tried to help the reader follow these connections with parenthetic guides; for instance, "(5.3)" would indicate that the issue in play will be, or was, discussed further in section 5.3. Finally, we have provided "Further Reading" sections at the end of chapters 2–15 to introduce and orient newcomers to the literature.

So we hope this book is accessible to both philosophers and biologists without previous experience of the other area. Our third intended audience is, of course, our peers. This book is not a view from nowhere. It's an introduction to philosophy of biology from our own perspective on the discipline. So it contains our own assessment of what matters and what does not—of what is central and what is peripheral. That perspective is not widely shared, for we are products of a hybrid zone (displaying, we hope, hybrid vigor rather than hybrid sterility). Our take on evolution integrates important elements of the adaptationist, gene-centric conception of evolution associated with the likes of Maynard Smith, Williams, and Dawkins with elements of the pluralist, hierarchical conception associated with Gould, Lewontin, and Eldredge. At the very least, we would like to convince others in the field that the space of viable options is larger than they might have supposed.

This book took a lot of writing, and we got a lot of help with that writing. First we would like to thank David Hull and Susan Abrams for their initial enthusiasm and continued support for the project. Second, we owe a lot to the opportunity to talk biology and philosophy of biology, over many years and on many occasions, with the following: Russell Gray, Peter Godfrey-Smith, Susan Oyama, Geoff Chambers, David Hull, Karen Neander, Michael Hannah, David Braddon Mitchell, and Lenny Moss. They helped to provide the intellectual matrix from which this book has grown. Third, Peter Godfrey-Smith, Geoff Chambers, David Hull, Elliot Sober, Richard Francis, and two University of Chicago Press reviewers read and commented extensively on the semi-final manuscript. To them we owe much: thanks (not enough; if you're lucky you might get a beer as well). We thank Dan McShea, Susan Oyama, David Sloan Wilson, Alan Musgrave, James Maclaurin, Mike Dickison, Karola Stotz, Werner Callebaut, and Annemarie Jonson for reading and commenting on sizeable chunks of that same draft. We have three more specific thanks to give. First, we shamelessly borrowed, though with permission, the title of chapter 1 from R. D. Gray and J. L. Craig, "Theory really matters: Hidden assumptions in the concept of habitat requirements" (1991). Second, chapter 4 owes a lot to Griffiths's collaborators on other publications, Robin D. Knight and Eva M. Neumann-Held. Third, chapter 12 owes much to David Braddon Mitchell.

There are some equally valuable nonintellectual inputs to acknowledge. Griffiths's thanks go to his former home, the University of Otago, for the outstandingly supportive research environment in the Department of Philosophy and for Richard Briscoe's valuable services as research assistant. Also to his present home, the University of Sydney, where he taught two courses based around this book and employed another indefatigable research assistant, Ross West, who prepared the illustrations.

Sterelny inflicted numerous extraordinarily rough drafts of various chunks of this book on students at Victoria University of Wellington in 1995 and 1996 and at the California Institute of Technology, also in 1996. He thanks them for suffering so patiently. He also thanks the Philosophy Department, Monash University for hospitality and support in 1994; the Philosophy and Law Program, RSSS at the Australian National University for similar hospitality and support in 1995, and Caltech for providing a home in 1996. His base institution, Victoria University of Wellington, has supported the project in many ways. It provided a grant for research assistance in 1997, which enabled him to employ James Mansell, who worked with great intelligence and enthusiasm in finding and tracking down references (thanks, James). It granted leave to visit the ANU in 1995 and, for a more extended period, Caltech in 1996. Most importantly, it remains a civilized and supportive environment in which to work. Finally, he would also like to thank Melanie Nolan for her (mostly) tolerant attitude to his various preoccupations with biology, preoccupations especially marked in the final burst of writing and rewriting this work.

Kim Sterelny, *Wellington, New Zealand*
Paul Griffiths, *Sydney, Australia*

I

Introduction

1

Theory Really Matters:
Philosophy of Biology and Social Issues

1.1 The Science of Life Itself

The results of the biological sciences are of obvious interest to philosophers because they seem to tell us what we are, how we came to be, and how we relate to the rest of the natural world. The media often report that "scientists have discovered" the original purpose of some common human trait—morning sickness during pregnancy is designed to prevent malformed fetuses (Profet 1992). Or a traditional but controversial claim about society is found to be a "biological fact"—boys are more prone to violence and in greater need of formal social training than girls. And the "gene for" this difference has been localized—the genes for good social adjustment are on the paternally derived X chromosome, which only girls receive (Skuse et al. 1997). In all these cases biology seems to yield clear factual answers to questions of enormous moral and social significance.

In the late nineteenth and early twentieth century many philosophers looked to biology for answers to basic questions of ethics and metaphysics. Herbert Spencer's evolutionary "synthetic philosophy" was the most influential philosophical system of its time. Friedrich Nietzsche, hero of today's "post-modernists," believed that Darwin's theory could demolish traditional views of humanity's significance in the overall scheme of things. In America, the pragmatist Charles Saunders Peirce investigated the implications of evolution for the nature and limits of human knowledge. But mainstream philosophy in the universities of the English-speaking world took a very different view. At the opening of the twentieth century Bertrand Russell declared that the theory of evolution had no major philosophical implications. The sciences that had something to teach philosophy were mathematics (particularly mathematical logic) and physics. Physics was to serve as a role model for the other sciences, and for the next fifty years philosophers nagged biology for its failure to live up to its example. The well-known philosopher of science and mind J. J. C. Smart compared the biologist to a radio engineer. Biologists study the workings of a group of physical systems that happen to have been produced on one planet. Smart thought that such a parochial discipline was unlikely to add to our stock of fundamental laws of nature (Smart 1963).

Mainstream philosophy has taken an equally dim view of the significance of biology for ethics. In the nineteenth century Darwin's theory was thought to have all sorts of moral implications. Darwin himself remarked that if "men were reared under exactly the same conditions as hive-bees, there can hardly be a doubt that our unmarried females would, like the worker-bees, think it a sacred duty to kill their brothers, and mothers would strive to kill their fertile daughters; and no one would think of interfering" (Darwin 1871). The view that our moral ideas are an accident of biology seems inconsistent with, for example, the Kantian idea that morality is binding on all rational beings. If human morality is an adaptation for survival in human ancestral conditions, perhaps we should not take it quite so seriously. Drawing very different lessons from evolution, Spencer and others identified social progress with the universal progressive tendency that they claimed to find in nature (Ruse 1996). Even at the time, some philosophers were skeptical about these claims. Thomas Huxley, for example, thought them wrong-headed (Paradis and Williams 1989). Many twentieth-century philosophers have been even more damning, seeing all such ideas as fundamentally misguided. Biology cannot settle ethical issues because it speaks to matters of fact, not value. According to this view, inferences from purely factual claims to moral ones commit the *naturalistic fallacy*. Normative claims about what ought to be true can never be validly inferred from factual claims about what is true. Debate about the naturalistic fallacy continues. But although some philosophers still try to derive ethical results from evolution (Ruse and Wilson 1986), the consensus is that this cannot be done (Kitcher 1994).

It has always seemed obvious to the wider community that biology has the potential to challenge our most treasured beliefs about ourselves and the way we should live. This view is probably correct. Even if moral principles cannot be inferred from purely factual biological premises, the biological sciences can discover morally relevant facts. Those discoveries can interact with existing moral principles to produce radical new practical policies. For example, early in the twentieth century, morality was connected to evolution via the supposed need to maintain the evolutionary pressures that have adapted humans to their environment. The result was a case for *eugenics*— policies intended to maintain or improve human fitness through selective breeding. The eugenicists put forward purely biological claims about the effects of the relaxation of natural selection on humans in technologically advanced societies. These claims were supposedly, in themselves, factual. But conjoined with standard moral ideas about the importance of human welfare, the resulting eugenic case seemed compelling to people of every moral persuasion, from socialists to liberal capitalists to fascists. Before the Second World War almost every advanced society had made some legal provision for eugenics (Kevles 1986). Only its enthusiastic adoption by the Nazis brought eugenics into disrepute. More recently, E. O. Wilson and other biologists have claimed that human economic practices are driving species extinct at rates comparable to the great mass extinctions of earth's history. They further claim that these extinctions have the potential to disrupt the ecological processes on which human life depends. They call for radical changes in social and economic policy (E. O. Wilson 1992).

There are many uncontroversial biological claims that are relevant to our moral and social views. Starving children stunts their growth and ruins their health, and that is one reason not to starve them. But biological claims that have *novel* social and moral implications are usually highly controversial. Media reports of "genes for" homosexuality or evolutionary explanations of female orgasm are followed the next day by contradictory claims by equally well qualified authorities. Controversy is possible because the exciting conclusion is usually linked to actual experiments and observations by complex, and far from obviously sound, chains of argument. This is one reason why there is *philosophy* of biology. Philosophers try both to disentangle these chains of reasoning and to evaluate the broader conceptual frameworks that make biological results yield these significant social lessons. In *Wonderful Life,* Stephen Jay Gould describes for the general reader the recent reclassification of a group of Canadian fossils. But he also draws from these fossils the lesson that human intelligence is an accidental product of history rather than an essential feature of the natural world (Gould 1989). In chapter 12 of this book we look at the arguments connecting the fossil data to this extraordinary conclusion and examine the broader views in biology and philosophy upon which these arguments rely.

So philosophy is important to biology because biology's exciting conclusions do not follow from the facts alone. Conversely, biology is important to philosophy because these exciting conclusions really do depend on the biological facts. *Biological determinism* is the family of views that share the idea that important features of human psychology or society are in some way "fixed" by human biology. Many moral and social philosophers would dearly love a guarantee that nothing like biological determinism could possibly be true. But philosophy cannot provide such a guarantee. We believe that most of the doctrines that go under the name of biological determinism are false, but they are false because of the facts of evolutionary theory and genetics. It is true that some defenders of these views suffer from philosophical confusions, but these confusions cannot be diagnosed without coming to terms with the biology involved. The role that genes play in evolution and development is the subject of part 2.

Another reason philosophers are interested in biology is that, like much of science, it expands our sense of the possible. We think that far too often metaphysics and philosophy of science have been dominated by models drawn from physics and chemistry. An impoverished list of possible answers will often lead to an invalid conclusion. For example, a standard distinction in our culture is that between "learned" and "innate" behavior. Thus many parents are worried that young boys' delight in weapons is innate. Moreover, this distinction has played an important role in philosophy (Cowie 1998). One of the great divides in the theory of knowledge has been between empiricists, standardly regarded as thinking that very little is innate because almost everything is acquired from experience, and rationalists, standardly regarded as supposing that we come equipped with much that is innate. We think it would be very unwise to attempt to resolve this debate without understanding how modern ethology has transformed the concept of learning and why many biologists consider the concept of innateness to verge on incoherence. These issues are discussed in many parts of this book, but particularly in chapters 13 and 14. To choose another example, the concept of *biological species* figures extensively in ethical discussions of our obligations to the environment. Most philosophers learned Ernst Mayr's definition of a species in high school: a species is a group of organisms potentially capable of interbreeding with one another. They will cite this definition when asked what species are, despite discussing in the next breath plants and asexual species, neither of which fit the interbreeding criterion. The nature of species is one of the most hotly disputed areas of biology (9.2), and the alternative definitions have very different implications for environmental ethics.

The aim of this book is to introduce the major areas of discussion in philosophy of biology, not to directly address the broader philosophical questions to which these discussions are relevant. In this introductory chapter, therefore, we sketch some of the links between the issues discussed in later chapters of the book and some broader philosophical questions, namely:

- Is there an essential "human nature"?
- Is genuine human altruism possible?
- Are human beings programmed by their genes?

- Can biology answer questions in psychology and the social sciences?
- What should conservationists conserve?

These questions have both empirical and conceptual strands, and it is this mixed character that makes philosophy of biology relevant to them.

1.2 Is There an Essential Human Nature?

What makes someone a human being? The idea that each human being shares with every other human being but with nothing else some essential, human-making feature goes back at least to Aristotle. He thought that each species was defined by an "essence"—a set of properties found in each individual of the species, but only there. That essence makes it the sort of creature that it is. Today most people suppose this essence is genetic, and that the job of the Human Genome Project is to reveal the genetic essence of humans.

In reality, however, there is no such thing as the "genetic essence" of a species. A central aspect of modern evolutionary theory is *population thinking* (Mayr 1976b; Sober 1980). Each population is a collection of individuals with many genetic differences, and these differences are handed on to future generations in new combinations. Populations change generation by generation. In many contemporary views of the nature of species, there is no upper limit to the amount of evolutionary change that can take place within one species. Over many generations a species may be transformed in appearance, behavior, or genetic constitution while still remaining the same species. Diversity is normal, and perhaps even functional, for lack of diversity makes a species vulnerable to parasitism and to extinction due to environmental change. So uniform populations in the natural world are unusual. Such populations do exist in the laboratory. For experimental purposes, biologists often want, and have generated by inbreeding, "pure" strains of fruit flies and mice. These strains are "standard" in the sense that they are the same in every laboratory, not in the sense that they are the "normal" or "correct" genome of the fly or the mouse. These invariant strains have to be carefully constructed by selective breeding; nature does not supply them for free.

It is not easy to repair Aristotle's idea in the face of this variation within species. That may seem surprising, for anyone familiar with field guides, identification keys, or floras will be familiar with the idea of "identifying traits." A *Field Guide to the Birds of Australia* will appeal to the characters of voice, plumage, and behavior to distinguish, say, one babbler species from another. But these identifying features are rarely truly universal at any time, let alone across time. A statistically atypical white crowned babbler is still a white crowned babbler. It may be the forerunner of the typical babbler of the future or a survivor of the typical babbler of the past. So from the fact that we can *reliably recognize* many species it by no means follows that there is an *invariant essence* of a species. Even more to the point, as we shall see in chapter 9, there is no good reason from biology to try to repair Aristotle's idea. Contemporary views on species are close to a consensus in thinking that species are identified by their histories. According to these views, Charles Darwin was a human being not by virtue of having the field marks—rationality and an odd distribution of body hair—described (in Alpha Centaurese) in *A Guide to the Primates of Sol,* but in view of his membership in a population with a specific evolutionary history.

The implications of this transformation of our view of species have been much discussed in philosophy of biology, although they have been surprisingly neglected in ethics. David Hull, in particular, has argued that nothing in biology corresponds to the traditional notion of "human nature" (Hull 1986). This idea is significant, for the concept of human nature has been historically important. It has underwritten the view that there is some way that human beings are supposed to be, and that other ways of being are deviant or abnormal. This view is still central to the thought of some contemporary moralists (Hurka 1993). Biology is often supposed to provide some backing for this notion of normality: that there is a way that members of any species—including *Homo sapiens*—are meant to be, and that deviations from this are abnormal. But Darwinian species are continually evolving clusters of more or less similar organisms. There is nothing privileged about the current statistical norm.

So no general biological principle suggests that human moral feelings, mental abilities, or fundamental desires should be any more uniform than human blood type or eye color. On the contrary, human cognitive evolution seems likely to have involved an evolutionary mechanism that produces variation within a population, called *frequency-dependent selection*. In frequency-dependent selection, the fitness of a trait depends on the proportion of the individuals in a population that have that trait. In a classic thought experiment to illustrate this idea, John Maynard Smith invited us to consider the interaction between two types in a population: an aggressive, hard-fighting "hawk" and a timid, quick-to-retreat "dove." Hawks win any contest against doves, and so succeed wonderfully well when most of the population are doves. But in hawk-dominated populations, hawks bear the severe cost of frequent fights, and doves do not (10.6). So in many circumstances both types will survive indefinitely in the population (Maynard Smith 1982). In general, *maintenance of variation by freq.-dep.-selection*

frequency-dependent selection often gives rise to the coexistence of distinct types within a population. The evolutionary psychologist Linda Mealey has argued that psychopaths may represent one "minority strategy"—a variant form of the human species that can reproduce as effectively as the other types as long as it remains a small minority (Mealey 1995). According to this picture, if there were more psychopaths, there would be stronger selection against psychopathy than there is now. Of course, Mealey's particular idea is speculative, and we are not endorsing it here. Our main point is that the amount of morally and cognitively significant variation in the human population is an open empirical question. The fact that we recognize one another as members of a single species neither establishes that there must be some enormously significant characters distinctive of humans nor excludes that possibility.

Just as our species, like other species, consists of a varied population of individuals, so too do groups within a species. Human beings form overlapping pools of genetic variation, not distinct races, each with its own distinctive genome. Because our genetic material dates back to the beginning of the evolutionary process, and because human populations have typically been separate for only tens of thousands of years, only a small proportion of variation is distinctive of particular human populations. It can be argued that the average genetic distance between two individuals within a population is typically larger than the average genetic distance between two populations (Lewontin 1972, 1982a; Cavalli-Sforza, Menozzi, and Piazza 1994). Phenotypic differences may follow the same pattern. So we should not assume that the "races" that have been so important in human ethnic politics correspond to well-defined biological populations. They may instead be illusions generated by a focus on features that are more common in some geographic location or social group than in others, so creating a stereotype that is more applicable in that group. If we look only at these specially chosen features and ignore the exceptions to the stereotype, the members of another race seem to be a single, different type of human being. But even if these races are well-marked subpopulations with distinctive local adaptations—if, for example, Inuit facial structure really adapts them for life in the cold—we should not suppose that such subpopulations are invariant. The Inuit will only sample the full range of human variation, but they will still be a varied and evolving sample. Modern technology probably has to some extent eased selection pressure for adaptations to cold, so their facial features may well be in the process of change. In sum, the only real subdivisions of the human species are its many populations: groups that have been genetically isolated from one another for a longer or shorter time. These populations often do not fit traditional "race" categories very well. The people of Finland are very historically distinct from other Northern European populations, but they share with those populations the socially prominent feature of white skin, so the differences are ignored.

Much of this book will help make clear how central the doctrine of the ubiquity of variation is to modern biological thought. In chapter 13 we discuss some recent evolutionary psychologists who have argued that the human mind is an exception to the rule. They have argued that all healthy humans inherit the same mental potential. Other parts of chapters 13 and 14 examine approaches to human evolution that fully embrace the ubiquity of variation and the possibility that distinct types coexist in a single human population.

1.3 Is Genuine Altruism Possible?

Richard Dawkins's *The Selfish Gene* (1976) has been one of the most successful works of popular science. It argues that people, like other organisms, are "survival machines" built by their genes. These survival machines have no function in life but to produce as many copies as possible of the genes that built them. Dawkins was not the only person to advocate this view. The 1970s were the decade of *sociobiology*—the attempt to extend evolutionary explanations to human behavior. Like Dawkins, many sociobiologists saw humans as survival machines for genes. Most of them also agreed with him that all evolved human behavior must be designed to benefit those genes.

The idea that organisms are survival machines for genes rests on the view that genes are the only things that are passed on when an organism reproduces. So everything that one generation inherits from the last must pass down this genetic highway. An organism can inherit its mother's long neck or its father's knowledge of what is good to eat only if these characteristics are somehow stored in the genes. If individuals with long necks or sensible food preferences become more common in future generations, it is only because the underlying genes are surviving and proliferating. This view is known as *gene selectionism,* and the arguments for it are assessed in part 2, chapters 3–5.

Dawkins thought there was an important connection between gene selectionism and another important debate in biology—the debate over the evolution of *altruism.* An altruistic act is an act performed by one individual to benefit another. The question that biologists have debated is, if organisms exist only to benefit their genes, could evolution create altruistic organisms?

Gradually biologists came to realize that there was a problem in supposing that particular behaviors—for example, warning others of the presence of a predator—were altruistic adaptations. An *adaptation* is a feature of an organism whose presence today can be explained by the fact that it served some useful purpose in previous generations. A cat's claws, for example, are adaptations for catching prey. How could evolution lead to adaptations that were costly to the animal engaged in the behavior but beneficial to other individuals? This problem was initially masked by a failure to distinguish clearly between adaptations that assist the survival and reproduction of individual organisms and adaptations that assist the survival and reproduction of the species or group of which that organism is a member. Because early evolutionists did not make this distinction, they were quite happy to explain some fact about an organism by pointing out its value to other individuals of the same species. For example, when a bird calls out a warning about a predator, it draws attention to itself. Surely a bird that stayed silent would do better in the struggle for existence, and so warning behavior could not evolve. One solution is to say that this behavior benefits the whole population of birds. Groups of birds that warn one another survive longer than groups in which birds sacrifice one another, and the superior survival of altruistic groups explains the warning behavior we see today. This is a "group selective" explanation.

George C. Williams is famous for his rejection of group selective explanations (Williams 1966). He argues that evolution cannot build an adaptation that is good for the group because of "subversion from within." Organisms within a group are in competition with one another. Suppose that there are two kinds of organisms in a group: those that act for the good of the group, and those that do not. The "selfish" individuals would get all the benefits that occur because of the "altruistic" behaviors, but would bear none of the costs. So evolution would favor the selfish individuals. Therefore, a feature cannot evolve because it is good for the group, only because it is good for the individual.

Williams developed a second argument that connects the debate about altruism to Dawkins's idea of gene selectionism. Williams argued that an adaptation can evolve only if it is reproduced in each generation. But only genes are passed on intact across the generations. So an adaptation can evolve only if it is produced by some underlying gene or genes. If this is true, then an adaptation can evolve only if it favors the gene(s) that produce it. Suppose that the bird giving warning calls makes its own "warning genes" less likely to be passed on, and the "silence genes" of other birds more likely to be passed on. Under those circumstances, we would expect the "warning genes" to become rarer and rarer in the population, and "silence genes" more and more common. Eventually, calling would disappear from the population.

The ideas sketched in this section suggest that insofar as human behavior is a product of evolution, it is created by certain underlying genes, and is designed solely to assist the reproduction of those genes. The message that many social scientists have taken from these theories is that if they are to respect the biological facts, they face a dilemma. They must either insulate a large part of human behavior from biological explanation, or they must explain all human behavior in terms of individual self-interest. Both of these unattractive alternatives have been extensively explored.

However, the original biological theory is subject to much debate. The dilemma may well be false, for the case against group selection has been revisited. Perhaps the problem of "subversion from within" is not fatal. A potentially similar problem arises in the evolution of adaptations that are for the good of the organism. Organisms are groups of cells, and each cell carries groups of genes. Building an organism is a community project. So why isn't it undermined by selfish struggles between genes and cells to get into the cell lineages that become the gametes and perhaps ultimately new individuals? As it happens, such struggles sometimes do happen, and they are bad news for the organism. But usually they do not. Organisms possess features to guard against subversion. For example, in many animals a particular cell lineage— the *germ line*—is fixed as the source of all future gametes early in the growth of the embryo. A human female is born with a fixed number of potential eggs already in place. This phenomenon is known as the "segregation of the germ line." A mutant "selfish" cell that is outside the germ line cannot hope to survive the death of the individual organism. We are familiar with just such "selfish" cells, which replicate freely without regard to the interests of the organism as a whole. They are known as cancers, and they have a very limited life span. If a cell is outside the germ line, its only reasonable strategy is to contribute to the general welfare of the organism in the hope of re-producing those copies of the genes within it that *are* in the germ line. By this means, most cells are forced to act for the good of the whole organism (Buss 1987).

Recent advocates of group selection have argued that groups, like individual plants or animals, possess a mechanism for enforcing cooperation and preventing subversion from within. This mechanism is population structure. Subversion from within relies on the fact that a selfish individual can associate with altruistic individuals and derive benefits from their altruism. If the distribution of individuals in a population makes altruists likely to associate with altruists and nonaltruists with nonaltruists, subversion from within may not be effective. We focus on altruism in chapter 8.

1.4 Are Human Beings Programmed by Their Genes?

Some human psychological characteristics are nearly universal. Almost all humans speak some language or other. Other features vary widely across cultures. Food taboos, for example, are often quite uniform within cultures, but not across them. Most people with a European cultural background find the thought of eating insects and their larvae repellent, though many will happily scoff raw oysters by the dozen. These preferences are reversed in other cultures. Still other features vary even within a culture. What explains these patterns, both of variability and of invariance?

A central problem within many contemporary debates on this subject is the relationship between human psychology and human genetic endowment. There are those who think both that our genetic endowment plays a central role in the development of many of our most important characteristics, and that this central role of genes in development implies that these characteristics are resistant to change by the manipulation of the developing individual's social environment. Sometimes this view is framed as an explanation of invariant, or allegedly invariant, features of human cultures. We all possess, say, genes for aggression, and hence aggression is found in all human cultures and will be found in all possible human cultures. Sometimes it is framed as an explanation of differences. In some views of intelligence, certain genes predispose their bearers to a lower IQ not just in statistically typical environments, but in all possible environments. *Genetic determinism* or *biological determinism* are labels for views of this general character.

A caricature version of biological determinism is the view that there are biological factors (usually genes) whose presence in an organism means that, no matter what other factors are present, a certain outcome will result. Thus, for example, a gene linked to the production of certain hormones in males might be thought to guarantee that its bearers will be aggressive, no matter what upbringing they are given. There are no biological determinists in this extreme sense. With the exception of mutations that are lethal no matter what, it is universally acknowledged that no feature of an organism will develop unless suitable environmental inputs are present. No one supposes that a plant will grow in just the same way no matter what sort of light or nutrients it receives. So the term *biological determinism* is often applied to more moderate, and often vaguer, views. Such a view might be that some trait will emerge in any organism that has the right gene and that has a "normal" environment. A "normal" environment might be defined as one suitable for producing viable organisms of that sort. So to create an organism that has the gene but does not have the trait, it would be necessary to interfere with its development so severely that the resulting organism would be abnormal and probably not viable.

There are bodies of scientific literature that defend some version of biological determinism about some human characteristic or other. The sociobiological literature of the 1970s gave the impression that large swaths of human behavior were the expressions of genes specifically selected to produce those behaviors. It conveyed the impression that only the most drastic alterations in other developmental factors could prevent the production of these behaviors. The second-wave sociobiology of the 1990s has switched its focus from human behavior to the psychological mechanisms that produce it. It proposes that human psychology contains "Darwinian algorithms" selected for their fitness-enhancing effects in our ancestral environments. These mechanisms emerge in humans the world over, whatever their upbringing. No environment that produces a functional human psychology can avoid producing them. Proponents of the "language instinct" maintain something similar about the psychological mechanisms that allow us to learn language (Chomsky 1980; Pinker 1994).

The reverse of biological determinism is *environmental* or *social determinism*. Naturally, no one believes that the environment will produce a certain outcome no matter what genes an organism has. No hothousing program will get a chimpanzee into Harvard Law School. Instead, social determinism is the view that biology provides only a broad constraint on the range of outcomes that can be produced by environmental factors. Our genes prevent us from becoming Superman at one end and chimpanzees at the other. Within those constraints, however, only social factors affect what is produced. The striking variation we see in actual human cultures is the result of variation in social environment, and even greater variation is possible—indeed, likely— as novel cultures come into existence. There is no relationship between the variety of human cultures and any genetic variation that may exist in the human species. All actual human cultures, and all the many possible cultures that have not been tried, can be supported by any genome capable of producing a working human being. This view has been expressed by innumerable authors in the social sciences. A typical statement of it occurs in Moira Gatens's book *Philosophy and Feminism:* "This is not to say that human being is not constrained by . . . rudimentary biological facts but rather that these factors set the outer parameters of possibility only. Within these constraints, if they can be called that, there is a variety of possibilities" (Gatens 1991, 98).

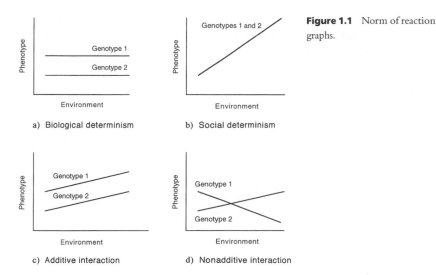

Figure 1.1 Norm of reaction graphs.

The observable features of an organism—its body and behavior—are jointly known as its *phenotype*. If there is one thing all biologists agree on, it is that the phenotype is the product of the *interaction* of genetic and environmental factors (Kitcher, in press). The weakest form of *interactionism* is one that even the strongest proponents of biological and social determinism can accept. The social determinist accepts that a human genome is needed if the environment is to produce a human individual, and the biological determinist accepts that no organism can develop without a suitable environment. A more substantial form of interactionism admits not only that both genetic and environmental factors are needed to produce a finished product, but also that *changes* in either can produce *changes* in the finished product. This is the sort of view often represented in a norm of reaction graph (figure 1.1).

A norm of reaction graph shows the pattern of variation in the kind of organism produced as genetic or environmental factors change (6.2). The two deterministic views just discussed can be seen as very extreme norms of reaction (figures 1.1A and B). Each of these views proposes that a change in one of the variables has no effect on the outcome, except at the extremes, where it has catastrophic effects. Figures 1.1C and D represent increasingly radical forms of interactionism. In the first, the relationship between the two variables is seen as purely *additive* (figure 1.1C). This means that a particular change in one variable has the same sort of effect no matter what value the other variable is set at. If norms of reaction are additive, then a change in a certain developmental factor will always produce a certain sort of difference in outcome. So if we were testing two genetic varieties of wheat, the influence of fertilizer would be additive if a little more fertilizer always produced, for both varieties, a slightly greater yield. A genetic difference between two organisms would produce the same sort of difference between their phenotypes in any environment that the two share. So one variety of wheat would have a higher yield than the other at each rate of fertilization. This view allows the author of *Male Dominance: The Inevitability of Patriarchy* to argue that, although the effect of the hormone testosterone depends on other factors, assuming that men and women share the same social environment, the fact that men have higher levels of this hormone than women means that they will always be, on average, more aggressive, no matter what social environment we create (Goldberg 1973).

The assumption that interaction is additive also underlies a piece of reasoning often used to dismiss the idea that genes affect human behavior. This argument starts from the premises that there are many radically different forms of human societies and that genetic differences across these societies are minimal. It concludes that the genes have no important role in the production of the features that differ across those societies. The intuitive idea is that difference is explained by difference, so if there are large differences in human behavior but only small genetic differences between cultures, then genetic differences can be playing no significant role. But this argument assumes that interaction is additive. In fact, as Richard Lewontin has argued, gene/environment interactions typically are not additive (Lewontin 1974). So, in the right environment, a small genetic difference can make a large phenotypic difference (figure 1.1D). Jared Diamond has suggested that African American populations have genes that make them vulnerable to hypertension and similar diseases, but that the phenotypic effect of these "salt-thrifty" genes in the very different cultural context of their recent past was very different (Diamond 1991). Some Asian populations have high frequencies of genes that now act to protect their carriers from alcoholism. Their phenotypic expression—if any—in other contexts might have been be quite different. So gene differences often contribute to widely varying phenotypes through naturally occurring environmental variation. But even if certain genes are correlated with certain phenotypes in all natural environments, we cannot "bracket off" the environment as a mere constant background factor, playing no important role in producing those phenotypes. For we cannot extrapolate the invariance of that gene/phenotype correlation to new environments. A novel environment may well produce a novel phenotype. Introduce alcohol into the diet, and a wholly new phenotypic effect occurs.

We suspect that dissemination of the idea that genes and other developmental factors may interact with the environment in multifactorial, nonadditive ways to produce outcomes would greatly improve the debate over the role of genetic factors in determining human behavior. It is probably fair to say that many people assume additivity when discussing gene/environment interactions simply because they overlook the full range of the possible. We hope to expand this sense of the possible, particularly in chapters 5–7 and 13–14.

1.5   Biology and the Pre-emption of Social Science

Many biological theories seem to threaten the independence of the social sciences: in Rosenberg's phrase, biological sciences "pre-empt" the social sciences. Biology offers explanations for the very same characteristics of human beings and human societies that psychology and sociology claim to explain. Are these explanations rivals? Sometimes one explanation does displace another. Clever Hans was a horse believed by his owner to understand simple arithmetic. When given a simple sum (in German), Clever Hans could stamp out the answer with his hoof. It was discovered, though, that his owner had accidentally conditioned Hans to start stamping and continue until cued to stop. The stop signal was an unintentional change in his owner's body posture when Hans had counted to the right answer. It was his owner's response that Hans had been conditioned to detect. The explanation of his behavior in terms of reinforcement displaces its explanation in terms of arithmetic understanding. But one good explanation does not always drive out another. An explanation of a riot that appeals to the frustration, poverty, and alienation of the rioters, and an explanation in terms of an igniting incident of police brutality, may be mutually illuminating. So one central question is whether biology and the social sciences offer competing explanations, of which only one can be true, or complementary explanations that mutually illuminate one another (6.1).

Biology seems to have the potential to pre-empt the social sciences in two ways: by constraining the range of admissible social scientific hypotheses, and by displacing those hypotheses. We begin with the idea of constraint. The effect of the group selection debate on social science is typical of the way in which social science seems to be constrained by the findings of biology. In the last few decades there has been a great deal of pressure on social theorists to account for human behavior in terms of individual self-interest. This has given a sort of automatic credibility to some theories and created skepticism about others. According to traditional economic theory, individuals act so as to maximize their individual income, and any cooperative activities are produced as a side effect of this pursuit of self-interest. But in many situations, common sense suggests that this assumption is flawed. Members of university departments, for example, do not refuse to work unless supplied with financial incentives. Despite the fact that their incentive structure primarily rewards individual research, they are often remarkably concerned about the smooth functioning and relative standing of the teaching units of which they are part. If we are convinced that humans are fundamentally selfish, then we will think up ways to explain this behavior in terms of self-interest. Perhaps a person who volunteers to run a graduate placement scheme in their spare time is trying to please the head of the department. If they do this even when their superiors disapprove, perhaps they are worried about the capacity of the department to attract students in the future and are acting to ensure their job security. Explanations of this sort are produced even when they seem rather forced, because it seems somehow disreputable to suppose that the persons simply values the success of the larger unit of which they are part. But if group selection has been a significant force in human evolution, then there is no reason to rule out this possibility. Group selection would select for individuals whose psychology allows them to sacrifice individual advantages for the good of the group. One obvious mechanism that might be selected would be the capacity to feel emotions such as loyalty, pride, and guilt. A person might experience these emotions in a way that motivates them to act for the good of the group.

The importance of the revival of group selection is not that it proves that mechanisms of this sort exist, but that it removes the assumption that they do not. It allows social scientists to concentrate on how people actually think and behave, rather than being constrained by ideas about how they "must" think and behave. In fact, the findings of social science about human motives may provide just the evidence biology needs to decide whether group selection has been an important force in human evolution. We shall return to this issue in chapter 8.

We now turn to the idea of displacement. A second threat to social scientific explanations seems to come from the suggestion that social and psychological traits are the products of evolution. The social sciences have traditionally assumed that only "human universals"—traits found in all or most human societies—can have evolutionary explanations. Culture is left to explain all those traits, such as clothing, family structure, or aesthetic preferences, that display a pattern of within-group similarity and between-group difference across human populations. This pattern is thought to result from cultural transmission in which individuals pass on mental representations by imitation and inculcation. Since most human characters of interest to social scientists do vary across cultures, this division of territory—biology gets to explain the invariant features, and the social sciences get to explain the variable ones—suits social scientists just fine. No pre-emption here. But John Tooby and Leda Cosmides have pointed out that evolutionary biology has no reason to cede varying traits to the social sciences. Instead, varying traits may be the result of a disjunctive developmental program that responds to local environmental conditions. Tooby and Cosmides, and the program of evolutionary psychology for which they are standard-bearers, offer evolutionary explanations of many psychological characteristics that are both important and varied: family patterns, mate choice, and much else (Tooby and Cosmides 1992). If an evolutionary explanation of, say, mate choice or the distribution of resources to children really does displace one from the social sciences, then the social sciences are indeed threatened with pre-emption.

However, we have our doubts about the contrast, on which this debate depends, between biologically and socially produced traits. First, evolutionary and cultural explanations may be mutually illuminating rather than inconsistent with one another. One of the founders of the evolutionary analysis of behavior, Niko Tinbergen, distinguished four explanatory projects: (1) the evolutionary history of a behavior; (2) the current use of the behavior in the life of the animal, which may involve a change from (1); (3) the development of the behavior over the life of the organism; and (4) the psychological and other mechanisms used in the control of the behavior (Tinbergen 1963). Given Tinbergen's distinctions, it's quite plausible to suspect that evolutionary theorists and social scientists may be engaged in different explanatory projects.

Second, even when we are considering the evolution of human behavior, there is a problem in contrasting biology with culture, for humans have co-evolved with their culture. Humans have had a culture since before they were human. This culture is one of the resources that feeds into both the evolution of human traits over time and individual human development. Social environment must be an essential aspect of both our evolution and our development, so the contrast between the biological and the cultural looks shaky. The cultural plays a deep role within biology, and vice versa. This suspicion is reinforced by the "developmental systems" approach that we discuss in chapter 5. We take up these issues further in chapters 13 and 14.

1.6   What Should Conservationists Conserve?

Ecology refers to both a biological science and the increasingly popular values espoused by the environmental movement. The scientific discipline of ecology is the study of organism/environment interactions. The environmental movement draws on the science of ecology. Moreover, its agenda poses many of the most difficult questions that scientific ecologists are trying to answer. These questions include the effects of environmental changes on a species, both changes that are the direct result of human action and those that are the result of the invasion or retreat of other species. In the 1960s, ecologists hoped to provide a general theory that could be used to predict the effect on ecosystems of, for example, the introduction of a new species. More recently, critics have stressed the historical nature of ecology. Predicting the fate of an ecosystem may be as difficult as predicting human history: particular facts count for too much and general principles for too little (Kingsland 1985). This issue is explored in chapter 11.

One conceptual issue with major implications for ecology, and for the environmental movement, is the nature of species (9.2). Species are the focus of conservation efforts all over the world. But many of the types of organisms that people try to conserve do not count as species under most scientifically well motivated definitions. New Zealand's black stilt and North America's red wolf are often cited as examples of "mere varieties" that are the subject of expensive conservation programs. Whether this matters depends on the source of concern for the environment. If conservation is seen as a human-centered activity, then we can justify our concern for a favorite color morph on aesthetic grounds. If we want to spend the conservation dollar to preserve biodiversity in some more objective sense, then we will be more concerned with the proper definition of species.

A connected debate concerns the proper measure of biodiversity. Intuitive conceptions of biodiversity seem to be sensitive to two different factors: first, how closely species are related, and second, how different a species is from its closest relatives. Relatedness is relatively easy to measure; divergence is more difficult. The degree of relatedness between two species can be expressed as the number of speciation events between them. This is the evolutionary equivalent of being, successively, sisters, cousins, second cousins, and so forth. A species represents more biodiversity the less closely it is related to its closest living relative. But this measure does not capture the intuitive notion of biodiversity very well. For example, the closest living relative of the Chatham Island black robin is a not too dissimilar robin. The closest living relative of the kakapo is another parrot, either the kea or the kaka, but neither of these is a large, flightless, highly sociable, nocturnal parrot. Many people have an intuitive sense that losing the kakapo would mean losing more biodiversity than losing the robin, even if the number of speciation events separating each species from its nearest relative were the same. This second aspect of biodiversity seems to concern whether a species has evolved into a new and different ecological niche, and whether it has changed physically in important ways. In section 12.3 we take up the idea of physical divergence; in chapter 11 we ask whether there are really such things as ecological niches, and if so, what they are.

Many conservationists argue for a move away from preserving species and toward preserving whole *ecosystems*. The basic idea behind this change in strategy is that species are not viable as isolated things, but only as parts of a larger whole. No one doubts that ecological communities are very complex, and that each species interacts strongly with many others. But many doubt that these communities are very systemlike. The idea of a system suggests a relatively stable set of relationships, rather than a continual state of change. The popular image of an ecosystem as a rich, diverse community that tends to return to its original state after small perturbations may be as much the result of wishful thinking as of observation. Some ecologists have even claimed to show that diverse ecosystems are less stable, more changeable, than simple ecosystems (although the arguments connecting their data to this exciting conclusion have been criticized: Mikkleson, in press). A major source of the ecosystem concept is undoubtedly the ancient idea of "the balance of nature," an idea that has its roots in the intrinsic order of a universe created by God, but for which it is difficult to find scientific justification (Egerton 1973). We discuss these ideas in chapter 11.

We hope that these short discussions are enough to convince you of both the intrinsic interest of philosophy of biology and its practical importance. On with the show.

Chapter

2

The Received View of Evolution

2.1 The Diversity of Life

As we saw in chapter 1, our conception of the living world is important both in itself and in its implications. If we are to understand that living world, evolutionary biology must explain three fundamental phenomena. One is life's variety. The world is rich in living things, yet that richness is limited in important ways. So we need to explain both why there are so many kinds of organisms and why there are not more. A second is adaptation. Organisms typically seem very well suited to their environments; they are *adapted* to their world. A third is development. Organisms "breed true": sparrows give rise only to sparrows, not to eagles. Furthermore, they do so through a long and complicated process of development from an apparently simple egg into a complex, organized, and differentiated adult organism.

In this chapter we introduce the main ideas—the "received view"—of contemporary evolutionary theory and its explanations of adaptation and diversity. Until recently, development played a less central role in evolutionary biology, and hence it is a less central element of the received view of evolution. Many commentators think that this relative neglect of development is itself significant, so we return to this issue in chapters 5 and 10. In the meantime, we focus on variety and adaptation.

Diversity and Its Limits

The world of life as we know it is fabulously diverse. Somewhere between one and a half and two million species have been described and named. There are no very reliable estimates of the number of living species still to be discovered, but one recent estimate is ten million (Minelli 1993, 129). Moreover, the life that now exists is only a fraction—quite likely only a tiny fraction—of the total historical diversity of the tree of life. Perhaps a quarter of

[handwritten margin note: 3 phenomena to explain. which do I find most interesting?]

22

a million fossil species have been described, and they must be only a minuscule sample of all the species that have been and gone.

Yet though life is so diverse, there are gaps in that diversity. To explain the notion of a gap in diversity, we shall hijack Daniel Dennett's metaphor of "design space" (Dennett 1995). Dennett thinks of design space as a vast library containing the exact specifications of all the ways organisms might be—of all the actual and possible creatures. So it includes specifications not just for all the actual dinosaurs, but for all the possible ones as well; not just the formidable enough real *Tyrannosaurus rex,* but also the intelligent, arms-building descendants it might have had.

An appreciation of life's actual diversity is important, for that diversity has generated important controversies in evolutionary theory. It's easy for human beings to overlook much of the actual diversity of life because we tend to think of ourselves as typical of the organic world. But our idea of life's workings should not depend on such unrepresentative exemplars of the living world. We are highly atypical. The vast majority of organisms are not vertebrates like us. Most creatures are single-celled organisms. Probably the most fundamental division in the history of life is within the category of simple single-celled organisms called *prokaryotes.* Prokaryotes have relatively simple genetic systems, and their genetic material is not segregated into a nucleus. They are by far the most numerous of organisms: almost every organism that has ever been alive is a prokaryote. They come in two basic kinds, the *eubacteria* and the *archaebacteria,* which diverged in the very ancient past (Ford Doolittle and Brown 1995). The archaebacteria comprise a diverse group of bacteria-like organisms with weird metabolisms. These are the organisms that live in extreme environments and in extraordinary ways; for example, by breaking down sulphur compounds in superheated water from deep ocean hydrothermal vents. The eubacteria are, relatively speaking, the "standard" bacteria—the kind that live in us and in our food. Branching off from one of the prokaryote lineages, probably the eubacteria, are the *eukaryotes.* Eukaryotic cells are complex, with a nucleus containing most of their genetic material and some other molecular machinery separated from the cytoplasm by a nuclear membrane. We and all other multicellular organisms are eukaryotes. We are assemblages of eukaryotic cells. So we are offshoots of an offshoot: we derive from one of the three branches of single-celled life.

So our size and our cells make us atypical. But we are atypical in other respects as well. Plants are physically robust in ways we and most animals are not, often recovering from being mostly eaten or mostly burned. Plant life cycles involve astonishing physical transformations. For example, in many plant lineages, the equivalents of eggs and sperm exist for some time as complex, independent individuals before producing cells that fuse with those produced by another plant to begin a new life cycle. In particular, in sea-weeds and ferns, individual plants exist in two forms. The *sporophyte* form is *diploid;* that is, it has two copies of each chromosome in its cells. It propagates by producing spores. These spores are *haploid;* that is, they contain a single copy of each chromosome, formed by combining genes from the two copies in the sporophyte. If all goes well, these spores germinate and grow into haploid *gametophytes*. A gametophyte can be physically quite different in size and shape from its diploid ancestor. The gametophytes produce haploid ga-metes, and the fusion of two gametes is the origin of a new diploid sporo-phyte and a new generation. This complex cycle is known as *alternation of generations*. It is a central feature of the plant reproductive cycle, though the haploid stage is very reduced in seed-bearing plants, and it is only in more ancient lineages that we find both generations living as distinct organisms (Niklas 1997, 157–162).

Alternation of generations is not the only respect in which plant repro-ductive habits are different from ours. Plants often reproduce *vegetatively* as a chunk or bud grows into a new individual. Even for animals, our life cycles are atypically simple. Many invertebrates undergo great physical transfor-mations across a single life cycle. For example, parasite life cycles often in-volve an organism traveling through a number of different hosts, in each of which its body form is very different. It's worth tracking through one of these cycles just to show how tame human development is by comparison. The trematode parasite *Dicrocoelium dendriticum* has a life cycle that takes it through three separate hosts. Adult flatworms live in livestock; they lay eggs in the livestock's dung. These eggs are eaten by snails, in which they hatch and in which they reproduce asexually for two generations before forming a mucus-covered larval mass, which the snail excretes. This mass of several hundred parasites is eaten by an ant, the parasites' next host. At this stage one of the larvae invades the ant's nervous system and changes the ant's behavior so that it spends much of its time on grass tips, thus greatly increasing the chances that the ant will be eaten by livestock along with the grass. Should this happen, the brainworm dies, but promotes the completion of the life cycle by the other larvae (Sober and Wilson 1998).

Some colonial organisms also have bizarre life cycles. The Siphono-phora—jellyfish-like colonial hydrozoans such as the Portuguese man-of-war—are so integrated that it is hard to say whether they consist of many cooperating organisms or a single organism. The various cells (the *zooids*) within the man-of-war are specialized: there are floatation specialists, pro-pulsion specialists, killer cells, and sex cells. In this respect, the man-of-war

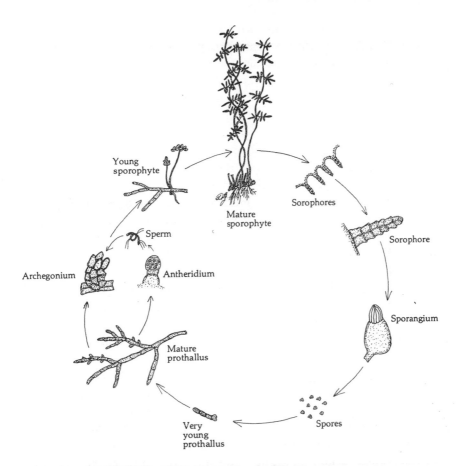

Figure 2.1   The haploid phase of a life cycle need not take the form of short-lived sex cells. The life cycle of *Lygodium,* like that of many other ferns, alternates between of a haploid gametophyte (prothallus) and a diploid sporophyte. (From Jones 1987.)

seems to be a single organism. On the other hand, each cell within the colony has an independent origin in a fertilized egg. In contrast to the zooids that jointly form a man-of-war, cellular slime molds spend most of their life as independent cells. But when food runs out, they aggregate into a single body, which develops specialized parts. Some of the cells form a stem, ending in a group of cells that specialize in making spores. So these cells too seem to spend part of their life as individual organisms, and the rest as parts of an organism.

So vertebrates like us are unusually huge and unusually fragile, have relatively simple life cycles, and are built from the least common type of cell.

The fact that we are not the standard mode of life is important to remember, for some think contemporary evolutionary theory is too influenced by the vertebrate paradigm of the organism (see, for example, Dawkins 1982; Gould 1996d); we take up this issue in section 3.4.

The variety of life is the focus of important debates within evolutionary biology. Gould distinguishes between *diversity*—the number of species in existence—and *disparity*—the extent to which evolution has manufactured organisms that are genuinely different in their basic organization (Gould 1991). He argues that evolutionary theory needs to explain not just life's great diversity, but also its disparity—the extent of design space that life occupies. One important aim of evolutionary biology is to explain why some regions of design space are occupied and others are empty. We seek an explanation of the actual richness of the organic world, as well as an explanation of why that richness is not more impressive still.

One group of evolutionary theorists, the *process structuralists,* think that design space is highly constrained, and that many of the organizational features of organisms are explained by intrinsic physical constraints on life's possibilities (Goodwin 1994). There is no doubt that some imagined organisms are not really possible. The constraints of gravity and limitations on the power that can be delivered through standard metabolisms make it unlikely that winged pigs could fly. Other real limitations are less obvious. There may well be constraints on how an organism can grow and work, constraints that make some apparently possible organisms really impossible. Some frogs glide, but there are not, and as far as we know never have been, any flying amphibians. The amphibian metabolism might simply not deliver enough energy for powered flight. If not, there can be no true flying frog. However, even if there are many hidden constraints on design space, many varieties of organisms that are surely possible nevertheless are not to be found. Most possible organisms have never become actual. Though there are plenty of vegetarian lizards, there are no grass-eating snakes. Asexual and two-sex regions of biological space are occupied, but no species has ever required three or more sexes for reproduction. What is so special about two? In some lineages, we find the extremes of sociality known as *eusociality.* In eusocial bees, wasps, ants, aphids, a weird marine crustacean, and the equally weird naked mole rat, some animals have given up reproduction entirely. They live as sterile workers in extended families, and they are often physically quite different from their fertile siblings. African wild dogs are a less extreme example, but here too a single dominant female suppresses reproduction by the other females while herself having large litters. In contrast, while there are plenty of birds that temporarily forgo reproduction to help their parents raise their siblings, none are irreversibly committed to sterility. So is the lack of eusocial birds living in avian hives a sad accident of history, or is there something about avian genes, avian bodies, or avian lifestyles that makes them unlikely or impossible?

We remarked above that evolutionary theorists seek to explain both the extent of and the limits on life's richness. One fundamental issue is how much of an explanation we should expect. It is no mystery that not all possible species are actual. The space of all the possible organisms there could be is so large that there has not been world enough or time for most of them to evolve. So often the explanation of a missing organism will turn on small accidents of history; no deep principle of biological organization will be involved. We have cheetahs, but no marsupial predators similarly adapted for running. Perhaps there was a marsupial lineage that might have gone down the cheetah path, but the right genetic variation never came together in the right organism at the right time. If so, the fact that there are no cheetahlike marsupials is just an accident of history. Given that many birds do breed cooperatively, it's quite likely that the absence of extreme sociality among the birds is just an accident of history. Eusocial birds didn't happen to evolve, and that is all there is to it.

However, there are other patterns in life's richness that are not likely to be historical accidents. Sex—one, two, but no more—is a good candidate for such a pattern. We shall see others. Gould, for example, has argued that almost no fundamentally new morphological organization has evolved since the Cambrian. The basic body plans all evolved in a short but exuberant burst a bit more than five hundred million years ago (Gould 1989). If this is true, he is surely right in thinking this shutting down of the machine that generated disparity requires explanation. There are also intermediate cases. Richard Francis has pointed out that we might expect many vertebrates to be able to change sex, as some fishes do. In many vertebrate species there is great variation in male reproductive success. Who has not seen documentaries of bull elephant seals fighting for supremacy? The few who succeed and become beachmasters will sire many pups. But most sire none. So if you were a slightly undersized male, or even a beachmaster beginning to feel your scars, surely you would be better off switching to the female role and settling for the modest output of a single pup. In any species in which male success is very uneven, and in which success is not just a lottery, we would expect "likely loser" males to turn female. Yet among vertebrates, only in fishes do we find a capacity to change sex (Francis, personal communication).

G. C. Williams wonders why there are no viviparous turtles (a *viviparous* animal gives birth to live young rather than laying eggs). Any of us who have seen a sea turtle's laborious and dangerous struggle up a beach to lay her eggs, the dangers to the clutch, and then the hatchlings' desperate race to the water will agree that a viviparous turtle would make excellent adaptive sense (Williams 1992). Moreover, other reptiles are viviparous. There are plenty of viviparous snakes. Is the missing turtle just an accident, or does she signal an unexpected constraint on the power of evolution to build genuinely different turtles?

There is one fundamental pattern in disparity that is very unlikely to be a historical accident. Organisms come packaged into species. The existence of clusters of similar organisms is an obvious, pervasive, and almost certainly important feature of life as we know it. It is so obvious and so pervasive that it's easy to overlook the need for its explanation. Yet we can certainly imagine life without species. Consider Carnivore Hall, that region of design space where we find dogs, cats, bears, Tasmanian devils, weasels, snakes, and all the various extinct and merely possible carnivores. As it is, Carnivore Hall is occupied in patches. But it's not hard to imagine its denizens varying seamlessly in size. We can certainly imagine a world in which wolves, coyotes, hunting dogs, and the rest vary smoothly from mouse-size poodles to bear-size wolves. On top of this gradation in size, we can add smooth variation in other carnivore characters—for example, from the rather omnivorous bears and foxes to the more meat-specialized cats and snakes; from the runners to the stalkers to the hide-and-wait specialists. So we can imagine a world with carnivores, but no carnivore species. Yet that imagined world is very different from our world.

Thus evolutionary theory must explain why and how life came to be organized into species. Explaining the existence and importance of species might seem a particularly challenging problem for evolutionary theory, for that theory is committed to the idea that one species can be an ancestor of others. All marsupials, for example, are descended from a single ancestral species. Furthermore, in this view, there is no sharp distinction between species and varieties. So there is controversy over whether some population is a species or just a variety: whether, for example, the crimson rosella group of Australian parrots (the crimson, yellow, Adelaide, and green rosellas) is a cluster of very closely related species or a single species with a number of well-defined subspecies. And the varieties within a species—say, the North and South Island subspecies of the New Zealand robin—are themselves potential, incipient species. So the evolutionary theorist has to take on the tough job of defending both the idea that species are important units in the biological world and the idea that they evolve from one another. It has often been argued that evolutionary theory is committed to some version of an "anti-realist" view of species; that is, to the idea that species only seem to be real objective units to us because of temporal limits on our perspective. That is not our view; we say why in chapter 9.

In sum, the living world presents us with both an array of species and an array of organizations. Life is *diverse;* there are numerous species. Life is *disparate;* those species manifest a considerable variety of adaptive structures and body plans. But though life is disparate, it is not endlessly so. There are adaptive structures and body plans that we might expect to see and do not. Evolutionary biologists diverge over what they take to be most problematic about these phenomena. The received view has focused on explaining how species are made—that is, on explaining the diversity of life. In contrast, some contemporary evolutionists take restrictions on disparity to be the most striking problem posed by life's richness. Mollusks, sponges, bivalves (oysters and the like), arthropods, and many kinds of worms first appear in the fossil record over five hundred million years ago, but the basic layout (the "body plan") of these organisms remains unchanged. In the eyes of these biologists, this stability in modes of bodily organization over hundreds of millions of years requires explanation (Raff 1996). They think that evolutionary processes have been surprisingly conservative. In their view, we would expect to see more change than we do. A problem with evaluating these ideas is that, despite the intuitive plausibility of the distinction between diversity and disparity, we shall see in sections 9.3 and 12.3 that it faces very serious challenges.

Adaptation

The structured complexity of organisms, and their adaptation to their environment, is every bit as obvious as the diversity of organisms. Perceptual systems are classic examples of such complex, fine-tuned adaptation. One striking example is the extraordinary and interconnected set of mechanisms that jointly compose bat echolocation systems. Echolocating bats have mechanisms that enable them to produce high-energy, high-frequency sound waves. They have mechanisms that protect their ears while they are making such loud sounds. They have elaborately structured facial architectures to maximize their chances of detecting return echoes, together with specialized neural machinery to use the information in those echoes to guide their flight. There are many other equally wonderful examples of adaptation.

Darwin and some other evolutionary biologists have emphasized the *imperfection* of adaptation. Tree kangaroos have many adaptations for life in the trees; their paws, pads, and tails are in many respects unlike those of their

Figure 2.2   Like many other cave dwellers (troglodytes), the salamander *Typhlomolge rathbuni* has vestigial eyes and has lost its pigmentation. (From Barr 1968.)

ground-dwelling relatives. They do surprisingly well. But no biological engineer building a tree-dwelling mammal from scratch would come up with the tree kangaroo. Organisms exhibit design compromises, vestiges, and the accidents of their history. These imperfections, vestiges, and accidents are important to evolutionary biologists because they reveal so much about the organisms' histories. The tree kangaroo is instantly recognizable as a kangaroo because it is carrying so much of its history with it. The different designs of bat and bird wings reflect the separate evolutionary histories of these mechanisms of flight. A famous example of an accident of history is the strange design of the retina in vertebrates like ourselves: light has to go through the cell to the photosensitive pigment at the back. The eyes of squid and octopus, which evolved independently but are in most respects very similar to our eyes, have their photosensitive pigment at the end of the cell nearest the light source. History is perhaps most obvious in vestigial organs. Many cave-dwelling creatures, for example, have vestigial eyes, even though their current environments contain no light with which to use them. Evolutionary theorists are right to point to design compromises and vestiges, but the existence and importance of adaptation is not in serious dispute.

except perhaps in the American South...

2.2   Evolution and Natural Selection

Darwin and his successors have constructed a "received view" of the pattern of evolution and of the mechanisms that allegedly explain that pattern. One characteristic formulation of this view derives from the work of Ernst Mayr, who has played a triple role as its architect, historian, and philosopher. He sees contemporary evolutionary theory as a complex of five separate elements:

1. The living world in general is not constant; evolutionary change has occurred.

2. Evolutionary change has a branching pattern. The species now alive are descended from one (or a few) remote ancestors.

3. New species form when a population splits and the fragments diverge. More specifically, most new species are formed by the isolation of subpopulations at the periphery of the ancestral species' range.

4. Evolutionary change is gradual. Very few organisms that differ dramatically from their parents survive to found populations.

5. The mechanism of adaptive change is natural selection.

Mayr's elements of evolutionary theory

These ideas are related, but they are no package deal. The first two were almost universally accepted in the biological community by the end of the nineteenth century. Darwin, Wallace, and others rapidly convinced the scientific community of the fact of evolution. Mayr's distinctive contribution to the received view is its conception of species and speciation through isolation and divergence. While most accept that there is an important insight in this conception, many think it needs some kind of revision. As the nature of species and speciation remains controversial, 3 is a good deal more controversial than 1 or 2.

In general, claims about the mechanism of evolution have remained controversial since Darwin and Wallace formulated them. In the first decades of the twentieth century, the Darwinian view of natural selection was seen as inconsistent with the developing science of genetics. For it was thought that Darwinism was committed to continuously varying traits, whereas genetics showed that trait differences were fundamentally discrete. It was not until the synthesis of genetics with evolutionary theory by Fisher, Wright, Haldane, and others in the 1930s that there was any consensus about the importance of natural selection in driving evolution. These scientists constructed mathematical models to show that genes inherited according to the patterns discovered by Mendel could replace one another in a population if they were associated with very small differences in the capacity of organisms to survive and reproduce. Since Mendelian genes and their mutations were known to be real, and the mathematics of the new "population genetics" was demonstrably correct, earlier worries about the power of natural selection were laid to rest. However, more subtle questions about the nature and role of selection persist. While very few deny the importance of selection, its nature and role remain the key controversy in evolutionary theory. So the consensus on the role of selection remains incomplete. But to understand these controversies, we must first explain the received view of selection.

Natural selection has often been presented as the inevitable result of the interaction of three general principles. It is the consequence of

> phenotypic variation
> differential fitness
> heritability

Organisms in a population vary. Some variants will be better suited to dealing with the problems presented by their environment than others. These variants are more likely to survive to reproduce, or to reproduce more fecundly, than their fellows. If the characteristics that promote survival and reproduction are in part heritable, subsequent populations will be biased in favor of these advantageous traits. Thus the distribution of traits in the population will change. It will keep changing if the mechanisms that produce heritable variation add new traits to the evolving population over time.

Imagine the population ancestral to the superbly camouflaged Australasian bittern. Suppose that it too lived in reeds adjacent to wetlands, and sought to escape predation by crouching motionless when a threatening creature was near. It's quite likely that feather patterns varied in this ancestral population. If so, some birds had plumage that made them somewhat harder to see when they froze among the reeds. Those birds were more likely to survive to breed. This advantage need not have been dramatic. Perhaps the advantage held by our proto-camouflaged bittern ancestors was slight. Perhaps it helped only at dusk or dawn, or when a predator was at the very edge of its effective visual range. But a small edge is an edge nonetheless. That marginal advantage would sometimes make the difference between success and failure. So if bittern chicks tended to inherit their parents' plumage patterns, then the plumage patterns of the descendant generation would be somewhat different from those of the ancestor generation. Perhaps, for example, the harder-to-see birds of the ancestor population had more bars and fewer spots on their feathers than the average of their generation. If their chicks tended to be more heavily barred, then the average of their chicks' generation would

Box 2.1 Variation

In discussions of evolutionary theory, variation is often said to be blind, undirected, or even "random," though this last expression is misleading. The forces that give rise to new genetic variation in a population may or may not be deterministic. They are, however, insensitive to the adaptive demands on a population. If a rat population on an island is under extreme hunting pressure from snakes, so that only the very fastest rats have a chance of survival, that fact about the selection pressures on the rat population does not make variation in the direction of high rat speed any more likely. If new mutations arising in the population are relevant to speed at all, they are as likely to produce slower rats as faster ones. Indeed, they are more likely to produce slower rats, for there are more ways of slowing a rat down than of speeding one up. Of course, in such a situation, the slower rats will not be around for very long, but that is a matter of the retention of variation, not of its generation. Natural selection does not require that the mechanisms that generate variation be nondeterministic, or even that they be independent of the adaptive needs of the population. Rather, selection can produce adaptive change in a population even though the mechanisms that produce variation are insensitive to the direction of selection.

be pushed just a little in the direction of bars. The genetic mechanisms in the population responsible for variation would rebuild population variation anew. But not only would the average be edged toward more bars, fewer spots; in all likelihood, so would the limits. The most spotty bird of the chick generation would be less spotty than the spottiest bird of the parent generation. The most barred of the downstream generation would be more barred than any member of the previous generation. Over time, the patterns characteristic of the population would have changed.

In our imagined example we have mentioned only one feature of the plumage: bars versus spots. But color and orientation (and, of course, much else) in the birds' plumage will vary too. So if these traits make a difference in a bird's visibility, and if they tend to be passed on to the chicks, then feather color and the orientation of the spots and bars will change a little in the next generation. The average color will be shifted toward the color of the background vegetation. The average orientation of the bars will line up just a little better with the reed stalks among which the bird crouches. Thus with the color, orientation, and other elements of the feather patterns gradually

Figure 2.3 The Australasian bittern *(Botaurus poiciloptilus)* is superbly camouflaged. (From *Reader's Digest Complete Book of Australian Birds,* 1977.)

shifting, generation by generation, over time, the superbly camouflaged contemporary bittern evolves. Natural selection selects fitter organisms, and the heritability of their traits ensures a changed descendant population. Organisms are selected; populations evolve.

It is very important to see that this change depends on more than just variation, heritability, and fitness differences. The adaptive shift depends on *cumulative selection.* The adaptive shift to good camouflage took place gradually over many generations. Innovation is the result of a long sequence of selective episodes, not just one. For, as creationists endlessly tell us, the chances of a single mutation producing a new adaptation are very low. It is vanishingly unlikely that a single mutation could take us from a poorly camouflaged bittern to a well-camouflaged one. Mutations with large effects are almost always disastrous. If we take any well-functioning mechanism and make large, random alterations—if we double the size of one component, shrink another, and change the shape of a third—we are most likely to produce junk, not an improved machine. For similar reasons, major mutations are almost always lethal.

The extraordinary power of cumulative, as opposed to one-step, selection can be seen in the two ways to open a combination lock. To hit on the

[handwritten margin note: but could big changes still be a major force in evolution?]

Box 2.2 A Caution on Heritability

Selection works only on traits with some degree of heritability. There is no point selecting parents with good qualities if their offspring will not share those qualities. It is a common mistake to think that high heritability means that a trait is genetically determined—a matter of nature rather than nurture (1.4). In fact, heritability has very little to do with how traits are built in the growing organism. Selection cares about whether your children resemble you, but it doesn't care why. Heritability is purely a measure of how well the state of the parent predicts the state of the offspring.

To measure the heritability of a trait, we need a population of individuals, some of whom have the trait and some of whom lack it. If the only offspring with the trait are those whose parents had it, and all offspring whose parents had the trait also have it, then the trait is perfectly heritable. If the presence of the trait in offspring is unrelated to whether it was present in their parents, then the trait has zero heritability. In between lie the different degrees of heritability.

One way to make a trait highly heritable is to make the environment the same for everyone. By controlling other causes of variation, we can make heritable variation a higher proportion of total variation. For example, IQ scores will be more heritable if we provide equality of educational opportunities. Conversely, genetic uniformity in a population will reduce heritability. In a commercial forest of *Pinus radiata* in which every tree is a clone, most differences in height will be accounted for by differences in microenvironment. Just as it would be a mistake in the first case to infer that intelligence is affected only by genetics, it would be a mistake in the second to conclude that height is affected only by the environment.

combination of the lock in figure 2.4 by chance, you would have to get every wheel in the right place simultaneously. The chances of that are very low. But as every safecracker knows, if you can hear a faint click when each individual wheel falls into the right position, the problem disappears. It will take an average of five random tries to get the first wheel right, and then you can go on to the second. On average, fifty random trials will find the right combination. Natural selection works like the safecracker, by variation and selective retention. Natural selection, of course, does not involve any agent listening for the "clicks." An organism with "one wheel right" will be the basis for the next set of variations because it will have more offspring than

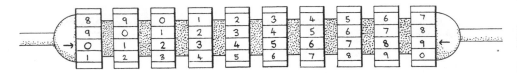

Figure 2.4 Like a safecracker, evolution cheats by solving complex problems one step at a time. (Adapted from Simon 1969.)

organisms with no wheels right. Its offspring with an advantageous mutation will have "two wheels right." It is this absence of any overseeing agent that makes natural selection natural.

So cumulative selection is the only realistic way in which natural selection can produce adaptive shifts in a population. But for selection to be cumulative, some additional conditions are needed on top of the triad of variation, differential fitness, and heritability. Most obviously, if the direction of selection is not fairly constant over a number of generations, no new adaptive traits will be built. In nature, the direction of selection is not always stable. For example, selection on a certain population of Galápagos finches is not stable; wet and dry seasons select for different beak shapes (Weiner 1994). Another requirement for cumulative selection is a relatively low mutation rate. If the mutation rate is very high relative to the strength of selection, then the mechanisms that generate variation will swamp the effects of selection. But just as too much variation swamps selection, too much selection drives out variation. Intense selection pressure reduces the genetic variation in a population. Since most variation in the physical form or behavior of organisms (so-called *phenotypic variation*) is the result of shuffling of the existing genetic variation, rather than of new genes being created by mutation, reducing genetic variation within the population reduces the differences selection has to work on. Artificial selection by human breeders is typically very intense, as all the organisms with any unwanted character are culled. So artificial selection often produces significant change quickly, then runs out of steam. In the nineteenth century, the experience of animal breeders was sometimes thought to show that there are limits to how much each species can be changed, and hence that evolution by natural selection cannot make a new species. But natural selection operates over a far longer time scale, and so can afford to wait for more mutations to come along.

There is another important restriction on the process of cumulative selection: each intermediate stage must be fitter than its predecessor. Think of an evolutionary change from (say) white herons to streaky dark ones. Suppose, for the sake of argument, that a very dark heron would be fitter, because better camouflaged, than any other color, but that a slightly dark heron would be easier to see than a wholly white one. If so, natural selection will not drive an evolutionary shift from white to dark herons because slightly darker variants will typically do worse than white ones. In the jargon of the trade, white herons are at a *local optimum*. That is, no small change in their current characteristics (with respect to color) will improve their fitness. Evolutionary theorists illustrate facts like these with geographic images called *adaptive landscapes*. In these diagrams, height represents fitness and the other dimensions represent features of the organisms. In an adaptive landscape, white herons would be represented by a hill, and very dark ones by a higher hill some distance away. Between the two hills would be a valley. Selection cannot drive a population off the top of a hill and down across a valley, even if a much higher hill is in the vicinity.

Thus variation, heritability, differential fitness, and the conditions admitting of cumulative selection result in selection on organisms, and this produces gradual change in populations over time. Such gradual change, continued over long periods of time, results in both adaptation and differentiation as distinct populations become adapted to distinct environments. Populations live in somewhat different local environments and thus face somewhat different challenges. Natural selection enables populations to respond to these different environments, and as they do so, the populations come to be different from one another. Thus one population of herons becomes small and cryptically colored as it becomes adapted to foraging within shallow marshy swamps; another becomes long-legged and white as it becomes adapted to foraging in open estuaries. As populations become increasingly distinct, their members become less able to treat one another as potential mates. Eventually the two populations will be reproductively isolated from one another, and hence speciation will have taken place. The differences between the two populations will have become permanent, leading to new species: to diversity. That diversity reflects both variation in the environment and the reaction of populations under selection to that variation.

In some views, speciation is wholly a by-product of the divergence of populations that have somehow become isolated from one another. In other views, as divergence takes place, there will be selection for reproductive isolation. The idea is that as two populations diverge, any hybrid matings that do occur will be penalized by selection. The issue from these hybrid matings may be sterile, nearly sterile, or suited to neither environment. So selection will start to favor, in both populations, any trait that makes its bearers less likely to accept a mate from the other population. Selection thus entrenches the differences between the two populations. The idea that isolation is critical to the formation of new species is an important part of the received view. There are, however, differing views on whether selection is important in reinforcing and entrenching isolation or whether new species arise simply as a by-product of isolation (Butlin 1987a,b).

Whatever stand is taken on this question, according to the received view, the two most striking features of life—diversity and adaptation—are both explained by natural selection. Natural selection is a constrained process. It is slow, and it is bounded by both previous history and the availability of variation. Nonetheless, the history of life is the history, essentially, of changes in populations of organisms as a consequence of natural selection.

2.3 The Received View and Its Challenges

Much in contemporary philosophy of biology revolves around the received view and its challenges. It's worth disentangling three strands in these debates. One focuses on the nature of natural selection itself—on what is being selected. A second concerns the place of selection within evolution. A third turns on the role of evolutionary theory within biology.

The Units of Selection

The received view conceives of natural selection as the result of competition between individual organisms in a population. Differences among those organisms result in their differing success, and those differences in success cause generation-by-generation shifts in the character of populations. The received view identifies some very familiar participants in life's history: organisms, populations, and species. In this view, the natural kinds of evolutionary biology are also the kinds identified by common sense. Adherents of the received view see no radical disjunction between our naive folk inventory of the biological world and that world as described by evolutionary biology. That is not always true of science. Color, for example, appears to us to be a simple objective property of material surfaces, but it turns out to be a very complex property. Indeed, perhaps color is not an objective feature of the world at all. The received view faces challenges, and among them are views that see a greater gap between the folk conception of the biological world and that derived from evolutionary theory.

One such challenge comes from the "gene's eye" conception of evolution, introduced by George C. Williams and developed and popularized by Richard Dawkins (Williams 1966; Dawkins 1976, 1982). According to this conception, when we think of the tree of life, we should think not so much of organisms as of genes. In section 2.2, we emphasized that evolutionary change depends on cumulative selection, and this is the gene selectionists' point of departure. Individual organisms, they say, are unique. When organisms reproduce, their offspring are not copies of either parent. Genes, in contrast, are copied when organisms reproduce. Most of a child's genes are copies of parental genes. In the terminology developed by Dawkins (1982) and Hull (1981), genes are *replicators*. Since organisms cannot be copied, they cannot form chains or lineages in which each link is a copy of the one before it. But since genes can be copied, they can form such lineages: chains of copies, with each link being a copy of its predecessor. Gene lineages can sometimes be many copy-generations deep. They can vary in bushiness, too, for, depending on the number of offspring in each generation, a gene may be copied many times, and the copies may form an increasingly broad lineage as well as a deep one. Alternatively, a gene lineage may be narrow, with only a few copies existing in each generation.

Individuals within a population are typically in competition with one another because resources are limited, and not all will secure enough. In many species of birds, for example, a third or more of the population starves over the winter. The gene selectionist view of evolution takes this notion of competition and applies it to competing gene lineages. Collectives of genes replicate by constructing a *vehicle* or *interactor*—that is, an organism or something like an organism—that mediates both their interaction with the environment and their further replication. So gene replication is typically a very indirect process and thus demands scarce resources. According to this conception, the life or death of an organism has its evolutionary consequences indirectly, by influencing the copying success of the genes within it. Well-built organisms mediate more effective replication of the genes within them, the replicators that help to build them. So selection acts through organisms to target some genes rather than others by virtue of those genes' differential influence on their probability of replication. This differential influence is typically exercised through the gene's organism-building role. Usually genes have high replication capacities because they build effective organisms. So the consequence of selection is the differential growth of lineages of replicators, and hence, indirectly, the differential production of interactors of various kinds. According to this "gene's eye" conception of evolution, the received view is in the grip of the wrong picture of evolution. It fails to make the distinction between replicators and interactors, and to see the fundamental importance of replication in cumulative selection.

As we shall see in the next three chapters, these ideas have generated fierce controversy. Critics of the gene selectionist view think that it requires a very

Box 2.3 Replicators, Interactors, and Lineages

In the gene's eye view, there are three fundamental kinds of entities that play a role in evolution. *Replicators* are copied into the next generation: their pattern survives intact. So they may give rise to a potentially unbounded sequence of descendant copies. However, the materials and energy for construction of those copies must come from somewhere. Replicators must therefore interact with their environment, and they do so with differential success.

One means by which replicators compete is by constructing special-purpose entities. These entities aid replication by mediating the interaction of the replicator with the rest of the world. Hence replicators are usually assumed to carry information that is used in the construction of these entities. These special-purpose entities are *interactors,* or, in Dawkins's roughly equivalent terminology, *vehicles*.

Genes are the paradigm replicator: they are copied across the generations, and are usually thought of as comprising a recipe or a program used in the construction of organisms. Organisms are the paradigm interactor, but perhaps not the only kind. Perhaps, for example, a termite nest is a single interactor, jointly constructed by all the genes in the nest. In any case, the typical replicator codes for characteristics of interactors—organisms, colonies, populations. These interactors reproduce, and their differential reproduction results in the differential copying of replicators. If a particularly aggressive type of ant founds new colonies at a greater rate than others, that differential reproduction will cause a differential replication of the gene(s) that code for that increased aggression. The lineage of copies of the aggression gene will become bushier than those of its more peaceable rivals.

So this picture of evolution recognizes three basic kinds: replicators, which are usually, perhaps always, genes; interactors, which are usually, but perhaps not always, organisms; and lineages, which are chains of copied replicators.

simple relationship between an organism's genes and its traits. They argue that gene selection would occur only if there were something like a one-to-one relationship between genes and traits. Sober, for example, has argued that there is selection for one gene over its rivals only if an organism in the relevant environment is *always* fitter—more likely to survive, reproduce, and replicate its genes—by virtue of carrying that gene (Sober 1984b). In turn, it seems that the only way a gene could guarantee such a fitness advantage is if it produced a specific trait. So we could think of selection as acting on lineages of bittern camouflage genes if there were particular genes, each of which gave its carriers a distinctive feather pattern. We would then have a cluster of feather pattern genes, and the one responsible for the most cryptic pattern would win out over its rivals coding for different patterns. Sober's interpretation of the debate has the merit of giving the gene selection hypothesis a clear empirical interpretation. But everyone agrees that the relationship between bittern genes and bittern plumage is anything but simple. So the critics conclude that gene selection is fatally flawed, while the advocates of gene selection deny that each competing gene needs to code for a distinctive trait.

A different, and perhaps less radical, challenge to the received view comes from defenders of "hierarchical" conceptions of evolution. They think that the received view has locked onto just one aspect of evolution. Populations of organisms do evolve under natural selection, just as the received view claims. But they argue that organisms are not the only biological entities that form populations. Recall the conditions for evolutionary change under natural selection. If a varied population of entities gives rise to descendants like themselves, and if those entities differ in fitness, selection will generate evolutionary change in that population regardless of the type of entity in question. Suppose, for example, that bat colonies form a population of colonies, in which individual colonies vary one from another. Perhaps some roost in hollow trees and others prefer caves. If colonies found others like themselves, so that daughter colonies tend to share their founders' roosting preferences, and if those colonies differ in fitness (for one type of roosting site may be safer than another), we should expect to see selection change the colony population: the proportion of colonies with the safer roosting preference should rise. Eusocial insect communities, and perhaps many other groups, seem to meet these conditions. In recent work, Wilson and Sober have drawn just this conclusion. They argue that we should regard hives and other cohesive animal societies as subject to selective forces (Wilson and Sober 1989, 1994; D. S. Wilson 1992). We consider these ideas in chapter 8.

Species, too, appear to form populations. Consider, for example, New Zealand's sadly extinct moa species. This group formed a population of closely related species. In the process of speciation, ancestral species give rise to similar descendant species, so we would expect features of species to be heritable. The exact ecological interactions among moa species are unknown, but we can assume that they affected one another and, in doing so, divided the New Zealand habitats of their time among themselves. This division would have left some species confined to smaller, less productive, or more unstable habitats, and those species would be more likely to face extinction as a result of natural environmental fluctuations. The species would not just vary; they would differ in fitness. So the moa species group seems to meet the conditions that would generate selection among species. Hence we should expect to see natural selection at work in a pool of species.

Just as selection on populations has had its defenders, so too has selection on species. It has been thought to be part of the explanation for the prevalence of sexual reproduction. Sexual reproduction is a major puzzle in evolutionary theory. After all, why waste all that time, energy, and risk on finding a mate when you could reproduce asexually instead? (Sexual reproduction has subtler costs as well.) One answer goes something like this: Sexual reproduction, by mixing organisms' genomes, increases genetic diversity and hence the evolutionary flexibility of a species. Sexual reproduction allows the combination of two independently favorable mutations within a population. If Abu the baboon carries a dominant mutation that makes him more tick resistant, and Belle one that makes her more tolerant of dehydration, and they mate, some of their offspring may benefit twice. Organisms that reproduce by cloning would have to wait for one of the clone lineages to repeat the luck of the other—for example, for one of Abu's descendants lucking upon Belle's mutation. So sexually reproducing species respond better to environmental change. They are more likely to resist extinction if their environment changes in ways that demand that they, too, change. As it happens, there are different, more recent theories of the adaptive advantage of sex to the individual organism. Moreover, sex can act as a brake on evolutionary change. Migration between populations followed by mating can homogenize them, stopping either population from adapting to its specific local conditions. So this particular idea has slipped from favor. But there are other features of the pattern of life that species selection may explain. Lloyd and Gould (1993), for example, have recently argued that variable species are more resistant to extinction than more homogeneous species, and that species selection has acted to preserve variability within species lineages. We return to this issue in section 9.4.

According to the hierarchical view, selection can be operating simultaneously at many levels. An individual organism competes in a population of organisms. If that organism is, say, a wolf, she may be part of a pack competing in a population of packs. It may even be that wolves, coyotes, and dogs form a population of species that are in competition, one with another. Perhaps there are features of the coyote species itself that have enabled it to adapt better to human-altered environments and hence replace the wolf in most of its American range. When selection operates at different levels at one time, we should expect to see conflicts between the levels at which selection occurs. What's good for General Motors is not necessarily good for GM's office cleaner. Equally, what's good for, say, a baboon troop is not necessarily good for Abu, a baboon currently low on the totem pole in his troop but wanting to rise. Group selection for internal cohesion will push baboon organization one way. Individual selection favoring those who insist on getting a piece of the action will push it another way.

The hierarchical conception of selection does not reject the received view outright. Rather, it proposes that the received view mistakes the most important case for the only case. Those skeptical about hierarchical conceptions of evolution usually do not think it impossible in principle for individuals collectively to form metapopulations that are selected. But they think that, in practice, selection at the level of individual organisms so dominates all super-individual processes that the latter are of no evolutionary significance. We will consider whether they are right in chapters 8 and 9.

Selection and Evolution

So one set of issues within evolutionary biology revolves around the units on which selection acts. But others turn on the effects of selection, and on its role within the total matrix of forces that jointly explain the evolution of life. These issues became prominent in evolutionary theory through the work of Gould and Lewontin (Gould and Lewontin 1978; Lewontin 1985a). Their attack on "adaptationism" initiated a fierce debate on the role of selection within evolution. In this debate, at least four issues are tangled together.

First, does the received view overstate the importance of adaptation? At first glance, it would seem that there is no room here for deep disagreement. No evolutionary theorist denies the existence of complex adaptation, or denies that natural selection plays some important role within evolution. Equally, no one denies the causal importance of other factors in evolution. Selection can operate only on the variation that is available, and history, development, and genetics determine the range of variation. Everyone accepts, for instance, that some genetic changes sweep through a population because they "piggyback" on others. If there is a favorable mutation in a mitochondrial gene, and as a consequence, that mutated gene becomes more frequent in the population, all the other genes in that mitochondrion will share in its good fortune, because all the mitochondrial genes are copied as a whole and are transmitted as a whole. So an advantage to one is an advantage to the whole mitochondrial genome. Moreover, there is no question that

does everyone really recognise this?

chance is important too. Fitness advantages are propensities; they make one outcome *more likely* than another. Better camouflaged bitterns have a better chance of surviving than less well camouflaged ones. But, especially in a small population, a greater chance of success need not translate into actual success. If a fair coin is tossed four times, we might well get three heads. If it's tossed a thousand times, however, we are extraordinarily unlikely to get 750 heads. So the smaller the population, the more likely it is that the actual success of organisms with some trait will vary from the expected fitness of organisms with that trait. Yet the received view itself—especially Mayr's version of it—suggests that most evolutionary change takes place in small, isolated populations, in which chance—both lucky and unlucky accidents—can play an important role. So the potential importance of both selection and other factors seems accepted on all sides, and there seems room for disagreement only on the mix of factors that explain particular cases.

But the impression that there can be no global disagreement on the importance of selection is misleading. Lewontin and other critics of what they call *adaptationism* claim that those in the grip of the received view pay only lip service to the possibility of evolutionary explanations in which selection plays a secondary role. As Lewontin and his allies read the situation, adaptationist evolutionary biologists, despite conceding the potential importance of chance and history, tacitly assume that virtually every striking feature of an organism somehow contributes to its survival and reproduction, and that that contribution explains why the creature has that feature. In other words, adaptationists suppose that most traits have functions. Bats, for example, often have elaborate facial architectures that play an important role in echolocation. A splendidly bizarre example is the large-eared horseshoe bat *Rhinolophus philippinensis,* which can broadcast directional ultrasound through its elaborately structured nose, so that it can eat and navigate at the same time. To claim that its facial structures function in echolocation is to say that they exist because of their role in the ancestral bat's echolocation. Those ancestral bats got to be ancestors in part because their facial architecture gave them an echolocating edge vis-à-vis their contemporaries. Thus that architecture's role in helping the ancestors of today's horseshoe bats echolocate explains its existence in bats today. This much is uncontroversial. At issue, though, is the role selection must play in this history for these functional claims to be true. Those skeptical about adaptationism think that an adaptive claim about bat facial structure depends on a very strong claim about the role of selection in its evolution. They think that adaptationist hypotheses are committed to an "optimality hypothesis" about bat facial structure: that bat echolocation is the best it could possibly be. As we shall argue in chapter 10, we very much doubt that those who think that the effects of selection

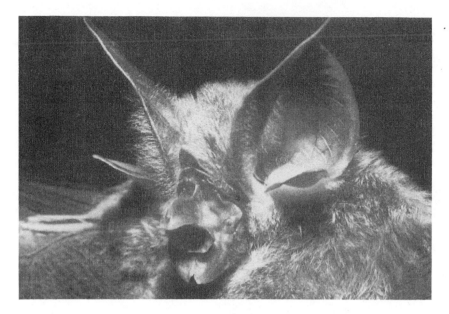

Figure 2.5 The large-eared horseshoe bat *(Rhinolophus philippinensis),* found in Australia and the Philippines, has impressive facial structures that assist echolocation. (From Strahan 1983, 297.)

are pervasive, and who think that most complex traits of organisms have functions, are committed to the idea of the best possible bat.

So the first thread in the adaptationist debate focuses on the importance of selection. The issues are partly empirical and can be addressed only on a case-by-case basis: How much variation do natural populations have? How often are populations small enough for chance events to be important? How often do deleterious genes spread by being physically linked to advantageous ones? But they are also partly conceptual: What role must selection play in the evolution of a trait for that trait to have a function and be an adaptation?

A second thread in the adaptationism debate focuses on the relationship between selection and the other factors that help to explain evolutionary change. How, if at all, can we compare these different factors? In thinking about the evolution of a trait, history matters. Adaptation through natural selection never redesigns from the ground up, but instead tinkers with the results of earlier history. Consider a feral cat hunting in the Australian bush. In semiarid areas of Australia, feral cats are considerably larger than their domestic counterparts, and are light tabbies. Size and color are apparently adaptations of that particular cat population—dark kittens are vulnerable to eagle predation. Some other aspects of their biology are the result, no doubt, of earlier selective processes that shaped the species. Others are inheritances from their felid, mammalian, and even earlier ancestors. These inheritances constrain the future trajectory of this population of feral cats.

We can make good sense of the idea that the moon has a greater influence on the earth's tides than the sun, for we can sum those effects and compare their relative magnitude. But it is unclear how to make sense of the idea that feral cats' biology is more the result of their history than of selection. History and selection conspire together to drive the evolutionary trajectory of the population. Both are necessary; neither is sufficient. Those skeptical of adaptationism often speak of historical constraints on natural selection as if the history of the cat lineage somehow blocks or prevents the operation of selection. There is something strange about this idea, for it is only the history of previous cat evolution that makes possible the adaptive shifts we now observe. Yet there is something right about this idea too. As we have noted, tree kangaroos, despite their adaptations for arboreal life, are obviously terrestrial mammals jerry-built for or press-ganged into life in the trees. We return to these problems in chapter 10.

A further problem is that some evolutionary patterns seem to be independent of selection. In 1972, Eldredge and Gould put forward the hypothesis of *punctuated equilibrium*. In this view, most species, over most of their life spans, do not change in body or behavior. Evolutionary change and speciation occur in brief—geologically speaking—bursts. Species come into existence quickly, remain phenotypically the same throughout most of their life spans, and then disappear, either through fragmentation into descendant species or extinction. More recently, in his *Wonderful Life* (1989), Gould argued that the standard image of life's history is quite mistaken. We typically think of life as becoming both more diverse and more disparate over time, viewing the history of life as a change from few, simple, and comparatively similar organisms to many, complex, and highly differentiated ones, like the pattern in figure 2.6a. Not so, he argued. Though *diversity* has increased, animal life was maximally *disparate* at the time of the Cambrian explosion, and has become less disparate since then, a pattern more like that in figure 2.6b. This is an idea about macroevolutionary patterns on the grandest possible scale.

How do these ideas about the history of life relate to the received view? While no one now suggests that the hypotheses of punctuated equilibrium or of maximum disparity in the Cambrian are inconsistent with the received view, equally, the thought goes, nothing in the received view predicts them. It is silent about that whereof it should speak. The received view is incomplete: there is at least one important feature of the history of life—the

a)

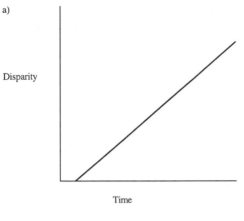

Disparity

Time

Figure 2.6 (a) The received view would lead us to expect a more or less uniform increase in disparity over time. (b) Gould suggests that disparity reached a plateau early in the history of animal life.

b)

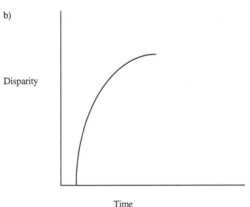

Disparity

Time

patterns in diversity
& disparity ⇒ Ch 12

plateau in disparity—that it does not explain. We shall explore these issues in chapter 12.

A third strand in the debate over the role of selection in evolution is methodological. Few will deny that bat echolocation is the product of natural selection. These mechanisms exist because they enable bats to navigate and hunt in the dark. But adaptation is very often less obvious than this. Are psychopaths a "minority strategy" that has been selected during human evolution to take advantage of the relatively benign behavior of most other humans? Is menstruation an adaptation to rid female monkeys of sperm-borne *eh?* disease organisms? How can we test these ideas? There has been a debate about the strength of evidence needed to establish an adaptationist hypothesis. Gould and Lewontin (1978) refer disparagingly to "just so stories" told by some theorists. A "just so story" is an adaptive scenario, a hypothesis about what a trait's selective history might have been and hence about what

[handwritten: human psych & social traits — biological or cultural evolution? ⟹ Ch. 13 & 14]

its function may be. Gould and Lewontin think that it is just too easy to think up such adaptive scenarios. Hence we should not accept an adaptive hypothesis just because it sounds plausible. So how do we show that a trait is the product of selection, and how do we show that it was produced by a particular set of selective forces? We deal with this question in chapter 10.

Finally, enmeshed in all these other disputes are questions about the proper scope of evolutionary explanations. One arena in which this issue has surfaced with great bitterness is the debate over evolutionary explanations of human psychological and social organization. In 1975, E. O. Wilson published *Sociobiology,* an adaptationist account of social evolution among animals. In a speculative final chapter, he extended these ideas to humans. This work was followed by many others that were even more speculative, often without emphasizing, or even admitting, their speculative nature. Moreover, they often advanced adaptationist explanations of features of human life that many people wanted to reform: differences in sex roles, xenophobia, rape, and the abuse of stepchildren, to name just a few. The response was often savage, but in the heat, confusion, and rancor, two somewhat different critical strategies emerged.

One response has been to accept that, in principle, evolutionary explanations of important aspects of our psychology and society are possible. According to this line of criticism, evolutionary theories of human nature are potentially promising. But human sociobiology as actually practiced has typically consisted of poor and ill-supported evolutionary hypotheses. Kitcher (1985) typifies this critical response. The second response has been to argue that sociobiological explanations of human social and psychological characteristics are mistaken in principle. Human psychological and social traits are the result of human culture. Evolutionary theory ought to explain how the human lineage evolved the capacities that enable us to be an encultured species. But ever since we acquired our cultures, it is those cultures that explain the distinctive and interesting features of us and our societies (Levins and Lewontin 1985). We explore these issues in chapters 13 and 14.

Evolution within Biology

Finally, in thinking about the received view, we need to think about the role of evolutionary theory within biology generally. Challenges to the received view can have important ramifications for our conception of the place of evolutionary theory within biology. For instance, they force us to reconsider the relationship between evolutionary biology and ecology. One of the virtues of the received view is the elegance and simplicity of its picture of that relationship. It perceives evolutionary change as driven by the demands the environment imposes upon organisms. Selection shapes organisms to their environment. Thus evolutionary theory is linked to, and depends on, ecology, because ecology provides a principled analysis of the environment and the selective pressures it generates. Ecology provides this analysis paradigmatically through its depiction of a biological community as a set of interrelated niches. These niches specify the different roles, or ways of making a living, that organisms have in their environments. Since organisms can vary in the degree to which they fit their niche, selection will prefer those that fit their niche well over those that fit it less well. In one of evolutionary biology's most vivid metaphors, one of the architects of the received view, Theodosius Dobzhansky, articulates this view of the relationship of ecology and evolution:

> The enormous diversity of organisms may be envisaged as correlated with the immense variety of environments and of ecological niches which exist on earth. But the variety of ecological niches is not only immense, it is also discontinuous. One species of insect may feed on, for example, oak leaves, and another species on pine needles; an insect that would require food intermediate between oak and pine would probably starve to death, Hence, the living world is not a formless mass of randomly combining genes and traits, but a great array of families of related gene combinations, which are clustered on a large but finite number of adaptive peaks. Each living species may be thought of as occupying one of the available peaks in the field of gene combinations. The adaptive valleys are deserted and empty.

> Furthermore, the adaptive peaks and valleys are not interspersed at random. Adjacent adaptive peaks are arranged in groups, which may be likened to mountain ranges in which the separate pinnacles are divided by relatively shallow notches. Thus, the ecological niche occupied by the species "lion" is relatively much closer to those occupied by the tiger, puma, and leopard than to those occupied by wolf, coyote, and jackal. The feline adaptive peaks form a group different from the group of canine "peaks." But the feline, canine, ursine, mustelid and certain other groups form together the adaptive "range" of carnivores, which is separated by deep adaptive valleys from the "ranges" of rodents, bats, ungulates, primates, and others. . . . The hierarchical nature of biological classification reflects the objectively ascertainable discontinuity of adaptive niches, in other words the discontinuity of ways and means by which organisms that inhabit the world derive their livelihood from the environment. (Dobzhansky 1951, 9–10)

relationship of ecology & evolution

If the received view is mistaken or incomplete in important ways, especially in its conception of the role of selection in evolution, the relationship between ecology and evolution needs to be rethought. In chapter 11, we sketch out some of the options for this rethinking.

So one problem concerns the relationship between evolution and ecology. But we also need to integrate our picture of evolutionary biology with developments in the other major domains of biology. The "molecular revolution" in biology blasted off after the discovery of the structure of DNA by Crick and Watson in 1953, and it is clear that many thought that our new understanding of the molecular mechanisms of inheritance would make evolutionary biology redundant. E. O. Wilson's autobiography, _Naturalist_ (1994), gives a graphic account of the atmosphere of this period, and of the apparent institutional imperialism of molecular biology from the perspective of one who felt threatened by it. (Some will find this ironic, for Wilson, as chief instigator of human sociobiology, is himself suspected of writing redundancy notices for those employed in anthropology, sociology, and psychology departments.)

Tinbergen and Mayr, each in somewhat different ways, showed that it would be a mistake to suppose that a molecular understanding of the mechanism of inheritance would replace an evolutionary one (Tinbergen 1952, 1963; Mayr 1961). As we noted in section 1.5, Tinbergen distinguished four questions we could have in mind in asking why a bittern stands still with its bill pointed directly at the sky. (1) We could be asking for a _proximal_ explanation: an explanation of the hormonal and neural mechanisms involved in triggering and controlling this behavior. (2) We could be asking for a _developmental_ explanation: an explanation of how this behavior pattern emerges in a young bittern. (3) We could be asking for an _adaptive_ explanation: an account, that is, of the role this behavior currently plays in the bittern's life. (4) Finally, we could be asking for an explanation of how and why this behavior pattern evolved in this bittern population or in its ancestors.

The first of these projects is specific to the biology of behavior, but the others apply to any trait. However, as we shall see (especially in section 6.1), identifying different explanatory projects only partially resolves the relations between different domains of biology. For while these projects are distinct, they are not independent. Our views on development, for example, affect our views on evolution, and vice versa. What is developmentally possible influences what is possible in evolution through its effect on the range of variation available to selection. At the same time, developmental mechanisms have themselves evolved. But while no one denies that there are connections between developmental and evolutionary questions, there is a good deal of controversy about the nature of those connections. For there is an influential line of thought within both developmental and evolutionary biology that suggests that the received view has failed to include developmental biology, and perhaps even is tacitly inconsistent with it. Those who suspect the received view of adaptationism typically think that it greatly understates the importance of developmental constraints in explaining why organisms are the way they are. Perhaps primates have two arms rather three simply because it is impossible to grow an organism with that asymmetry. Perhaps unsuccessful male elephant seals cannot turn female because it is impossible to finesse internal changes of such complexity in an organism that must continue to lead a relatively normal life throughout its metamorphosis. Indeed, the only possible counterexample to the claim that the significance of adaptation is universally accepted is the view of the process structuralists, who suggest that the constraints on possible organisms are so severe that natural selection plays only a minor role in evolution (Goodwin and Saunders 1989; Goodwin 1994). We discuss these issues in chapter 5 and in section 10.5.

More generally, as G. C. Williams (1992) notes, evolutionary theory operates on the assumption of mechanism; that is, on the assumption that causal processes in biology involve no occult forces. The mechanisms involved in development, inheritance, selection, and speciation are, or are composed of, standard physical and chemical processes. Inheritance, for example, might be physically or chemically very complex, but it does not involve fundamental processes found only in living matter. The mechanistic hypothesis in biology moves molecular biology onto center stage, for, prima facie, molecular biology vindicates mechanism, and perhaps even stronger views about the relationship between biological theories on the one hand and those about physical and chemical processes on the other. These issues are the focus of chapters 6 and 7, where we take up the surprisingly complex problems of saying what a gene is and of describing the relationship between genes as they have been conceived in evolutionary theory and the structures of DNA and the associated cellular components currently being revealed by molecular genetics. In the next part of the book we look at gene selection and the problems it poses.

Further Reading

2.1   The idea of design space is introduced in Dennett 1995; the idea is adapted from Dawkins 1986. For some insights into the extravagance of the actual biological world, see E. O. Wilson 1992. For some sense of the weirdness of organic form, browse in Margulis and Schwartz 1988. Most of the genuinely weird forms are to be found among the invertebrates; one standard survey of these is Brusca and Brusca 1990. Gould's essays exhibit his wonderful feel for the strange; see Gould 1996e for a very recent celebration of the intricate complexity of parasite lifestyles, and Gould 1996c and 1996a for even more vigorous statements than usual of the dominance of bacteria in the history of life. J. T. Bonner discusses the world of the slime molds in most of his books, most recently in Bonner 1993. For some of the constraints on possible design, see McMahon and Bonner 1983, which discusses the relations between body size and possible body design. Vogel (1988) discusses the effects of differing physical forces on organic design more generally.

2.2   Evolutionary theory is so hotly debated that it's not surprising that there are no uncontroversial introductions to the received view and the controversies it has generated. Perhaps the closest is Depew and Weber 1995, a work with the added advantage of detailed reading guides. But though it is very fair-minded, it is also a very long, detailed historical treatment of the development of Darwinian ideas. Another, less terrifyingly long, historical introduction is Mayr 1991, but it is by a partisan of the received view. Peter Bowler has also produced a number of historically based introductions to evolutionary thought; the one with the broadest scope is Bowler 1989. Cronin (1991) blends historical and contemporary material, concentrating on two problem cases for evolutionary theory: sex and altruism. This is a clear and lively read, but it is very partisan. She writes from the "gene's eye" perspective without making it clear just how controversial that perspective is. Richard Dawkins's introductory books are even more lively, more readable, and more partisan, but they never pretend to be anything else (Dawkins 1986, 1989, 1995, 1996). However, the clarity and vigor of Dawkins's account of the importance of cumulative selection, and of the difference in power between cumulative and single-step selection, is unsurpassed, and this aspect of his thought is not controversial. A more understated account of the same basic picture is Ridley 1985. Defenders of a hierarchical conception of evolution have written plenty of books, but none really intended as general introductions to their views. Perhaps Eldredge 1985a might be the closest. For a very good account of the development and, from his perspective, fall from grace of the received view, see Gould 1983a. Futuyma 1998 and Ridley 1993b are much more technical surveys of evolutionary biology. For good accounts of the critical concepts of the received view, adaptation, fitness, and heritability, see the relevant entries in Keller and Lloyd 1992.

2.3   Suggested reading on these issues will be given at the end of the chapters in which these problems are discussed more fully.

II

Genes, Molecules, and Organisms

3

The Gene's Eye View of Evolution

3.1 Replicators and Interactors

In a widely quoted passage, Richard Dawkins sums up the message of his book *The Selfish Gene:*

> What was to be the fate of the ancient replicators? . . . Now they swarm in huge colonies, safe inside gigantic lumbering robots, sealed off from the outside world, communicating with it by tortuous and indirect routes, manipulating it by remote control. They are in you and me; they created us, body and mind; and their preservation is the ultimate rationale for our existence. They have come a long way, those replicators. Now they go by the name of genes, and we are their survival machines. (1976, 21)

Stripped of its purple prose, this is gene selectionism, the "gene's eye" view of evolution.

This view of evolution sees it as consisting of two fundamental processes: replication and interaction. As we have seen, significant evolutionary change results only from cumulative selection (2.2). So one fundamental feature of evolution is replication: the process of copying from generation to generation, ensuring that successive generations are similar enough for selection to be cumulative. The other fundamental process is ecological interaction. Interaction between organisms and their environment—including the other organisms in that environment—biases the copying process and causes differential copying from one generation to the next. To each of these two fundamental processes there corresponds a special sort of entity. *Replicators* are things that are copied into the next generation: they form lineages of things with the same structure. *Interactors* (or *vehicles*) are entities that exist in each generation of a copying cycle and interact, more or less successfully, with the environment.

Box 3.1 Terminology: Interactors and Vehicles

Hull's term *interactor* and Dawkins's term *vehicle* are often treated as equivalent, although there are significant differences in the definitions the two authors give. Dawkins's term nicely expresses his view that organisms are vehicles of their genes in the sense of being under their control—"lumbering robots," as he puts it. The term *interactor* is more neutral, and more suited to our discussion of potential higher levels of selection, in which the interactor may be a local population or a species.

Once we draw the replicator/interactor distinction, we can see that there is room in evolutionary biology for two sets of debates about the units on which selection operates. One is about replicators. What are replicators? How do they influence the world in ways that make themselves more, or less, likely to be copied? How are replicators related to other replicators and to interactors? The other debate, obviously, is about interactors, and the adaptations that promote successful interaction.

Many of those who accept the gene's eye view of evolution have taken questions about replicators to be straightforward. Replicators are genes, and genes are chunks of DNA. Long chunks of DNA are broken up when gametes are formed (by a process known as *crossing over*), but short chunks can reappear unchanged through many generations of organisms. So replicators are relatively short chunks of DNA. These replicators are the beneficiaries of interaction and adaptation, because the ultimate effect of that process is to promote the replication of favored genes. In contrast, this idea goes, questions about interaction and the bearers of adaptation—what is adapted rather than what benefits from adaptation—remain open. Are hives or wolf packs interactors, or do they merely consist of interactors? Are they themselves adapted, or do they just consist of adapted organisms? As we shall see in section 8.2, the gene's eye conception of evolution arose partly in response to empirical debates about the evolutionary explanation of altruism, debates that are best seen as debates about interactors. However, while questions about the nature of interaction are certainly pressing, and are central to chapters 8 and 9, we deny that questions about replication are unproblematic. We shall argue in sections 3.3, 5.2, and 5.6 that genes are not the only replicators. Moreover, the question "What is a gene?" turns out to be very difficult indeed.

Gene selectionism, then, begins with the distinction between the evolutionary roles of two kinds of entities: replicators and interactors. Genes—

replicators—can be copied to form lineages or chains of identical copies. In some organisms, there is an important distinction between two kinds of genetic replicators. In those organisms, there is an early developmental differentiation between the *germ line*—cells that will be the ancestors of the organism's sex cells—and the *somatic line*—cells fated to form the tissues and other components of the organism's body. Genes in germ line cell lineages are paradigmatic replicators. They are potential ancestors to indefinitely long lineages of copies. Although the individual copies of a gene—gene *tokens*—have a limited life, the lineage—the reproductive family of copied gene tokens—may persist indefinitely. In this sense, but only in this sense, germ line genes are, as Dawkins says, potentially immortal. The somatic line genes in the cells that form, say, a koala's ear lack this potential for immortality. They may still be ancestors to a long lineage of copies, but not to indefinitely many. The lineage cannot outlive the individual koala. However, this germ/soma distinction does not exist in all animals, and is not applicable to plants.

Some replicators are loners. *Outlaw genes* are replicators with characteristics that promote their own replication at the expense of the other genes with which they share an organism's genome. Outlaw genes are relatively uncommon, but not unknown. One set of examples is sex ratio distorters. In most circumstances, selection on individual organisms favors a sex ratio of 50/50. But not all genes are equally likely to end up in each sex. Most of the genetic material of complex animals like humans is organized into *chromosomes*. In our somatic cells, we have 46 chromosomes, organized into 23 pairs; these are our diploid cells. Each gene, then, exists in two copies, one on each chromosome of a pair. Sometimes the two copies in a chromosome pair are identical, in which case the organism is *homozygous* with respect to that gene. Sometimes they are different, in which case the organism is *heterozygous*. When the sex cells (the *gametes*) are formed, by a process called *meiosis,* the chromosome number is halved. In meiosis, each chromosome in the sex cell draws its genetic material from the two paired parental chromosomes. So our sex cells have 23 chromosomes. They are haploid cells, in contrast to our diploid somatic cells with their 23 chromosome *pairs*. A fertilized egg results when sperm and ovum fuse, each contributing 23 chromosomes and thus building a new diploid cell with 23 chromosome pairs.

In most cases, when a 23-chromosome cell is made from a 46-chromosome cell, any particular gene in the parental cell has a 50/50 chance of making it to a sperm or an egg. But not in every case. Mammals like us have a sex-determining process that depends on the nature of one of the 23 chromosome pairs. A mammal with two X chromosomes is female; a mammal with an XY pair is male. No gene on the Y chromosome ends up in a female, because females do not have Y chromosomes. So there is selection at the gene level for any mutation on the Y chromosome that biases the sex ratio toward males, even if that mutation reduces the fitness of the organism. Males have an XY sex-specifying chromosome pair, so they produce some X-carrying sperm (daughter-makers) and some Y-carrying, son-making sperm. Now imagine a mutant gene on the Y chromosome that biased the sex ratio of a male's offspring in favor of males. Suppose it produced fast-swimming sperm, or sperm that somehow poisoned X-carrying sperm. There would be selection at the gene level in favor of speedy-Y sperm or sabotaging-Y sperm genes, even if the resulting males reproduced less successfully (being too numerous) than females, or even if speedy-Y males were less fit than other males. The proportion of speedy-Y males in the population would depend on the balance between sperm-level interaction, which would favor the speedy-Y gene, and organism-level interaction, which would penalize it. If speedy-Y males were very weedy, there might not be many in the population at all. But there would still be more speedy-Y males than one would expect from their fitness *as an organism*. Their numbers would still be boosted by sperm-level interaction.

There can be female-biasing outlaws, too. For though most of our genetic material is organized into those 46 chromosomes, not all of it is. Eukaryotes are thought to have originated in a symbiotic union of prokaryote cells, one living within the other. One legacy of this evolutionary history is the existence of *mitochondria*. These are structures outside the cell nucleus that contain their own genetic material. They are believed to be the reduced remnant of formerly independent prokaryotes. Mitochondria are typically inherited maternally: your mitochondria are all inherited from your mother. So there would be selection in favor of any mitochondrial mutation biasing the sex ratio toward females, even if there were deleterious consequences for individual fitness. So sex ratio distorters are examples of genes that do not have the same chance of replicating as all the other genes in an individual organism. A gene that (1) makes an individual more likely to have female offspring and (2) is copied into all of that individual's female offspring has a higher fitness than the individual's other genes, and a much higher fitness than any gene in the same individual that is copied only into its male offspring.

A second class of outlaw genes is *meiotic drive* or *segregation distorter genes*. As we noted above, when the gametes are formed in a sexually reproducing organism, meiosis halves the normal complement of chromosomes, and normally each allele on each chromosome has a 50/50 chance of being copied into the gamete. Segregation distorter genes bias this lottery in their favor in a variety of chemically complex ways. Such genes improve their chances of

Box 3.2 Gene Selectionism and Genetic Determinism

Gene selectionism is committed to the idea that genes make a distinctive contribution to building interactors, but it is not committed to genetic determinism—the view that only genes matter. As we shall see in chapters 4 and 5, there is considerable controversy over what commitments gene selectionism must make to the role of the genes in development. Its critics claim, and its advocates deny, that gene selectionism is committed to the existence of a relatively simple and regular relationship between genes and phenotypic traits. But no one thinks that gene selectionism is committed to the view that the effects of genes on traits are insensitive to the environment. Whatever problems gene selectionism faces, genetic determinism is not one of them (see Dawkins 1982, chap. 2).

making it to the gametes. Hence a segregation distorter gene will be fitter than the corresponding gene on its paired chromosome (unless matched with an identical twin). But in increasing their own fitness, segregation distorters often decrease the fitness of the organism carrying them. Organisms homozygous for segregation-distorting genes are often sterile (Godfray and Werren 1996; Hurst, Atlan, and Bengtsson 1996).

A comparatively benign example of an outlaw is *junk DNA*. Plenty of genetic material exists that is never transcribed into protein, nor, as far as we know, does this noncoding DNA make any other contribution to development. We cannot rule out the possibility that some or all of this allegedly junk DNA will turn out to play some role in development. But there is another option: perhaps this DNA is passively parasitic, hitching a ride, generation by generation, on the replication of functional DNA.

One avenue through which the idea of outlaw genes has been explored is a thought experiment about a "green beard" gene. This hypothetical gene has two phenotypic effects: it causes its bearer to grow a green beard, and it causes its bearer to be willing to sacrifice its own reproductive prospects if in so doing it can sufficiently boost the reproductive prospects of another green-bearded individual. Thus the green beard gene acts to sacrifice its own replication prospects, and the replication prospects of all the other genes with which it shares a home, if by doing so it can secure still greater replication of another copy of itself. Such a gene could increase in frequency in a population despite its potentially adverse effects on its own interactor.

But outlaws are the exception rather than the rule. Most replicators are organism builders, and succeed or fail because of their contribution to the organisms they build. Replicators do compete with one another, but not usually as lone wolves. The typical replicator's strategy is to participate in concert with others in the construction of interactors that promote their copying. These replicator combinations are often thought of as programs, or as carrying information about interactors that guide their construction. In this view, the typical replicator codes for characteristics of interactors— organisms, colonies, populations. These interactors reproduce, and their differential reproduction causes the differential copying of the replicators. A replicator is more likely to be copied if its presence makes an interactor more likely to reproduce.

The success or failure of one replicator lineage has implications for the success or failure of others. Rabbit gene lineages in Australia are, in a sense, in competition with sheep, kangaroo, and wombat gene lineages. There is limited ecological space for grazing animals, and so limited ecological space for their genes. Thus the success of a rabbit-building gene lineage may blight the prospects of gene lineages that produce wombat fleas. It's a crowded and interconnected world, and the triumphs of one lineage will send causal ripples into many corners. Flea genes and rabbit genes, though, are not irrevocably fated to be competitors. Quite often their fates will be independent of one another, and sometimes, over evolutionary time, different gene lineages can become allies. Many fungus genes are allies of tree genes, because there are many *symbiotic* associations—associations of mutual benefit—between fungi and trees. Such associations are also common between ants and trees, so many ant genes are allies of tree genes. Equally, genes that travel together in the same organisms, and hence which typically succeed or fail together, are natural allies. As noted in the discussion of outlaw replicators, such genes are not always allies, but they usually are.

However, there are cases in which competition is inescapable. Some gene lineages in organisms that reproduce sexually are in direct competition with one another. The different alleles of a gene are the different DNA sequences within a species that one can find in the same slot—known as a *locus*—on a chromosome. These alternative alleles are rivals for particular loci on the chromosomes of that species. Triumph for one allele means extinction for the others. For example, Australian magpies (a crowlike species quite unlike European magpies) are cooperative breeders, and live in extended families that defend their territory with some vigor. If a particularly aggressive type of magpie community founds new colonies at a greater rate than other types, that differential reproduction will cause differential replication of the gene(s) that codes for that increased aggression. The lineage of copies of the aggression gene will become bushier than that of its more peaceable rivals. The alternative alleles for that chromosome slot—alternatives that, in that slot, in that environment, result in less aggressive magpie communities—will come to exist in a smaller proportion of the magpie population. Their lineages will become thinner.

So the struggle between rival replicators is carried out by collectives of replicators constructing interactors—often organisms—that mediate both their interaction with the environment and their further replication. Well-built interactors cause more frequent copying of their replicators. It's important to emphasize this dual vision of replication intersecting with interaction in any discussion of gene selectionism. It is easy to read Dawkins's *The Selfish Gene* as a defense of the idea that only genes matter in evolution. Indeed, it is probably true that defenders of gene selectionism have vacillated between the idea that genes are the primary agent in evolution and the idea that evolution is a dual process of replication and interaction. But the actual doctrine of gene selectionism need not emphasize replication at the expense of interaction. Interactors—organisms—are a vital part of the gene selection process, for it is interactors that cause differential replication. The idea that the gene is the unit of selection does not deny the reality or importance of organisms. Organisms are not epiphenomena of the genes. (This more balanced perspective is clear in Dawkins 1982; Hull 1988; Williams 1992.)

Gene selectionism is an important challenge to the received view of evolution. In this chapter we outline three arguments in favor of the view that it is really genes that are selected. In the following chapter we look at a critical counterargument that tries to return the organism to center stage. We then consider, in chapter 5, an alternative to both these ideas, and also consider whether we really have to choose between these different visions of the evolutionary process. An alternative is to regard them as different frameworks in which to represent the same facts.

3.2   The Special Status of Replicators

The most direct argument for gene selectionism is that it follows from the distinction between replication and interaction and from the point that selection must be cumulative. Both Williams and Dawkins take this view:

> The natural selection of phenotypes cannot by itself produce cumulative change, because phenotypes are extremely temporary manifestations. They are the result of an interaction between genotype and environment that produces what we recognize as an individual. . . . Socrates consisted of the genes his parents gave him, the experiences they and his environment later provided, and a growth and development mediated by numerous meals. . . . however natural selection may have been acting on Greek

phenotypes in the fourth century B.C., it did not of itself produce any cumulative effect. The same argument holds for genotypes. (Williams 1966, 23)

An individual body seems discrete enough while it lasts but alas, how long is that? Each individual is unique. You cannot get evolution by selection between entities when there is only one copy of each entity. Sexual reproduction is not replication. Just as a population is contaminated by other populations, so an individual's posterity is contaminated by that of his sexual partner. . . . Individuals are not stable things, they are fleeting. Chromosomes too are shuffled into oblivion . . . but the cards themselves survive the shuffling. The cards are the genes. (Dawkins 1989b)

Two threads can be distinguished in these arguments. First, complex adaptation evolves only through cumulative selection; hence, it involves persistence. Only replicators persist. Only they, by forming lineages of nearly identical copies, are exposed again and again to selection. Organisms, Williams and Dawkins argue, are not replicators: they are not copied. This seems particularly clear in the case of sexually reproducing organisms. They may have offspring, but they do not have copies. But not even asexually reproducing organisms are really copied. A change in the phenotype—for example, an aphid accidentally losing a leg—does not reappear in that individual's descendants. The information flow across the aphid generations will not preserve information about that aphid's body, except insofar as that information is contained in the aphid's genome. The aphid genome is a replicator, but the aphid itself is not (Dawkins 1982, 97–98).

The second thread that can be distinguished in Williams's and Dawkins's arguments is the idea that because organisms do not form lineages, they cannot be the beneficiaries of adaptation. Such benefits accrue only to a lineage. A trait is an adaptation if it has promoted the copying of the genes that code for it—the replicators that invented it. Evolutionary benefit is just an effect of interaction that projects the replicator lineage into the future. The winners and losers in evolution are replicator lineages. Though organisms are sometimes the bearers of adaptations, they are not the beneficiaries of adaptations.

We think there are powerful and persuasive arguments for the gene's eye view of evolution, though neither of us would accept the whole package without considerable modification. But we do not think that gene selectionism is a simple corollary of the replicator/interactor distinction and the cumulative nature of evolutionary change. We think this argument miscarries in two ways.

First, selection can be cumulative even when the entities that are being selected do not form lineages of copies. Elliott Sober, a longtime critic of gene selectionism, thinks that selection acts on the traits of organisms in a population. So, for example, there must have been selection for a certain array of brown streaks in Australasian bittern plumage patterns (see fig. 2.3). For this pattern to be created, fine-tuned, and then made universal in a bittern population, selection must act in roughly the same way generation after generation. Certain traits must be present in the population again and again. Yet these traits are not replicators, and do not form lineages. A plumage pattern does not make a copy of itself. Plumage patterns recur again and again because what meiosis breaks up, it also puts back together. The gene combinations of streakily plumaged birds would be dissolved by meiosis when they bred, but selection would make those combinations more frequent in the next generation. Gene combinations for streaky plumage would be put back together with increasing frequency, most often in the offspring of birds with those very combinations. So though we cannot trace ancestor/descendant links between phenotypic trait tokens in the way we can between gene tokens, there can be selection for a particular pattern again and again through the generations, until that pattern is characteristic of the species (Sober 1984b, 9.1).

The selection of a "lineage" of traits in the manner just described can be remarkably independent of selection of any underlying lineages of replicators. Frederick Nijhout and Susan Paulsen have shown in some detail how this could work when several genes are involved in making a butterfly wing pattern. In their model, there is such strong selection for a certain wing pattern that eventually every individual in the population has it (it reaches *fixation*). There is selection for some genes in the early stages of this process, but against them in the later stages. The mechanisms underlying this apparent paradox are quite simple. For example, if the wing pattern requires enough, but not too much, of a certain chemical, genes that reduce the rate at which this chemical is metabolized into other chemicals can enhance the pattern early on when the chemical is scarce. However, the very same genes may disrupt the pattern later, for selection for the pattern may result in genes that synthesize the chemical becoming more common in the population, and hence a gene that blocks its metabolizing can cause too much of a good thing (Nijhout and Paulsen 1997).

The second way in which Dawkins's and Williams's argument for gene selection miscarries is that it begs the question. Dawkins denies that whole organisms are replicators because they fail to pass on their *genetic structure* intact. But this assumes the primacy in replication of the genome. Parents pass on many of their traits to their offspring. Even sexually reproducing organisms standardly pass on morphological, physiological, and behavioral structure. So the idea that genes, but not organisms, are potentially immortal and hence the potential beneficiaries of selection merely assumes that the copying relation between gene tokens is identity-preserving in a way the parent-offspring relation is not. Dawkins's rejection of the idea that asexual organisms are replicators is even harder to defend. It is true that *most* alterations to the phenotype of an asexual organism are not passed on, but neither are *most* alterations to the genetic material! That is what the extensive proofreading mechanisms built into the machinery of DNA replication are for. Most genetic mutations are deleterious, and there has been strong selection for mechanisms that reduce the rate at which mutations occur. Moreover, some alterations to the *phenotypes* of asexual organisms *are* passed on. A well-known example involves an experiment in which a portion of the cortex of a paramecium was cut out and pasted back in so that its cilia faced in the opposite direction. This trait was inherited by the offspring of the paramecium (Maynard Smith 1989b, 11).

So even if we assumed that selection must act on lineages of copies, it would not follow that only genes can be selected. Lineages of genes persist and are subjected again and again to selection, but organisms and phenotypic traits form lineages too, which persist through time and are pruned by the vagaries of selection and chance. So a defender of gene selection rejecting the idea that organisms are replicators must argue that preservation of genetic structure has a significance unmatched by parent-offspring similarities. At the very least, the argument from cumulative selection and the conceptual distinction between replication and interaction to the idea that germ line gene lineages are the units of selection needs an extra premise.

This extra premise is often supplied by a view of evolution called *molecular Weismannism*. It has a standard diagrammatic representation: we see causal arrows that lead from genome to genome, with causal side branches leading off to organisms (see, for example, Griesemer and Wimsatt 1989; Maynard Smith 1993, figure 8). These are the only causal arrows in the diagram. There are none that lead from organism to genome, or from organism to organism directly. Molecular Weismannism is therefore a denial that organisms form lineages. If there is no causal arrow from parent to offspring, an offspring is surely no copy of the parent.

There is an important truth that this diagram hints at: there is a regular relationship between the structural genes an organism carries and the proteins made in that organism. If there is a change in a structural gene, that change can change the proteins made in the organism. But if there is a molecular accident that changes a protein in that organism, that change in the

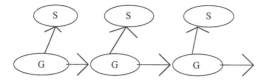

Figure 3.1 A diagrammatic representation of molecular Weismannism. G = genes; S = soma, or body (organisms).

protein will not cause a change in that organism's structural genes, nor in the structural genes of its descendants. So there is an asymmetry of some kind between an organism's genes and its proteins. Even so, this diagram is puzzling. There are very many ways in which parents affect their offspring so as to make them more like themselves. Some of these are recent scientific discoveries. Cell membranes, for example, cannot be constructed without a pre-existing membrane template: only membranes make membranes. Gene activation in the early development of the embryo is controlled by substances in the cytoplasm of the egg to such an extent that this phase has been described as being under the control of the mother's genome rather than the offspring's. And parents attach methyl groups to the DNA of their offspring, which block the expression of certain genes throughout the offspring's life. Other ways in which parents affect their offspring to make them more like themselves are far less recondite. Many parents feed and nurture their offspring, an activity that has an important role in making the offspring like the parent. So the puzzle is, why would anyone think that parents make their young similar to themselves only by passing genetic material on to them? Why would anyone think that this is even a useful simplification? We shall return to this problem in chapter 5.

So, drawing these two threads together: first, it is not obvious that cumulative selection requires lineages of copies. Second, despite the kernel of truth in molecular Weismannism, reproduction may be a copying relationship between interactors. As it stands, the simple and decisive argument for gene selectionism stated above seems to miscarry. Indeed, we do not think that there is any knockdown argument for gene selection. We do, however, think that there are persuasive arguments in its favor. These arguments, in different ways, turn on an appeal to generality. Their common thread is that selection on the traits of individual organisms, in a population of competing individuals, is only one variety of evolutionary change. So a picture of evolution with this kind of selection as its near exclusive focus encourages us to miss much that matters. Gene selectionists argue that if we accept their conception of evolution, apparently disparate phenomena can be grouped together—they can be seen as particular instances of a more inclusive process. This appeal to generality takes two distinct forms in the gene selection literature. One depends on the organization of the biological world into relatively distinct levels. The other questions the paradigmatic status of the organism. These arguments are the subject of the next two sections.

3.3 The Bookkeeping Argument

In section 2.3, we sketched a "hierarchical" conception of evolution that recognizes selection at distinct levels of the biological hierarchy. One version of gene selectionism's appeal to generality fastens onto the same idea, but suggests that gene selection gives us a common currency for representing, comparing, and explaining evolutionary changes. We can track evolution by tracking the fate of genes; gene fate "keeps the books" of evolutionary change. Whenever selection acts, it can be thought of as selecting replicators with particular phenotypic effects over rivals with different effects. However, as we shall see, these phenotypic effects are not always effects on the organism carrying the genes.

Junk DNA, for example, is untranscribed. It is never used as a template for constructing RNA. It plays no role in protein synthesis. If junk DNA is preserved by selection, that selection cannot be via its effects on organisms, for it has no effects on organisms. There can be selection for junk DNA only through the immediate chemical properties of the genes themselves. If magpie communities compete with one another, and if aggression is an adaptation of the community, the evolution of aggression would not be a consequence of selection on individual organisms. But magpie aggression is no problem for the gene selection framework. Selection is picking aggression genes in place of peaceable genes, where both genes express themselves as community traits. All selection is selection for genes by virtue of their phenotypic powers. The adaptations of cells, organisms, and groups are the means by which genes battle for their replication.

Those skeptical of gene selection have responded to this particular appeal to generality in three ways. One response, which we will consider next, is to give a counterexample: an evolutionary change not captured by gene selection. A second is to deny the heuristic advantage of the unified picture gene selection offers; that is, to deny that it provides a productive way of thinking about evolutionary problems. We will consider this response shortly. A third response is to argue that this alleged unified picture is wholly bogus, as genes do not really have the phenotypic powers by virtue of which they are supposedly selected. This idea is central to the attack on gene selection outlined in chapter 4.

Elliott Sober has chosen the counterexample response, though not exclusively. He argues that there is a well-known phenomenon that cannot be captured in the language of gene selection: *heterozygote superiority*. The textbook example is the sickle-cell allele found in some human populations. Homozygotes, with two copies of the sickle-cell allele, develop sickle-cell anemia. They rarely survive infancy, for they cannot produce normal hemoglobin. But in some African populations, heterozygotes, with one normal and one sickle-cell allele, survive better than homozygotes with two normal alleles. They produce functional hemoglobin, yet they can resist malaria better than normal homozygotes. So being heterozygous for the sickle-cell allele in malarial environments is an adaptation, but one that resists the gene's eye perspective. Only pairs of genes can be, or can fail to be, heterozygous. So Sober argues that there is no gene's eye story to account for heterozygote superiority.

Sterelny and Kitcher (1988) reply that at the level of the gene, heterozygote superiority can be seen as an ordinary case of *frequency-dependent selection*. The selective forces acting on a trait can depend on the frequency of that trait. The fact that many species produce 50% male and 50% female offspring is a result of frequency-dependent selection—not a sign of God's benevolence, as claimed in the seventeenth century by the pioneers of social statistics. If the sex ratio of a species were biased toward males, selection would then favor individuals that produced female offspring. Suppose, for example, that the current generation of kiwis somehow gives rise to a generation of chicks that is male-biased, either by statistical fluke or because the females suffer some sex-specific disaster. Imagine, say, that the chick generation sex ratio is 70 males to 30 females. Since each kiwi in the following generation— the "grandchick generation"—will have exactly one male and one female parent, it follows that the average female chick will have more offspring than the average male. So the kiwis of the current generation that produce female chicks will on average have more grandchicks than those that produce males. Having female chicks will be favored by selection until the sex ratio is restored to 50%. In general, if the sex ratio slips away from the 50/50 ratio, selection will favor the rarer sex. The point to note is that the adaptive value of having female chicks depends on the frequency of that trait. Similarly, Sterelny and Kitcher argue that the sickle-cell allele is favored by selection in malarial environments when it is rare, and penalized when it is common. For when it is rare, it will usually be advantageously paired with a normal allele, but when it is common, it will often be disadvantageously paired with another copy of itself.

In *The Extended Phenotype* (1982), Dawkins generalizes this response. He suggests that selection processes that appear to be operating at higher levels in the hierarchy of evolutionary units can often be understood as frequency-dependent processes at lower levels. Imagine moths with stripes on their wings, stripes resembling the grooved bark of the trees on which they rest. In some moths, the stripes run parallel to the moth's body. In others, they run perpendicular to it. So a moth is camouflaged only if it positions itself appropriately when it perches. Moths with parallel stripes must perch vertically; moths with perpendicular stripes must perch across the grain of the tree's bark, at right angles to the vertical. Let us assume that there is a gene complex responsible for parallel stripes together with vertical perching, and another gene complex responsible for perpendicular stripes together with right-angled perching. We could suppose that these complexes have been selected for as a unit. Among the pool of gene complexes, only those two have survived. Complexes that resulted in mismatches between perching behavior and stripe pattern have been weeded out by natural selection. We can think of selection as operating not on individual alleles, but on higher-order gene complexes as the result of a higher-order selection process. This view would parallel Sober's view of heterozygote superiority: there is no selection for the allele for vertical perching, because the effect on fitness of that allele is not uniform; it is sometimes good, sometimes bad.

Dawkins's response is to invoke frequency-dependent selection. If the allele for vertical perching finds itself in a local population in which the allele for parallel striping is rare, there will be selection against it. If, however, the parallel striping allele is more common than the alternative allele, there will be selection for the vertical perching allele. As the vertical perching allele becomes commoner, that will bring into play selection for parallel striping, which will increase the selection pressure for vertical perching, and so on. So frequency-dependent selection on initially accidental imbalances can produce harmonious gene combinations (Dawkins 1982, 241). This idea recurs repeatedly in the defense of gene selection. Selection pressures act relative to background conditions, including the genetic background. The context sensitivity of genes is no different from that of other proposed units of evolution.

The debate over the legitimacy of invoking context sensitivity and frequency-dependent selection is a complex one. Dawkins himself recognizes that some restrictions are required to avoid the conclusion that there are just four genes, the single nucleotides that all DNA is made up from, whose fitness (dependent, of course, on the particular background conditions in which they occur) explains all evolutionary change: "Have we, then, arrived at an absurdly reductionistic *reductio ad absurdum?* Shall we write a book called *The Selfish Nucleotide?* At the very least, this is not a helpful way to express what is going on" (Dawkins 1982, 90). We tackle these issues in section 4.1.

Thus we think that the counterexample response to gene selection fails. In his more recent work, especially with biologist David Sloan Wilson, Sober has emphasized a different response to gene selection, though one that has always been present in his writing (Sober and Wilson 1994, 1998; Wilson and Sober 1994). He accepts that, in a weak sense, gene selection is a general conception of evolution. All evolutionary change can be represented as a change in gene frequencies over time. But Sober thinks that gene selection purchases this generality at the price of neglecting critical information about *why* gene frequencies change. He thinks this information cannot be captured using only the language of genes and their properties. Therefore, though it's true that gene selection is a general conception of evolutionary change, this is an uninteresting truth. The critical problem in evolution is not identifying the replicators, but identifying the interactors.

We by no means agree that identifying the nature and role of replicators is a trivial problem. Indeed, we think gene selectionism clearly goes astray in assuming that nearly all replicators are genes. Not all replication is genetic replication; not all inheritance is mediated through gene lineages. One clear example is the nongenetic mechanisms by which symbionts are transmitted across generations. Many organisms depend on symbiotic relations with creatures that live on or in them. For example, no animal can digest cellulose, the material that makes up plant cell walls. Animals that eat cellulose depend on bacteria in their guts to digest the cell walls for them. Parents have various ways of transmitting such symbiotic organisms to their offspring. They pass across the generations whole functioning populations of symbionts, not just symbiont DNA. This is not just true of cellulose eaters. For example, many arthropods transmit in their eggs, sometimes by very precise mechanisms, microorganisms on which they depend for growth (see for example Morgan and Baumann 1994). The queens of leafcutter ants have special adaptations for carrying the symbiotic fungi they rely on for food when they found new nests. All of these cases involve nongenetic intergenerational copying, for more than symbiont DNA is passed across the generations. For these reasons, we think that gene selectionism should be generalized to "replicator selectionism."

Most of the conceptual innovations of gene selectionism would be preserved by a more general replicator selectionism, though there would remain an important empirical debate about the extent and importance of nongenetic replication. The main effect of this change would be on the rhetoric of the gene selectionist movement. Although Dawkins specifies the nature of replicator and interactor in ways that do not commit him to the view that genes are the only replicators, he is inclined to write as though they are. The only other replicators he has endorsed are *memes*—units of cultural replication. Hull is much less inclined to suppose that genes are the only replicators that matter. Genes are the paradigmatic replicators, but they are not the only ones. In asexual or genetically homogeneous populations, much larger units—chromosomes, genomes, or even the organism itself—qualify as replicators (Hull 1988). As we have seen in this chapter, membranes, cytoplasmic traces, DNA methylation patterns, and symbionts all have a case for being considered replicators as well.

3.4 The Extended Phenotype

The received view sees selection as acting on the traits of individual organisms in a population, rewarding some of those organisms and penalizing others. It places organisms at center stage in the evolutionary drama. So, as we noted in section 2.3, it is a very conservative, common-sense view of evolution. It takes a common-sense category—the concept of an organism—and uses it as the key *natural kind* in evolutionary biology. So if this category is seriously problematic, then the received view inherits those problems. Putting this idea in the most provocative way possible: if there is no such thing as an organism, then the received view cannot be true, but gene selection might be.

The received view is committed to the idea that the category of individual organisms is a natural kind. That is, it assumes that there is a cohesive characterization of what it is to be an individual organism, and that most living things on which selection acts fit that characterization. Yet, as Hull emphasizes, within evolutionary biology, "organism" does not name a natural kind. In addition to integrated multicellular animals like ourselves, there are, at a minimum, single-celled organisms like bacteria, colonial organisms like jellyfishes and corals, and plants, all of which challenge the conventional idea of an organism. From an evolutionary point of view, these are all quite different sorts of things. Gene selectionism, in contrast, is not committed to endorsing the common-sense categorization of the biological world. How well the categories of replicator and interactor map onto common-sense categories is an open empirical question. So we have here a negative argument for the gene's eye view: The received view is committed to the claim that organisms constitute a cohesive natural kind, but the gene's eye view is not. Similar considerations support two positive arguments as well. The first turns on Dawkins's idea of the *extended phenotype:* that the traits by virtue of which genes are selected need not be traits of the organisms in which they are contained. The second is the idea that the received view makes it harder to see some important evolutionary problems. In the rest of this section, we will flesh out these ideas.

Let us begin with the concept of the organism. The common-sense conception of an organism is that of a cohesive and integrated body. But many examples fail to fit this conception cleanly. Many plants can survive major losses of parts quite happily, and animals that metamorphose move through an undifferentiated stage. Moreover, slime molds, corals, and some medusae ("jellyfishes") have life cycles in which the cells live independently for some of the time, and come together into a single physiological unit for some of the time.

There have been two important attempts to revise this focus on organizational integration, but these have problems of their own. First, Daniel Janzen (1977) argues against emphasizing physical cohesion. He claims that we should see clones of genetically identical but physiologically distinct units as single organisms: for example, the clonal population of aphids that forms from one asexually reproducing mother in a single season. In some types of plants the genetic and the physiological unit come apart spectacularly. The *ramet* is the physiological unit; a *genet* is the cluster of ramets derived from one fertilized cell. A park full of dandelion ramets may contain only one dandelion genet. However, Janzen's criterion leaves many cases open, for obligatory sexual reproduction and invariable asexual reproduction are two ends of a continuum, not discrete alternatives into which all organisms can be unambiguously sorted (Templeton 1989).

Dawkins defends an alternative to both Janzen's idea and the traditional picture: An organism is that which persists through a developmental cycle from single-cell bottleneck to single-cell bottleneck. This cycle is important, for a mutation can make important differences to the whole organism when development is funneled through such a bottleneck. But developmental bottlenecks are a matter of degree, and can occur at various points in an organism's life history. Some organisms change structure dramatically during their lives. A caterpillar that pupates loses its structure before emerging as a moth or butterfly. A good deal of the internal physical complexity of the caterpillar's organization is lost before a new one emerges, and a genetic change acting during this period could surely result in an overall change to the emergent moth. This reduction of structure at an intermediate stage of the organism's life cycle can take a very extreme form in some parasites. In the life cycle of parasitic barnacles of the *Rhizocephala* group, an individual that attaches itself to a host crab must penetrate the crab's exoskeleton, and in doing so, some species reduce themselves to a single cell, from which the parasite develops anew once inside the host. It thus has two complete cycles of development from a single cell before reaching a reproductive stage (Gould 1996e). The "alternation of generations" in some plant groups also involves multiple developmental cycles between reproductive attempts, for in some plants gametes grow into complex structures before fertilization, rather than existing as single-cell propagules until they fuse into a fertilized egg. This pattern is particularly striking in mosses, in which the haploid phase and the diploid phase can involve structures of similar size and complexity (Niklas 1997, 161) (see figure 2.1).

Putting all this together, we get the following picture: Vertebrates are integrated units with clear physical boundaries. To a reasonable approximation, their germ line cells are genetically uniform. They develop from a single cell, and reproduce by constructing single-celled, genetically distinct gametes. These three properties—cohesion, genetic uniformity, and cyclical development—are each important. But they need not co-occur; indeed, their co-occurrence is not typical. Our paradigms of evolution—bat echolocation, bittern plumage patterns—are organized around vertebrate examples that are not typical of the living world. There seems no compelling reason to choose any of these three characteristics as definitive of the organism as an evolutionary kind. And even if we do choose one, the organism so characterized will not typify life. For each of the differing ways of defining organisms, there are many life forms that the definition will not clearly fit. This issue is difficult, important, and unresolved. We shall return to it in section 8.6.

But the received view faces yet another problem. Gene lineages are the beneficiaries of adaptation. The effect of a successful adaptation is to promote the replication of the genes that built that adaptation. But what is adapted? What carries or bears adaptations? The received view supplies its own answer to this question: Organisms are adapted, and by virtue of selection on organisms for their adaptations, organisms transmit, with greater or lesser success, their genes, and hence those adaptations, to successive generations. Setting aside worries about the concept of the organism, often this idea is right. But as Dawkins (1982) argues, it is not always right. There is an important group of cases in which the effect of a gene by virtue of which it is replicated is not an effect on the organism in which it resides. Genes have *extended phenotypes:* their replication-enhancing effects may be on organisms other than the one in which they are replicated (Dawkins 1982, 1989b).

Perhaps the most vivid and powerful examples of extended phenotypic effects involve the actions of parasite genes on host bodies. There are many gruesome examples of such gene actions. The fungus *Entomophthora muscae* infects and kills domestic flies. It causes dead females to develop features, such as a distended abdomen, that are sexually attractive to male flies, so that the necrophile males become infected with the fungus (Moeller 1993). So the adaptive effect of certain fungus genes—the effect that explains their prevalence in the fungus gene pool—is an effect not on fungus bodies, but on the behavior of male flies. These genes in the fungus cause flies to come to sites of infection. Of course, they have this effect only in the context of the fungus and its genes, the environment, and certain features of the fly. Their effect is context dependent in many and varied ways. Even so, this effect explains why those genes are around. Dawkins mentions a parasitic fluke that operates in a similar way, manipulating the behavior of its intermediate host (a snail) so that it is more likely to be eaten by its ultimate host (Dawkins 1982, 212–13). Parasitic barnacles of the *Rhizocephala* group take over the behavior of their crab hosts even more completely, suspending the crab's molt cycle (which might shed the parasite), biochemically castrating and feminizing it (if it is male), and subverting its brood care behaviors so that they support the parasite's eggs rather than the crab's own eggs (Gould 1996e, 15–16). In all of these cases the altered behavior and morphology of the parasites' hosts are rightly seen as adaptations of parasite genes.

A somewhat simpler example of an extended phenotype, and one that allows a direct comparison with a standard adaptation, is house construction by caddis fly larvae. These larvae typically live on the bottoms of streams, and glue together assorted debris to form a "house" in which they live. These houses plainly serve the same protective function for the larva that is served by a mollusk's shell. The replicator/interactor distinction enables us to see the evolutionary identity of house building by the caddis with shell secretion by a mollusk. The gene's eye view allows us to treat like cases alike; to see the genes for both as building adaptations for physical protection. We cannot treat like cases alike under the received view, for the caddis house is no trait of the caddis.

Defenders of the received view might reply by relocating the phenotypic effect of the gene. They might suggest that the phenotypic effects of the parasite genes are the chemical signals that subvert the host's behaviors, rather than those subverted behaviors themselves. These chemical signals are features of the conventional parasite phenotype. The caddis gene's adaptive effect is house-building behavior—a behavioral trait of the larva—rather than the completed house, which is not. We think this objection is unconvincing. There are many links in the causal chain from one replication of a gene to the next. These range from protein products through effects on the organism's social and physical environment. There are legitimate explanatory interests in each of these links. But there is good reason for singling out effects on the hosts' behavior. The adaptive value of, say, crab castration and feminization is insensitive to differences in the biochemical means through which those changes in the host are effected, but it is sensitive to variation in the effectiveness of castration and feminization itself. Incomplete feminization, for example, would be bad news for the parasite, since its own brood would then receive less effective care. Putting the point rather too metaphorically, the barnacle gene does not care how it changes the crab's repertoire; it does not care about the intermediate links in this chain, just so long as they are effective in securing the requisite alterations in the crab. The adaptive effect of the parasite's genes is their effect on the host's behavior.

We think that Dawkins's case for the extended phenotype and its importance is well made. Organisms are not uniquely privileged targets of selection. They are not the only interactors, and they are not the only bearers of adaptation. The parasitized crab body is surely most strikingly adapted, but not for crab genes. So it is not adapted in the sense that a parasite-free crab—with all its parasite-shedding behaviors—is adapted. Genes have extended phenotypes. Some of their jointly constructed adaptations are aspects of the organisms that contain them, and through which they replicate. But some are not.

In summary, genes have replication strategies that vary in two independent dimensions, and that is why the replicator/interactor distinction is not just a pretentious synonym for the gene/organism distinction. A few genes are outlaws, replicating themselves at the expense of others in the same genome, but most are not. So one dimension is the distinction between outlaw genes and cooperator genes. A second dimension is the site of interaction. Genes come with "arms" of very different lengths. Outlaws can have short arms: their replication-enhancing effects need not be effects on the organism through which they are replicated. Speedy-Y genes have effects on speedy-Y males, but these are not their replication-enhancing effects, so the speedy-Y gene is a short-armed outlaw. The hypothetical green beard gene is a long-armed outlaw: its replication-enhancing effects are effects on the reproduction of other carriers of the green beard gene. Standard cooperator genes are replicated by virtue of their contribution to the adaptive architecture of the organism through which they are replicated. They are not outlaws. If they benefit their own prospects, they benefit those of all the other genes of the organism that contains them. In these cases, the gene's eye and the received views more or less coincide, and if such cases were the only important ones, the replicator/interactor distinction really would be just a redescription of the gene/organism distinction. But some genes have long arms, and are not outlaws. If they enhance their own replication prospects,

but the adaptation is still the behavior, yes?

they also enhance the replication prospects of every gene in their organism. But their adaptive effect is outside the body they inhabit.

A final point that argues for gene selection and against the received view is that a focus on organisms in competition in a population makes certain evolutionary problems harder to see. One particularly striking example is the existence of the organism itself. The existence of organisms is much more likely to strike us as posing a genuine problem if we see evolution from the gene's point of view. From the perspective of the germ line cells, the construction of a body is an enormous investment of resources that might instead be allocated directly to replication. Why is it worth it? Suppose we see a solution to this problem, a reason why a set of genes that join together to construct an organism would each do better than any would do alone. That solves only part of the problem. How could such a collective evolve? What are the intermediate stages like, and what makes them superior to other arrangements for replication? Finally, once the organism has evolved, we still need to explain the stability of the gene alliance. Why are cancers—cells that replicate at the expense of the interests of the rest—not so common as to undermine the viability of indirect replication through massive organismal machines? Dawkins has developed an interesting response to some of these problems, but his particular answers are not central in this context (Dawkins 1982, 1989b). Our point here is that seeing life through the lens of the received view makes us less likely to see the need for such answers.

Further Reading

3.1, 3.2   Williams, Dawkins, and Hull are the central defenders of the gene's eye conception of evolution. Williams's most important statements are in his *Adaptation and Natural Selection* (1966), the first explicit formulation of gene selection, and his much more recent *Natural Selection: Domains, Levels and Challenges* (1992). This recent work defends the idea that selection acts on a variety of replicators. For Dawkins's views, see his *The Selfish Gene* (1976), *River out of Eden* (1995), and, especially, *The Extended Phenotype* (1982). For a more recent version of the same ideas, see Cronin 1991. Dennett (1995) defends a very similar conception of evolution. Dawkins (except in *The Extended Phenotype*) can often be read as emphasizing replication's importance at the expense of interaction; the same cannot be said about Hull (1981; 1988, chap. 11). For concise overviews of different interpretations of the question "What are the units of selection?" see Lloyd 1993 and Mayr 1997. The two leading philosophy of biology anthologies, Sober 1994 and Hull and Ruse 1998, both have good selections on the general issue of the units of selection.

For those who would like more on the empirical background of gene selection, most introductory texts in biology have a chapter that will serve. Two
good, accessible introductions to the basic genetic mechanics are Trivers
1985, chapter 5, and Futuyma 1998, chapter 3. Ridley 1993b is quite technical, but chapter 2 of Ridley 1985 is a short, nontechnical introduction to
the basic mechanisms of heredity. Similarly, Maynard Smith 1989b is technical, but chapter 3 of Maynard Smith 1993 is clear and helpful. An excellent
alternative to all of these, especially for those with no prior background, is
Moore 1993. Part 3 is devoted to the development of genetics, and walks the
reader through the history of genetic ideas to a near-contemporary view.
This book discusses the molecular biology of the gene at greater length in
chapters 6 and 7. Sarkar (1996) discusses the implications of modern molecular biology for what genes are.

Dawkins discusses outlaw genes in chapter 8 of *The Extended Phenotype*
(1982). For more recent, though technical, discussions of outlaw genes, see
Haig and Grafen 1991, Haig 1992, and Beukeboom and Werren 1993. For
a brief, nontechnical review of sex ratio distortion and its significance, see
Werren 1994. An excellent overall review of intragenomic competition is
Hurst, Atlan, and Bengtsson 1996. These papers are also relevant to the issue
of the evolution of the organism. The classic reference here is Buss 1987.
The parallel between a genome and a political collective is developed in
Skyrms 1996. The outlaw status of the green beard gene is evaluated in
Ridley and Grafen 1981. Our tale of speedy-Y genes is only a thought experiment, but there has been empirical work on genetic conflicts between
sperm. However, we understand that claims about sperm competition remain very controversial. For a review of this work by its defenders, see Baker
and Bellis 1995.

3.3, 3.4   The bookkeeping argument is found in Williams 1966 and Dawkins 1989. The criticisms raised in this section are found in Sober and Lewontin
1984, and in Sober 1984b and 1993. Sterelny and Kitcher (1988) reply to these
criticisms, as does Dawkins (1982, especially in chapter 5). See Godfrey-Smith
and Lewontin 1993 for a critique of these responses. Dawkins (1982) puts
forward the advantages of this viewpoint. The argument is carried further in
Sterelny, Smith, and Dickison 1996, which also discuss the prospect of extending the cast of replicators. A good introduction to the science of other
replication systems is Jablonka and Lamb 1995. The implications of these
findings are assessed in Maynard Smith and Szathmary 1995, which defends
a "first among equals" status for specifically genetic replicators.

4

The Organism Strikes Back

4.1 What Is a Gene?

In our view, the most important argument in favor of gene selection is that it offers a very general conception of evolution. That claim is rejected outright by its critics, who accept only a very weak interpretation of the "bookkeeping argument" described in section 3.3. They typically accept that all evolutionary change involves changes in the frequencies of genes (or some expanded cast of replicators). But they deny that we can explain evolution by explaining gene frequency changes. This thought can be captured in a simple analogy: The success or failure of lineages of human phenotypes is fully reflected in changes in the proportions of different surnames in the population. But the forces driving these changes will be largely invisible if we look at the forces acting on surnames: how likely certain names are to be misspelled by ignorant immigration officials, how embarrassing they are, or even the social prestige they confer.

The difference between surnames and genes, of course, is that while surnames play, we conjecture, only a minor role in determining phenotypes, genes play a very significant role. Hence we might expect that genetic change would track the success or failure of phenotypes. The critics of gene selectionism claim that while genes are connected to phenotypes, that connection is so indirect and variable that genes are actually "invisible" to selection. In explaining this idea, we will borrow a distinction from Sober (1984b). He distinguishes between *selection for* some trait and *selection of* that trait. Running water sorts sediments by mass: heavier sediment settles out at higher flow speeds than lighter sediment. So a given flow speed selects for a certain weight: the threshold at which particles will precipitate out rather than flow with the water. But particle weight is often correlated with color, for minerals of different densities have characteristic colors. That is one reason why sedimentary rocks often have bands of different colors running through them. So the flow speed of water *selects for* weight, but there is *selection of* color as well. Sediment color banding is a side effect of selection for particle weight. Similarly, those who think genes are invisible to selection think there is selection for organisms with certain traits, and as a consequence, there is selection of any genes that are statistically correlated with those phenotypes, however indirect their causal contribution to the phenotypes—indeed, whether they are causally connected to the phenotypes at all.

Arguments of this general style go back to Mayr, and they have many recent defenders. Some of these emphasize the variability of a gene's phenotypic expression, and others emphasize the variability of a gene's effect on fitness. But we assume that the only way a gene could have a relatively invariant effect on fitness would be by having a relatively invariant phenotypic effect in the world. So though these arguments have different formulations, we take them to be essentially equivalent.

Particular gene tokens often have determinate effects on the phenotype of the specific organisms carrying them. But in the evolutionary scenarios we sketched in chapter 3—the evolution of genes for magpie aggression, of fungal genes with effects on fly behavior, and the like—we wrote of the phenotypic effects not just of individual gene tokens, but also of *kinds* of genes—of *gene types*. So in discussing the visibility of genes to selection, there are two critical scientific issues on which debate turns. The first is which DNA sequences count as genes. The second concerns the complexity of the relationship between these genes and the phenotypes they help to build.

good question

We begin with a deceptively simple question: What is a gene? In most discussions of genes, a gene is thought of as a functional unit of some kind. *Codons*—sequences of three nucleotides—are the smallest functional units—the smallest meaningful units of the genetic code. Each codon corresponds to a single *amino acid*—the building blocks of proteins. But genes are usually taken to be larger functional units. The most common sense of *gene* in molecular biology is "a *reading sequence*"—that is, a sequence of nucleotides that is transcribed into a piece of messenger RNA that is either translated into protein or used directly in the metabolism of the cell. When we read that humans have 75,000–80,000 genes, this is how genes are being counted.

However, in their explanations of gene selection, Dawkins and Williams introduce an alternative conception of the gene that cuts this link between being a gene and having a specific function in protein construction. They introduce the so-called *evolutionary gene concept*. In this conception, a gene is *any* reasonably short sequence of DNA on a chromosome. The sequence must be fairly short, because long sequences are frequently broken up by crossing over in meiosis, and so are not "potentially immortal." Two DNA sequences are copies of the same gene if they have the same (or similar) sequences (by descent), no matter where they are in the genome. The important feature of the evolutionary gene concept for our current argument is that *any* DNA sequence counts as a gene:

> When I said "arbitrarily chosen portion of chromosome," I really meant arbitrary. The twenty-six codons that I chose might well span the border between two cistrons [functional units of gene action]. The sequence still potentially fits the definition of a replicator, it is still possible to think of it as having alleles. (Dawkins 1982, 87)

With this official definition of "gene" laid out clearly, it is easy to see why critics have argued that they are invisible to selection. If genes are just arbitrary DNA sequences, then most of them will have no more systematic relation to the phenotype than an arbitrary string of letters has to the meaning of a book. It is true that every change in the meaning of a book involves changing the letters in the book, but it does not follow that we can predict the change in meaning from the change in letters. What can you predict from knowing that we added the letter sequence "in" somewhere when preparing the final draft of this book, either as part of a word or on its own?

In *The Selfish Gene,* Dawkins defended the evolutionary gene concept with an analogy between the genes in an organism and the crew in a racing rowing eight. Coaches choose rowing crews by racing all possible combinations of rowers against one another. They choose the rowers who have, on average, had the most success. Dawkins suggests that we look back on all the organisms in which we can find a copy of some stretch of DNA—any stretch of DNA—and take the average fitness of all those organisms as the fitness of that DNA sequence: "a gene which is *consistently* on the losing side is not unlucky; it is a bad gene" (Dawkins 1976, 41; italics in original).

But this claim is mistaken: genes, like rowers, *can* be consistently unlucky. There are innumerable noncoding stretches of DNA that are characteristically found in unsuccessful groups of organisms. These stretches of DNA have no biological effect whatsoever, and so cannot be "worse genes" than similar noncoding stretches in successful species. *Introns,* for example, are parts of genes whose corresponding sections in the RNA transcript are cut out and discarded before a protein is assembled using the transcript as a template. So introns are DNA sequences that make no difference to the phenotypes of the organisms that carry them. Consequently, introns in the dodo were not "worse genes" than the slightly different introns in the corresponding sequences in cockroaches. Yet they were consistently on the losing side.

In effect, the rower/gene analogy breaks down because many possible gene combinations are never tried out. Dodo introns did not get tried out in cockroach genes!

Dawkins's averaging proposal fails because it ignores a central feature of scientific explanation: not every correlation is explanatory. If men born under the star sign Leo had more sexual partners than average, we could appeal to the "attractiveness of Leos" to explain this. The "attraction coefficient" of Leos could be obtained by averaging the attractiveness of individual Leos. The sexual success of Leos as a group could then be "explained" by their attraction coefficient. Yet in reality, the correlation would have to be either a fluke or the result of that date of birth correlating with some nonastrological property (perhaps Leos are the oldest members of their cohort in the education system). In the former case, nothing really explains the success of this group. In the latter case, it is the nonastrological property, the *confounding variable,* that explains their success. A good sign that a correlation is a mere coincidence is that it has no *counterfactual* force. If we cannot say of something that is not an A, that if it *were* an A it *would be* a B, then being an A does not cause being a B. See how this works with an uncontroversial causal claim: Smoking causes heart disease. Neither of us smokes, but if (say) Griffiths were a smoker, he would elevate his risk of heart disease. The fitness of genuinely arbitrary evolutionary genes does not pass this test. It is not true of cockroaches that if they had more dodo-like introns in their genes they would be at greater risk of extinction.

Dawkins recognizes that some absurd consequences follow from the definition of the evolutionary gene. One is that it becomes legitimate to say that there are precisely four genes—the four DNA nucleotides, A, T, G, and C—whose relative allelic fitnesses exhaustively explain the composition of all the DNA in existence. Dawkins calls this the "selfish nucleotide" theory. It works as follows: Each token of A, T, G, or C at a particular point in a particular DNA molecule has occurred in an environment of two other adjacent bases, making up a potential codon. Each of these potential codons has been part of a string of codons, and some of these strings have been real, functional genes. For each individual token of a base, any of the three other bases could be substituted. In some cases the result would increase the fitness of the organism in which the change occurred, in others it would decrease it; in the vast majority of cases it would have no effect on fitness whatsoever. Despite this variety of effects, we could derive an average fitness for a base such as thymine by averaging the fitnesses of the actual organisms in which each thymine token has occurred. Since the fitnesses of these organisms "determine" the number of copies of that thymine token that will be passed on to the next generation, the average fitness figure could be used to predict the proportions of bases in the next generation. Four such average fitnesses, for A, T, G, and C, would "explain" the composition of all the DNA in the world.

Dawkins discusses this "selfish nucleotide" theory in *The Extended Phenotype,* and in this discussion, the evolutionary gene concept is quietly, and we think rightly, buried. Nucleotides are not targets of selection because there is no meaningful sense in which they are in competition. They do not exert phenotypic power in the world that makes them, or copies of them, more likely to be replicated.

> The single nucleotide . . . cannot be said to have a phenotypic effect except in the context of the other nucleotides that surround it in the cistron. It is meaningless to speak of the phenotypic effect of adenine. . . . The case of a cistron within a genome is not analogous. Unlike a nucleotide, a cistron is large enough to have a consistent phenotypic effect, relatively . . . independently of where it lies on the chromosome. (Dawkins 1982, 91–92)

We think that this admission is a positive step for gene selectionism, though it is important to notice that it is not just sequence length that raises a problem for the evolutionary gene concept. A DNA sequence can fail to have a phenotypic effect that influences its replication prospects because it is too short, but that is just one way in which a sequence can fail to have this capacity. Some long DNA sequences are compressed into *heterochromatin,* a form of DNA that is generally not transcribed into RNA or protein. They too exert no phenotypic effect on the world. We will see many other examples in section 6.4.

If gene selectionism really did rely on the evolutionary gene concept and the averaging strategy Dawkins described in *The Selfish Gene,* then the allegation that gene selection does not capture the dynamics of natural selection would be justified. As we have seen, Dawkins tacitly abandons that strategy in favor of the idea that genes are DNA sequences with phenotypic power. But what is phenotypic power? After all, all gene action depends intimately on the action of other genes and on the cellular milieu. As we see it, in response to this question, the gene selectionist road forks. One route takes genes to be real, functional biochemical units—the "genes" that are being counted when you read that humans have fewer than 100,000 genes. These *molecular genes* are the genes molecular biologists make discoveries about:

[handwritten margin note: genes as biochemical units — "molecular genes"]

pieces of DNA with particular properties that cause other things to happen around them. Dawkins adopts something like this strategy when he defines genes as "active germ line replicators."

The second route available to the gene selectionist is to guarantee that genes have a consistent effect on the world by defining a gene in terms of its effects on that world! We have already used such definitions: "genes" for striped plumage in bitterns, "genes" for aggression in magpies, "genes" for hideous manipulations by parasites. Notice that there is nothing suspicious or circular about this idea: poisons are defined in terms of their effects, but there really are plenty of poisons, and they really do explain deaths. They pass the counterfactual test above: had Sterelny taken a poison in his wine, he would have died. Genes in this sense are "difference makers": they are DNA sequences whose presence or absence in a particular genetic and environmental context *makes the difference* between one phenotypic trait and another (Sterelny and Kitcher 1988).

So, in responding to the challenge that genes are invisible to selection, we have two options to explore. We begin with the idea that genes are DNA sequences that exert phenotypic power over their own replication prospects by playing a specific molecular role.

4.2 Genes Are Active Germ Line Replicators

Dawkins defines an active replicator as

> any replicator whose nature has some influence over its probability of being copied. For example, a DNA molecule, via protein synthesis, exerts phenotypic effects which influence whether it is copied. (Dawkins 1982, 83)

The invisibility argument depends on the variability of gene/phenotype relations. Those who deploy this argument think that it applies to molecular genes, not just evolutionary genes. Indeed, many of them clearly assume that gene selectionists are talking about molecular genes (Wimsatt 1980a,b). Their charge is that even gene sequences that make a specific protein do not have an identifiable phenotypic effect by virtue of which they are replicated. For the broader effects of having a particular gene for a given organism are very varied. A single protein can be put to very different uses. Lysozyme, for example, is a protein that is variously used as an antibiotic (in tears), in digestion, and in the production of milk (Geoff Chambers, personal communication). Moreover, a gene's effects will depend on the other genes that are present in that genome. The allele causing hemochromatosis leads to excess iron buildup in men, but not typically in women, who lose iron by menstruation and are typically in much more danger of anemia. It has even been speculated that this allele evolved by selection in women despite its adverse effect on men! As we have already seen in the case of sickle-cell anemia, the effect of a gene on the actual fitness of an organism will depend on many factors, including other genes.

Thus these critics interpret the idea of a "consistent phenotypic effect" in a way that excludes most genes. The general form of the invisibility argument, expressed in different ways by Gould, Sober, Lewontin, Wimsatt, and Brandon, is something like this:

Premise 1. There would be selection for genes if and only if gene action were robust and constant. If a gene G^* is to be selected, individual copies of G^* must have a consistent effect on survival and reproduction, or they will be "invisible" to selective pressures. G^* must have a reliable phenotypic consequence, or the replication rate of the copies of G^* will be a mere average of many unrelated processes. Some will do well for one reason, others for other reasons, and some will not do well at all.

Premise 2. "Beanbag genetics"—the idea that each trait is explained by the action of a particular gene for that trait—is false. There is nothing like a one-to-one correlation between genes and phenotypic phenomena. All traits require the action of many genes, and many genes contribute to the development of more than one trait.

So,

Conclusion 1. Genes do not have constant effects; there is no one thing that a given type of gene does by virtue of which it is visible to selection.

Conclusion 2. Except, perhaps, for certain special cases, the gene is not the unit of selection.

In replying to this argument, Sterelny and Kitcher argue that demanding "constant phenotypic effect" subverts the point of selective explanations. Selective explanations average out idiosyncratic scenarios to capture the central tendency in the dynamics of a population. Most mammal species, for example, include occasional albinos, and these usually have a tough life. There is selection against albinism in sugar gliders, and that is so even though albinos born in zoos are cosseted and protected because of their rarity and appearance. So it seems that even routine, uncontroversial adaptive hypotheses are ruled out by the demand that the feature selected have a constant effect across the range of environments in which it occurs (Sterelny and Kitcher

1988). We could avoid this consequence by subdividing environments ever more finely. One could say that there is selection against albinism in sugar gliders, except in zoos and in the snowy fringes of the species' range. But there is no explanatory point (yet) in regarding zoos, or any other tiny oddities and aberrations within the normal environment of a population, as bona fide environments that reverse the usual selection pressures. That would subvert the point of selective explanations, for as we shall shortly show, selective explanations are robust process rather than actual sequence explanations.

Actual sequence explanations seek to explain the nuances of the causal history of the world we find ourselves in. They explain the contrasts between our actual history and the histories of the nearby possible worlds. For such purposes, the more fine-grained the explanation, the better. But that is not true of all explanations. _Robust process explanations_ reveal the _insensitivity_ of a particular outcome to some feature of its actual history. Thus an explanation of World War I that appeals to the political divisions of Europe is a robust process explanation, seeking to show that some World War I-like event was very probable. The detailed unraveling of diplomatic and military maneuverings is an actual sequence explanation, showing how we got our actual World War I. These explanations are not rivals, and constraints appropriate to one are not thereby appropriate to the other. In particular, while detail is appropriate to actual sequence explanations, it subverts the point of robust process explanations. We wish to characterize all the possible World War I worlds, and since Grey might not be British foreign secretary in some of these, we would not want to identify these worlds by appeal to his doings.

Selective explanations are robust in the same way. Our description of selection for tabbiness on feral Australian cats (2.3) identifies a trend over many possible evolutionary trajectories of that population; it does not specify the precise character of the actual trajectory. Rather, the explanatory hypothesis is that many features of the actual history are not critical to the fixation of tabbiness, but that selection pressure—predation on kittens—is critical. The more we subdivide environments, the more we turn selective explanations into actual sequence explanations that depend on the nuances of the microhistory of the local population. That undercuts their point (Sterelny 1996a).

Moreover, ever finer subdivision of environments would undermine a distinction that is central to the empirical content of evolutionary theory: the distinction between actual and expected fitness. That distinction is important to rebutting the claim that "the survival of the fittest" is a tautology. The charge is that "the survival of the fittest" has no empirical content. Because fitness is defined by survivorship, it's true by definition. This

[handwritten margin notes: "contingency vs. inevitability ↓ level of explanation?"]

[handwritten note at bottom: "survival of the fittest" not a tautology (?)]

charge fails, however, because we have a conception of the *expected* fitness of an organism that is independent of its *actual* reproductive success. Sometimes accidents happen, and the unfit survive. But the tautology problem would become acute if "accidental" variation in reproductive success disappeared as more and more of the actual causal history of particular organisms were regarded as part of their normal, albeit unique, microhabitat.

Let us sum up the state of play. The evolutionary gene concept is an inadequate conception of the gene on which to base gene selectionism. It does not guarantee that gene tokens exert phenotypic power, let alone that each copy of a given gene type exerts a similar kind of effect on its replication prospects. So we need to replace or amend that notion with one that insists that if genes are targets of selection, then they must control phenotypic effects that influence their replication prospects. But what does it mean for a gene to have phenotypic control in this sense? We have just considered and rejected one proposal as too strong: that a gene type has a phenotypic effect, and hence there is selection for or against its replication, if it has a constant effect on an organism's fitness by virtue of having a constant effect on the organism's phenotype. We have seen that there are reasons quite independent of the debates over gene selection to reject this proposal.

We are left, however, with both a conceptual and an empirical problem. Given that the criterion of constant effect on phenotype and fitness is too strong, our conceptual problem is to replace it with another. Our alternative had better not be too weak. If we define phenotypic control of replication prospects too inclusively, we will revive the "absurdly reductionistic *reductio ad absurdum*" (Dawkins 1982, 90) of there being only four gene types: the four nucleotides A, C, G, and T. After all, as Dawkins notes, "substituting adenine for cytosine at a named locus within a named cistron" may have a phenotypic effect (Dawkins 1982, 91). On the other hand, if "having a phenotypic effect" is defined strongly enough to exclude single nucleotides, gene selection faces a serious empirical problem. Its defenders would need to show that most genes do have effects that explain their histories of replication success.

This empirical problem is not trivial. Suppose, for example, that we look for a robust process explanation to give content to the idea of a gene's exerting phenotypic power in the world. Some gene types play absolutely central roles in the life of the organism. Protein kinases, for example, transfer phosphate groups to enzymes and other proteins, a basic part of cell regulation. We can't get by without them. Hence there are a huge range of protein kinase genes in our genome and in that of other organisms. Chromosomes are wound around proteins called histones, so again, there are lots of histone gene tokens, and they stay almost identical across vast stretches of evolutionary time. If we focus on examples like these, one might think that there would be an irresistible robust process explanation for protein kinase genes and histone genes that appeals to their central role in making molecules essential to life. These genes exist, and exist in large numbers of copies, because of their effects. Their existence is *robustly* explained by their effects; they would exist in large numbers even if some details of life's history were changed.

another kind of robustness

So far so good. But it would be a big leap to the conclusion that there are robust process explanations that explain the presence of all the genes underlying each phenotypic adaptation. For the robust explanation is an explanation of the existence of (for example) the kinase gene *family,* rather than particular *forms* of the kinase gene. It is far from obvious that the particular forms of the kinase genes found in, say, humans have a robust explanation. So even if it's true of a particular gene type that its being a kinase gene of some kind is "visible to selection," it does not follow that its being this particular kinase gene is so visible. An analogy might make this clear. Rhinoceroses have horns, and for all we know, there may well be a robust selective explanation—perhaps defense, perhaps sexual selection—that explains why they have horns. But the African rhinoceros has two horns, while the Asian has one horn. Even if it's true that being horned versus hornless is visible to selection in a way that robustly explains the hornedness of the rhino, it by no means follows that having one horn versus having two horns is visible to selection in the same sense. The fact that the African rhinoceros has two horns may be an artifact of particular historical events in that lineage. Equally, even if the proliferation of *some kind* of kinase gene is robustly explained by an appeal to its adaptive effect, it by no means follows that the proliferation of the *particular kind* of kinase gene we have has such an explanation. So there may be a problem in treating even genes whose products play a fundamental role in metabolism as targets of selection by virtue of the phenotypic control they exert over their world. If so, then there are likely to be even greater problems in extending this idea to all the other genes that play a role in less fundamental adaptations.

In short, if gene selection takes this fork in the road, it faces a potential problem. It needs an account of phenotypic control that is strong enough to exclude individual nucleotides (and, in our view, other DNA sequences that are copied as a side effect of other processes), but not so strong that many molecular genes are also excluded. If we require that gene replication be robustly explained by its adaptive effects, then it is likely that many molecular

difficulties of defining genes as replicators

genes will be excluded. It is not at all obvious that there is a way of formulating the notion of phenotypic effect that meets our three conditions: (1) it counts molecular genes as having phenotypic effects; (2) it excludes impostors like individual nucleotides; (3) the phenotypic effects of genes (when they have them) explain their replication propensity.

4.3   Genes Are Difference Makers

The other way in which gene selectionists can respond to the failure of the evolutionary gene concept is to define genes in terms of the difference they make to their world. It is a commonplace in discussions of genes that there are not really any genes for traits, only genes for trait *differences*. This idea dates back to the earliest days of Mendelism. Alleles were the postulated factors that were supposed to make the difference between alternative phenotypes, such as blue and brown eyes. There were no Mendelian alleles for constant traits, for there could be no breeding experiments to see whether those traits would "Mendelize"—that is, to see whether the offspring of hybrids would express those traits in the ratios Mendel made famous. It is still a truism that there is no interesting sense in which a gene builds a trait. No gene is responsible for the capacity of kangaroos to make long hops. But genes do cause differences in hop length. A particular allele, G^*, might be causally responsible for its kangaroos having a longer hop length than average. That would be the case if substitutes for G^* on the relevant chromosome would lead, in the relevant environment (including the genetic environment), to a kangaroo that made shorter hops. Moreover, differences, and only differences, are visible to natural selection. A few doomed developmental disasters aside, the ability to hop is universal in the kangaroo population, so there is no selection for hopping over failing to hop. But if hopping stride varies, there may well be selection on hop length. So the aspects of the phenotype that we can attribute to the action of genes are the same aspects that vary in the population and are subject to selection.

Sterelny and Kitcher (1988) responded to the idea that genes are invisible to selection by treating genes as *difference makers,* and as visible to selection by virtue of the differences they make. In doing so, they provided a formal reconstruction of the "gene for" locution. The details are complex, but the basic intent of the reconstruction is simple. A certain allele in humans is an "allele for brown eyes" because, in standard environments, having that allele rather than alternatives typically available in the population means that your eyes will be brown rather than blue. This is the concept of a gene as a difference maker. It is very important to note, however, that genes are *context-sensitive* difference makers. Their effects depend on the genetic, cellular, and other features of their environment.

Most discussions of gene selection define the genes that are being selected in this way, via the phenotypic differences they produce. However, this way of defining genes is ambiguous. According to one way of understanding it, the phrase "gene for blue eyes" can be used to pick out all copies of the particular DNA sequence that is the blue-eyed difference maker in some specific organism. We might notice an unusual fruit fly—perhaps a mutant—with white eyes, and use the words "gene for white eyes" to mean the DNA sequence that made the difference in its having white rather than red eyes, along with all the other copies of that sequence. Those other copies are, in this sense, "genes for white eyes" even if in other individuals, in their standard environments, they would make no difference to eye color at all. If that is how we understand "gene for white eyes," then the difference-maker concept is simply another version of the idea that genes are active replicators, which we discussed in section 4.2. With this definition of the "gene for white eyes," we are guaranteed that the gene type—all the copies of this gene—has an underlying molecular unity. It comprises all the same or similar sequences. But this definition does not guarantee that there is any unity at the phenotypic level—that all the carriers of the gene have some propensity to develop white eyes. There may be such a phenotypic commonality in the carriers of the gene, but we would need to establish that empirically.

There is, however, an alternative, purely functional sense of "gene for white eyes." In this sense, only DNA sequences at loci where their presence actually makes them difference makers with respect to eye color are "genes for" eye color. Note that this sense does not simply define the gene for white eyes as all those sequences, and only those sequences, that actually cause white eyes. Recessive alleles that are not expressed in a particular individual, and alleles in individuals that fail to develop, will still be counted. Thus, in the human case, brown-eyed people and those born with no eyes may still have the (recessive) gene for blue eyes. These unexpressed genes are still difference makers with respect to eye color in humans.

With this functional interpretation, we can be sure that difference-making genes really are subject to natural selection, because we can be sure that they have a sufficiently constant phenotypic effect. Myotonic dystrophy, a hereditary muscle-wasting disorder, reduces fitness enormously. The "gene for" myotonic dystrophy is any one of a number of sequences of between 50 and 200 trinucleotide repeats. The normal form of the gene is any of several sequences of between 5 and 27 repeats (Brook et al. 1992). The difference-maker concept explicates the sense in which there are two genes here, not several dozen. We can be sure that gene sequences whose substitution for the wild type in real populations actually makes the difference between sufferers and non-sufferers are being selected against.

The disadvantage of difference-maker genes, defined in this sense, is that it is unclear that they have any reality as gene types independent of the phenotypes that are used to define them. The definition itself guarantees that all carriers of the blue-eyed gene will have a propensity to develop blue eyes, though this propensity may be blocked in various ways. But it does not guarantee that all the copies of the gene have any unity at the molecular level. It does not guarantee that those copies form a replicator lineage—that they are copies of one another. Unless the different sequences of the myotonic dystrophy gene are identical by descent, the facts about their fate given their phenotypic expression do not combine into an explanation of the fate of *a single gene lineage*. We have instead a classic example of the culling of many unrelated gene lineages through single-step selection. Yet gene selectionism was born out of a sense of the importance of cumulative selection (3.1). Cumulative selection requires gene lineages that have some form of underlying molecular unity and some form of similar phenotypic effect. Otherwise we can make no sense of the idea that the fate of phenotypes affects evolution only through its effect on gene lineages.

To see this, consider some of the extended phenotype examples we discussed in section 3.3. We described genes for camouflaged plumage, for aggressive behavior, and for the extended traits of parasites. We described a selective explanation for each phenotypic trait and the claim that it is really the "genes for" this trait that are being selected. The distended abdomen of the housefly is selectively advantageous to the parasitic fungus, and hence there is selection for the "genes for" the sexy abdomen—genes in the fungus. The "gene for" locution makes sense only if it is tracking a constant, underlying difference maker or set of related difference makers. Recall, for example, our aggressive magpies: a lineage of magpie aggression genes expands at the expense of other gene lineages by virtue of its behavioral effects on magpie families. To make sense of this scenario, we must suppose that if the gene that makes one magpie family aggressive were copied into another family, it would make (in the same context-sensitive way) that other family aggressive, too. The substitution of that DNA sequence for the alternative alleles that are actually present would cause that other family to be more aggressive. Only then does it make sense to talk of that gene lineage becoming deeper and bushier, and displacing rival lineages for the same chunk of magpie chromosome, by virtue of its effect of making magpie families more aggressive. However we interpret "a gene for blue eyes," for the gene selectionist, the different copies of a gene must have some form of underlying molecular unity. Similarly, the "gene for myotonic dystrophy" cannot just be all the sequence tokens that, if substituted into normal human developmental contexts, can make the difference between normal and dystrophic development. If there is a *single gene* for myotonic dystrophy, in the sense relevant to the gene selectionist conception of evolution, these sequences must all be part of a single gene lineage. If not, the gene selectionist will have to recognize a number of, perhaps many, "genes for" myotonic dystrophy. A purely functional notion of a gene, untied to anything constant at the molecular level, is not a definition suitable for gene selection theory, whatever its other uses might be.

Once again, as in section 4.2, gene selectionism seems to be placing an empirical bet. Even though there are two ways of interpreting the idea that genes are context-sensitive difference makers, gene selectionist evolutionary narratives are committed to a gene type having an underlying molecular unity: the individual tokens of the gene for white eyes must be copies of one another. The purely functional definition, in which individual gene tokens might have no historical connection with one another at all, turns out to be ill-suited for the gene selectionist idea. Suppose that a recurring phenotypic trait—aggressive magpie families—co-occurs with a gene lineage. The gene lineage explains (in a context-dependent way) the expression of that trait in those families. In turn, the ecological effects of aggression explain the evolutionary success of the gene lineage. This would count as a triumph of the gene selectionist idea. Suppose, however, that the phenotypic trait of aggression regularly occurs in the population, but that there is no gene lineage making a regular developmental contribution to the expression of that trait. Then this version of the gene selectionist story would fail. It would also fail—though perhaps it could be rescued—if the gene lineage had tendrils in peaceable magpie families, where it played a very different phenotypic role.

The fact that a serious empirical commitment is being made here is obscured by the fact that the "gene for" locution has its primary home in medical genetics, where in many cases the complexities of the gene/phenotype relation can be ignored. Medical genetics discovers "genes for" disease phenotypes. These involve some major defect in an evolved gene that, in its normal role, interacts with many other genes. Such disease genes include those for albinism, melanism, and other pigment changes caused by defects in pigment-making genes, and those for dwarfism caused by hormonal defects. They also include the most famous case of all: the genes responsible for phenylketonuria and its variants, caused by the absence of an enzyme or failure of its function (5.3). These pathological genes impair normal development to such an extent that they dominate variance in the phenotypic traits they affect under almost any background conditions. If we were interested only in the evolution of disruptions of existing phenotypes, we could perhaps assume that the difference-maker genes for such traits were real entities at both the molecular and the phenotypic level.

But biology is at least as interested in the evolution of new complex phenotypes as in such disruptions. So let's consider some examples in which gene effects are not overwhelmed by developmental catastrophe. Here it is far from clear that the difference-maker gene concept describes a real entity at the molecular level. As we noted in section 3.2, the relationship between phenotype and genotype can change over evolutionary time. There we discussed an example in which there is such strong selection for a particular butterfly wing pattern that the pattern becomes fixed. All the butterflies in that population have that phenotypic trait. But the gene for this pattern changes as this evolutionary transformation proceeds because the causal context changes. Since the pattern requires enough, but not too much, of a certain chemical, genes that reduce the rate at which this chemical is metabolized into other chemicals can enhance the pattern when the chemical is scarce. However, the very same genes may disrupt the pattern when the chemical has become abundant because other genes favoring the desired pattern are now common in the population. So over time, there is no single molecular gene type that is responsible for this pattern, no particular sequence whose copies make the difference between a butterfly having and lacking this pattern. Nijhout and Paulsen summarize their results on the changing roles of genes as evolution proceeds by noting that

> . . . whether a particular gene is perceived to be a major gene [a *major gene* for a trait is a gene that accounts for a significant fraction of the variance in that trait], a minor gene or even a neutral gene depends entirely on the genetic background in which it occurs, and this apparent attribute of a gene can change rapidly in the course of selection on the phenotype. (Nijhout and Paulsen 1997, 401–402)

This example and others like it are not decisive refutations of the gene selectionist claim. We should not exaggerate the empirical risks of the gene selectionist hypothesis. Gene selectionists can live with more than one gene for longer hops, more than one gene for magpie aggression, and more than one gene for butterfly wing patterns. And they can live with the idea of a gene lineage becoming bushy in one population for one reason and in another, for another. This is no more mysterious than a defender of the received view arguing that feral cats become tabby in Australia because of eagle predation on kittens, and in New Zealand to camouflage the adults while they are hunting. Moreover, distinct populations may be separated by time—by generations—rather than by space. So the fact that "the gene for blue stripes" picks out one sequence lineage early in a process of butterfly evolution and a different sequence lineage later, *by itself* presents no problem to gene selection. The worry is that the context sensitivity that these dynamic models reveal opens up the possibility that a gene lineage in the one population at the one time might be expanding or shrinking for reasons that vary widely from copy to copy. If that were the case, surely the replication success of a particular gene lineage would be an epiphenomenon of the success of organisms, not an explanation of it.

So here is the state of play as we see it. Gene selectionists, perhaps not surprisingly, have attempted to formulate their conception of evolution in a way that minimizes its vulnerability to empirical refutation. This strategy will re-emerge in the next chapter, in which we will consider Dawkins's attempt to insulate claims about the role of genes in evolution from claims about their role in development. The evolutionary gene concept should be seen as an attempt to insulate gene selection from controversial claims about gene/phenotype relations. According to this concept, gene types are sequences—any sequences—whose success is specified by the fitness, averaged, of all the organisms carrying them. We suggest that this attempt to insulate gene selection from all empirical risk cannot succeed. If different gene types merely covary with different levels of fitness—if all or most of the copies of a given gene play no systematic role in explaining what organisms with a copy of that gene are like—then they do not explain fitness. So the evolutionary gene concept does not support gene selection, because evolutionary genes need not have phenotypic effects by virtue of which they are selected. Gene selectionism cannot have it both ways. It cannot both propose such a weak notion of a gene's phenotypic effect that genes are bound to have phenotypic effects, *and* argue that the success or failure of gene lineages is explained by their phenotypic effects.

So gene selectionists need some alternative conception of the gene. One possibility would be to co-opt a notion from molecular biology: genes are sequences of DNA that code for specific proteins. A problem with this suggestion is that the effects of such genes on an organism's phenotype can be very variable, because the same protein can play very different roles in an organism. DNA sequences' effects on phenotype depend heavily on context; hence the same sequence in different contexts has different effects. We have considered an alternative approach: define gene types via consistent phenotypic effects. The "gene for red eyes" would then be all sequences that can (in an appropriate context) make the difference between possessing and not possessing red eyes. A critical problem for this idea is that the underlying DNA tokens may not form a gene lineage; they may not be related by descent. Hence the reproductive success of red-eyed phenotypes will be irrelevant to the expansion or contraction of most of the red-eyed gene lineages.

So gene selectionism seems to have placed an empirical bet that there exists some kind of well-behaved relationship between phenomena we can describe at the phenotypic level—magpie aggression, fruit fly eye color—and the genetic triggers for phenotypic differences. In this view, the continuing gene selectionist research program faces both a conceptual and an empirical challenge. Its conceptual task is to specify the genotype/phenotype relation to which gene selectionism is committed. Its empirical task is to confirm its existence.

[handwritten marginalia: gene selectionism relies on successful definition of genotype-phenotype relationship]

Further Reading

Different versions of the invisibility argument are given in Gould 1980a, Sober and Lewontin 1984, Brandon 1982, Brandon 1988, and Wimsatt 1980a,b. Sober (1984b, 1993) gives an argument from causal robustness. Sterelny and Kitcher (1988) reply to these arguments. Godfrey-Smith and Lewontin (1993) and Gray (1992) reply to Sterelny and Kitcher.

The distinction between robust process and actual sequence explanations is developed, under a different name, by Jackson and Pettit (1992). Sober (1983) draws a similar distinction, and the idea is exploited by Sterelny (1996a). The contrast between the genes of molecular and evolutionary biology is explored by Griffiths and Neumann-Held (in press).

5

The Developmental Systems Alternative

5.1 Gene Selectionism and Development

In chapter 3 we laid out the case for gene selection, and in the following chapter we discussed a composite "received view" reply. Here we turn to a radical alternative: a view that rejects the replicator/interactor framework itself. *Developmental systems* theorists claim that there is no privileged class of replicators among the many material causes that contribute to the development of an organism—that the entire replicator/interactor representation of evolution is refuted by the facts of developmental biology.

As we noted in section 3.2, Dawkins and Williams assume that genetic resemblances between parents and offspring have a significance that other resemblances do not. Dawkins tries to exclude nongenetic factors from evolutionary biology, as opposed to developmental biology, on these grounds:

> When we are talking about development it is appropriate to emphasize non-genetic as well as genetic factors. But when we are talking about units of selection a different emphasis is called for, an emphasis on the properties of replicators. . . . The special status of genetic factors is deserved for one reason only: genetic factors replicate themselves, blemishes and all, but non-genetic factors do not. (Dawkins 1982, 98–99)

We have already seen that the claim that nothing but genes are replicated in evolution is less obvious than it first seems (3.2). Developmental systems theorists argue that it is simply false.

The developmental systems critique is developed in two main stages. The first is to argue that the gene can be the unit of selection only if the gene plays some distinctive and privileged role in development. The second is to deny that genes play such a role. The main steps of the argument can be laid out as follows:

Step One: Organisms inherit a great deal more than their nuclear DNA. The epigenetic inheritance of nongenetic structures within the cell is a hot topic in current biology. Organisms also behave in ways that structure the broader environmental context of their successors. For instance, many birds inherit their songs through the interaction of their developing, species-specific neural structures with the adult songs to which they are exposed. So an organism inherits an entire *developmental matrix,* not just a genome.

Step Two: The orthodox view of development is that all traits develop through the interaction of genes with many other factors. So genes are neither the only things that are inherited nor the only things that help to build the organism. There is more to evolution than changes in gene frequencies. But genes might still be "privileged causes" of development, which control, direct, or act as an organizing center for everything else. If gene selectionism is to get off the ground, it must demonstrate that genes play some such privileged role.

Step Three: The notion of genetic information and its relatives cannot be made good in a way that singles out genes as privileged causes of development. Every reconstruction of the notion that genes contain information about the outcomes of development turns out to apply equally well to other causes of development.

Step Four: A range of further attempts to draw a distinction between the role of genes in development and the roles of other developmental factors fail. These attempts are either mistaken or overstated (for example, the idea that genes are copied "more directly").

Step Five: Developmental systems theorists conclude that for all their biological importance, genes do not form a special class of "master molecules" different in kind from any other developmental factor. Rather than replicators passing from one generation to the next and then building interactors, the entire developmental process reconstructs itself from one generation to the next via numerous interdependent causal pathways.

In this chapter we assess each step of this argument. We conclude that the argument as a whole has considerable force, and in the final sections we consider what this might mean for the debate over the units of selection.

5.2   Epigenetic Inheritance and Beyond

Developmental systems theorists agree with the normal emphasis on the cumulative nature of selection. But they point out that lineages of organisms show repetition of many important elements from developmental cycle to developmental cycle. In many species of birds, for example, the juveniles acquire their songs, their preferences for nest sites and nesting materials, and many other aspects of their behavioral repertoires from their parents. Their experience in the egg, as nestlings, and as juveniles is critical to the acquisition of the skills that are normal for their species. In any species in which learning, broadly conceived, is important, there is likely to be this type of flow of information across the generations. It need not involve anything like explicit teaching. Parents structure the learning environment of their young and provide them with information just through their normal, species-specific activities of daily life. So "cultural transmission" in this sense is not restricted to cognitively fancy animals. Indeed, as we shall see, there is an important sense in which we find this phenomenon among the arthropods.

Moreover, the idea that nuclear genes are all an organism inherits in the cells carrying the gametes is simply out of date. To develop normally, the egg cell must contain a great array of complex biochemical machines. Any account of the molecular details of how these machines work would take us well beyond the scope of this book (and of our competence), but they include basal bodies and microtubule organizing centers, cytoplasmic chemical gradients, DNA methylation patterns, and membranes and organelles, as well as DNA. Changes in these mechanisms can cause heritable variation that appears in all the cells descended from that egg cell. These elements of the cell have been labeled *epigenetic* inheritance systems (Jablonka and Lamb 1995; Jablonka and Szathmary 1995). For example, the so-called *DNA methylation system* has excited a great deal of interest recently. It has even been suggested that some behavioral differences between human males and females are due not to genetic differences, but to the inheritance of a methylation pattern. DNA methylation is the attachment of a series of additional chemical groups to a DNA sequence in a sperm or egg by the parent organism. These methyl groups block transcription of any genes to which they are attached. The methylation pattern is replicated by a special methylation copying system in all the cells descended from that sperm or egg. Some recent research suggests that human females methylate a sequence of the X chromosome, so that individuals who get only one X chromosome and get it from their mothers cannot transcribe the genes in that region. Hence certain gene products are denied to all males. Males demethylate that sequence in their sperm cells, so that females get a working X chromosome from their fathers (Skuse et al. 1997).

Developmental systems thinkers extend the idea of inheritance still further. The characteristics of epigenetic inheritance systems within the cell are shared by many extracellular structures. Some castes of the aphid *Colophina arma* require a growth spurt as part of their life cycle. These, and only these, castes inherit the microorganisms that make the chemicals on which this growth spurt depends (Morgan and Baumann 1994). The morphology of queens and the colony structures of the fire ant *Solenopsis invicta* differ radically between genetically similar lineages of the species because of stably replicated nest "cultures" mediated by pheromones (Keller and Ross 1993). Any queen raised in a colony with a particular culture will found a colony with the same culture, as can be demonstrated by moving eggs from one culture to another. Many parasites, both vertebrate and invertebrate, maintain associations with particular host species over evolutionary time through *host imprinting*. Thus insects of many kinds lay their eggs on the plant species whose leaves they tasted as larvae or caterpillars. Some parasitic finches lay their eggs in the nests of the host species that they imprinted on as chicks (Immelmann 1975). So *host switching* can occur when—once in a blue moon—something goes wrong and a moth, say, lays her eggs on a plant other than the one on which she fed. Usually those eggs are doomed, but occasionally they will survive (perhaps the plant is a new arrival in the region), and that same imprinting mechanism will then ensure that the moths that grow from those eggs return to the plant on which they, not their ancestors, fed. So parents pass on much to their offspring: genes, cellular chemistry, and other cell structures; features of their physical environment (burrow systems, nests, and the like); behavior patterns.

The developmental systems view argues that we should redefine *inheritance* so that every element of the developmental matrix that is replicated in each generation and which plays a role in the production of the evolved life cycle of the organism counts as something that is inherited (Gray 1992). Genes cannot be singled out as the unit of replication on the grounds that they, and they alone, persist through lineages long enough for cumulative selection to act upon them. Lineages can be selected for having good symbionts or being imprinted on a good host, and these features can persist for evolutionarily significant periods of time.

5.3   The Interactionist Consensus

Given that there are many different strands in inheritance, how do they combine to build a new organism? In section 1.4, we introduced the idea of genetic determinism. In its crudest form, genetic determinism is the view that a trait is genetically caused or innate; in contrast, other traits are environmentally caused or acquired. On this view, traits observed in all normal members of a species, such as mating rituals, would be regarded as innate, while traits that differ widely between individuals, such as preferred foraging sites, are acquired. However, no one accepts this crude division between genetically caused and environmentally caused traits. All traits have both genetic and nongenetic causes. The development of any trait can be blocked by some genetic modification. Equally, barring mutation-induced disaster, nongenetic modifications can stop any trait from developing. Social deprivation of young rhesus monkeys will prevent them from displaying their "innate" sexual behaviors as adults. Yet a rat and a bird will emerge from an identical program of conditioning having learned very different behaviors: their genetic endowment affects what is "acquired."

So it is universally accepted that all biological traits develop as a result of the interaction of genetic and nongenetic factors. But perhaps some traits depend more on genes and less on the environment. It is now common to read that homosexuality, for example, is "substantially genetic," or that schizophrenia may be "partly genetic." Often actual figures are cited. One study might suggest that homosexuality is 30% genetic, another that schizophrenia is 10% genetic. These figures are produced by a statistical technique called *analysis of variance* or *ANOVA*. To perform an analysis of variance, we need a population of individuals, some of whom have the trait of interest and some of whom do not. Some individuals that differ with respect to the trait will also differ with respect to some genes. The more often this is true, the more of the *variance* in the trait can be correlated with that variation in the genes. If every individual with the trait has certain genes and every individual without the trait lacks those genes, then the proportion of the variance accounted for by those genes is 100%. If possession of the trait is random with respect to possession of those genes, then the proportion of the variance accounted for by those genes is 0%.

In many people's minds, the discovery that a trait is "substantially genetic" means that it is substantially genetically determined. The more "genes for" complex human behaviors are reported in the media, the more genetic determinism seems true. But this interpretation is simply wrong. Measuring the amount of variance accounted for by genetic factors does not measure the degree to which a trait is genetically caused or genetically determined (Lewontin 1974). A trait would be literally genetically *determined* if it could not be altered by changing nongenetic factors, a situation that we can be sure never arises. More realistically, a trait may be said to be genetically determined when altering it by changing nongenetic factors is difficult or impractical (if, for example, such changes would always kill or severely deform the embryo). But high scores for genetic factors in an analysis of variance do not show that it is hard to alter the trait by nongenetic means, and hence do not show genetic determination. They show only that the actual environmental factors in the population under study do not alter the trait, not that no feasible set of environmental factors could alter the trait. One well-known example that illustrates this distinction is the disorder called phenylketonuria (PKU), which causes mental retardation. It is caused by a mutation that results in the bearer's inability to metabolize the amino acid phenylalanine. Under standard conditions, possession of the PKU mutation accounts for 100% of the variance between those who suffer PKU retardation and those who do not. However, PKU can be effectively treated by feeding people with the PKU mutation a special diet low in phenylalanine.

As we noted in discussing heritability in section 2.2, a uniform environment tends to increase the score of genetic factors in an analysis of variance. Conversely, genetic uniformity will increase the score of nongenetic factors (see box 2.2). Whenever a number of causal factors interact to produce an outcome, we should expect the effect of changing one factor to depend on what is happening to the other factors. To establish genetic determinism we would need high ANOVA scores for genetic factors across a wide range of values of all the other factors that typically play a role in development. Only if changes in those other factors had little effect on the relationship between genes and trait would it be proper to speak of the trait as genetically determined.

The points made so far are fairly uncontroversial—they make up the *interactionist consensus* in current biological thought. While nothing in the interactionist consensus makes genetic determinism (in the sense just described) impossible, there is plenty there that makes it unlikely. In the interactionist view, genes are "context-sensitive difference makers." They produce their effects by adding a physical product to a complex network of causes consisting of other genes and their immediate products, the other constituents of the initial cell, and all the inputs of materials and energy to the developing organism. The effect of one cause on the final outcome is mediated by all the others. Except in those cases in which having a nonfunctional gene is a disaster without remedy, it is unlikely that a change in an individual gene will produce the same effect no matter what changes occur in the other causes. The other causes, after all, include factors that affect *whether* the gene will be transcribed, *when* it will be transcribed, and *which* of the various possible final products will be made from its transcript (6.3, 6.4). In section 4.3 we saw that it is possible for a gene that is normally a "gene for" a trait to become a "gene for" its absence.

The argument so far creates a substantial challenge for gene selectionism.

Gene selectionism holds that evolution is nothing but the differential replication of genes. But genes are not the only things an organism inherits. Nor, as we have just seen, are they the only things that go into building an organism; on this, gene selectionists agree. So the gene selectionists need to show that the other elements that change over time through natural selection are somehow subordinated to the genes. They must demonstrate that, among the many inherited elements of the developmental matrix that combine to build an organism, the genes enjoy some special, privileged status. Otherwise, evolution will be the differential replication of the whole developmental matrix, not just the genes. The normal way of establishing this privileged status is to argue that while there are many *material causes* of development, genes are the only things that transmit *information* from one generation to the next.

5.4 Information in Development

In his later work, George C. Williams, the originator of the evolutionary gene concept, redefined evolutionary genes as units of pure information:

> DNA is the medium, not the message. A gene is not a DNA molecule; it is the transcribable information coded by the molecule. . . . the gene is a packet of information, not an object. (Williams 1992, 11)

This completes the drive to make the evolutionary gene concept independent of molecular biology, on which we commented in section 4.1. Williams's idea explains the sense in which it is widely thought that the organism gets nothing from its parents but its genes. The genes are the only things that contain information: they are the blueprint or program for building the organism. Genetic changes are changes in this plan and so constitute real evolutionary change. The other material causes of development are only building blocks, which are assembled according to the genetic plan (Lorenz 1965). Since changing the building blocks cannot alter the plan, nongenetic changes can only disrupt development by causing poor execution of the plan. Hence epigenetic inheritance is of no great evolutionary significance. The building blocks are not part of the evolving plan, so it is of no importance whether they are passed on by the parents or found in the wider environment.

Susan Oyama argues that the whole notion of developmental information that is transmitted from one generation to the next should be abandoned. Instead, she argues, the information manifested in an organism's life cycle is itself reconstructed in development; thus she speaks of the *ontogeny of information* (Oyama 1985). To understand Oyama's ideas, it is useful to see them as analogous to the theory of memory according to which a rat that has learned to run a maze does not have in its brain a map of the maze with the route marked out. Instead, the rat has learned cues that, in conjunction with the maze itself, suffice to reconstruct the route as the rat passes through the maze. Just as the rat constructs its route using information from cues in the physical world and traces in its own memory, development in the embryo relies on cues in the developmental environment working with traces in the embryo itself. There is no developmental plan within the embryo.

However, a much weaker position than Oyama's would be strong enough to defeat the view that genes are privileged causes of development because they alone convey information. Developmental systems theorists argue that in any sense in which genes carry developmental information, nongenetic developmental factors carry developmental information too. If they are right, then gene selectionists will either have to come up with an alternative ac-count of why the transmission of genes across the generations has special significance (we shall come to some suggestions shortly), or concede that both genes and other information carriers have this special significance.

So let's turn to the idea of the genome as a program. There are essentially two concepts of information, which we can label *causal* and *intentional*. *Causal* notions of information derive from the mathematical theory of communi-cation, the discipline originally invented to design efficient telephone sys-tems in the 1940s (Shannon and Weaver 1949). Mathematical information theory studies only the quantity of information in a physical system; it says nothing about what the information is about. The quantity of information in a system can be understood roughly as the amount of order in that sys-tem, or the inverse of the entropy (disorder) that all closed physical systems accumulate over time. However, there is a closely related causal notion of information content. Information flows over a *channel* connecting two sys-tems: the *receiver*, the system that contains the information, and the *sender*, the system that the information is about. There is a channel between two systems when the state of one is systematically causally related to the state of the other—when we can infer the state of the sender from the state of the re-ceiver. All scientific instrumentation is designed to ensure a reliable flow of information in this causal sense from sender to receiver. Thus there is a chan-nel connecting a barometer to the state of the atmosphere because the state of the barometer is reliably caused by the state of the atmosphere.

When the states of two systems are reliably related, but not directly caus-ally related, there is a *ghost channel* between them. There is a ghost channel between two copies of this book—you can reliably find out what is in our copy by reading your own. The channel between the barometer and the weather is also a ghost channel, because the barometer reading "rain" does not cause rain, and the rain does not cause the barometer to read "rain." Instead of causing one another, both are caused by a drop in atmospheric pressure.

The existence of channels depends on the factors that connect the sender to the receiver: the *channel conditions*. There is a channel between the television studio and the television screen whose channel conditions include the machinery at the studio, the relay stations, the atmospheric conditions, the antennae, and your TV set. So what you see on the read-out device of an instrument causally depends on the state of the source and the states of the channel conditions. Think of a very simple instrument, a doorbell. The silence, as distinct from the buzz, of a doorbell depends on (1) whether the buzzer has been depressed, (2) whether the battery is charged, and (3) the condition of the wiring. We regard the buzzer as the source and (2) and (3) as channel conditions. But that is a fact about *us*. The sender/channel distinction is a fact about our interests, not a fact about the physical world.

Channels, whether real or ghost, can contain noise. The ratio between *noise* and *signal* is a measure of how reliably states of the receiver depend on states of the sender. So as noise increases, the amount of information at the receiver about the sender goes down. Cheap barometers are noisier than expensive ones: many of their readings are noise rather than signal.

The idea of information as systematic causal dependence can be used to explain how genes convey developmental information. The genome is the signal and the rest of the developmental matrix provides channel conditions under which the life cycle of the organism contains (receives) information about the genome. If we hold the developmental history of organisms constant, then their behavior carries information about their genes. We can tell if someone has the dyslexia mutation by whether they become dyslexic given a normal education. But if this is the sense in which genes convey information, it does not single them out from other developmental causes. It is a fundamental fact of information theory that the role of signal source and channel condition can be reversed. In this conception of information, information is just covariation. So if we hold the other developmental factors constant, genes covary with, and hence carry information about, the phenotype. But if we hold all developmental factors other than (say) nutrient quantity constant, the amount of nutrition available to the organism will also covary with, and hence also carry information about, its phenotype. Biologists exploit this fact when they use a clonal population of plants planted across a landscape to measure variation in some environmental factor. Natural selection exploits this fact when different castes are produced in different conditions. A clone of genetically identical aphids is not necessarily morphologically identical: in some species, some individuals will develop into warrior morphs that protect the others. A constant genetic channel is used to transmit information from nongenetic factors to the next generation of organisms. So genes have no distinctive role as bearers of causal information.

Another way to see the parity between genes and other developmental causes is to return to the ideas of noise and signal. So far, our examples have relied on holding every factor but one constant so as to get a pure signal. But typically, many factors are changing at once. What is noise and what is signal depends on what you are interested in. When you see a white dot passing across your television screen, it may be a tennis ball (signal) or it may be atmospheric interference or the cat sharpening its claws on the aerial (noise). But nothing in nature dictates that one dot is signal and the other is noise. Typically, we want desperately to know what happened at Wimbledon and care little about what the cat is doing on the roof. So to us, dots caused by balls are signal and dots caused by the cat are noise. A television engineer, however, will tune the television to receive a constant "test card" transmission, so that irrelevant noise from Wimbledon will not interfere with the important signals from the guts of the TV that are being received by the screen. Similarly, a geneticist may want to raise monkeys under constant conditions so as to detect genetic mutants, but a developmental biologist may want to raise cloned monkeys to detect the effects of different maternal care or social interactions. In causal terms, information is covariation, and all the factors with which development covaries are sources of developmental information.

The importance of channel conditions has been underscored by recent developments in molecular biology. The DNA sequence of a gene corresponds to the sequence of amino acids in the proteins made from that gene. This is the famous *genetic code*. But this code operates through an intermediate stage: the DNA is first used as a template for an RNA sequence, *messenger RNA*. RNA, not DNA, is directly involved in the assembly of amino acids into proteins. It is normal for much of the sequence of messenger RNA transcribed from the gene to be cut out and discarded as *introns* before the messenger RNA is translated into a protein. Different proteins can be made from one gene by cutting out different introns, a phenomenon that turns out to be very common. Which protein is made from a gene at a given time in a given part of the body depends on the overall chemical state of the cell, which can be influenced by many elements of the developmental matrix. So even the fundamental idea that the series of bases in DNA is a linear "code" for a protein needs to be stated carefully; even this depends on channel conditions. Only a DNA sequence plus just the right cellular context contains enough information to specify the structure of a protein, let alone to specify a phenotypic trait (see 6.3, 6.4 for more detail).

The other concept of information is *intentional information* (sometimes called *semantic information*). Many of the thoughts possessed by intelligent beings like ourselves are about things with which they have only the most tenuous causal connection (e.g., thoughts about distant galaxies) or about things that do not exist (e.g., thoughts about phlogiston or Pope Joan). The relation between thoughts and things is called *intentionality* or *aboutness*. Thoughts contain intentional information *(intentional content)* about the objects of thought. Intentional information seems like a better candidate for the sense in which genes carry developmental information and nothing else does. If genes have intentional content, then they mean the same thing no matter what the state of the rest of the developmental matrix. When other conditions change, the content of the genes is merely misinterpreted. If other developmental causes do not contain intentional information and genes do, then genes do indeed play a unique role in development.

The idea that genes have meaning in something like the way that human thought and language have meaning is lurking in the background in many discussions of genetic information. For example, it is often said when an organism develops different phenotypes under different environmental conditions that the message of the genes is "Do this in circumstance A, do that in circumstance B" (a *disjunctive genetic program*). If genetic information is causal information, then this is just a quirky way of saying that changing the channel conditions changes the signal. A distinctive test of intentional or semantic information is that talk of error or misrepresentation makes sense. A map of Sydney carries semantic information about the layout of Sydney. Hence it makes sense to say of any putative map that it is wrong, or that it has been misread. Error and misrepresentation make no sense in the context of the purely causal notion of information. In the causal sense, a doorbell that rings because of corrosion in the wiring has not generated a false alarm. It is merely "reporting" a change in the channel conditions. Strikingly, genetic information is often described as if misinterpretation made sense. So no one says that the human genome encodes the instruction "when exposed to the drug thalidomide, grow only rudimentary limbs." This really would be the instruction if we were talking about causal information. When the channel is contaminated by thalidomide, human genes really do, sadly, contain this causal information.

To reiterate, according to the causal conception of information, there is no such thing as a channel that misinterprets the causal information in a signal sender. Any talk of the genes being misinterpreted, or of the information in the genes being ignored or unused, is a shift from the purely causal notion of information toward something like the intentional notion. So one way to make sense of the idea that some developmental pathways are programmed while others are misreadings of the program is to suppose that genes contain intentional information rather than causal information: information that remains the same when the channel conditions change.

Unfortunately, it is so hard to see how intentional information could be a property of physical systems that this has become one of the great stumbling blocks of contemporary philosophy of mind! The apparently magical nature of intentional information is one of the major objections to a materialistic account of thought. After all, how can a thought be about something that does not exist? Hence arguments for the special status of genes that rely on attributing intentionality to them face a very serious problem. The difficulties faced by attempts to "naturalize" intentional mental content form a vast and expanding literature, which is impossible to summarize here. But we will mention one such idea, for it shows that a successful attempt to remove the magic from intentionality might well restore the parity between genetic and other causes that the appeal to intentional information is being used to avoid.

One of the most popular attempts to explain intentional content in scientific terms appeals to the evolution of the mind. According to the *teleosemantic* theory of intentional content, a thought is about the things that evolution has designed it to be about. When a rabbit thinks PREDATOR, its thought may carry very little causal information about predators, because most such thoughts are false alarms caused by wind or shadows. The teleosemantic theory suggests that the thought PREDATOR has the intentional content that there is a predator here and now because it was produced by mental mechanisms selected for detecting predators. This theory can be applied to genes, yielding the conclusion that a gene contains information about the developmental outcomes that it was selected to produce. There are many possible objections to this idea. Many genes have important effects that they were not selected to produce. But these objections are not our concern here. We merely point out that many other means through which parents influence their offspring have selection histories too. These other elements of the developmental matrix have been selected for their developmental effects, hence they too can be said to contain information about the effects they were selected to produce. There seems to be a trade-off between defining a concept of information that is free of magic and defining one that applies to genes but not to other developmental causes. We return to this idea at the end of the chapter.

5.5   Other Grounds for Privileging Genes

Developmental systems theory argues for "parity" between genes and other developmental causes. It does not deny that nucleic acid sequences play a unique molecular role. It only denies that the differences between the role of DNA in development and the roles of other biological factors justify placing a distinction between genes and everything else at the heart of a theory of development. Nucleic acid sequences and phospholipid membranes both have distinctive and essential roles in the chemistry of life, and in both cases there seems no realistic substitute for them. However, the facts of development do not justify assigning DNA the role of information source and controller of development while inherited membrane templates, or methylation patterns, or pheromonal nest cultures get the role of "material support" for reading DNA.

Genes have been held to be unique on several grounds, the most important of which we considered in the last section. We cannot exhaustively survey the other possibilities, but here is a sketch of some of them, with brief indications of how developmental systems theory deals with them:

- *Genes are unique in their directness of replication.* Recent research casts doubt upon this claim. The accuracy of gene copying is purchased at the cost of a complex and mediated replication process. Now that the molecular process of gene replication is being described in detail, it seems at least as complex as many of the epigenetic inheritance mechanisms (Griesemer 1992a).
- *There is a causal asymmetry between the genes and other developmental factors.* The idea here is that every extragenetic element of the cell depends on the genes. There can be no membranes without genes for their constituents. Host imprinting events and maternal care also number gene products among their causes. So ultimately, everything depends on genes. This is one of the most popular responses to the idea of epigenetic inheritance. But the replication of the germ line genes is equally dependent on the reliable reproduction of a host of nongenetic factors. There can be no genes without membranes, for genes cannot exist without membranes, and gene products destined for membranes must be assembled using an existing membrane template. It is of no use to claim that all cellular conditions have genes among their causes, because every case of gene activation has cellular conditions among its causes. We cannot show that everything is in the genes by tracing the ramifying tree of causes back and stopping on each branch only if we reach a gene. We might equally arbitrarily decide to stop only at nongenetic causes and declare that developmental information is "in the environment"! It is possible, of course, that if we traced replicating DNA or

RNA back far enough, its replication would be the sole lifelike process. But, first, this is by no means certain to be true (15.3), and second, even if it is, those early replicators would bear little resemblance to current ones.

• *Causal responsibility for variance distinguishes the role of the genes.* Genes can be selected by virtue of their effects. Relativized to typical background conditions, the substitution of one gene for a rival allele may yield a boldly striped organism. So that gene is the "gene for bold stripes." But the same comparison between variants, relativized to a normal background, gives us incubation temperatures for traits, cellular chemicals for traits, and so on.

• *Replicators must be reliably re-made, generation by generation.* Genetic replication is high-fidelity replication. But fidelity does not single out the genes, for they are not alone in reappearing with great reliability. It is also worth noting that the fidelity of genetic replication is overestimated by looking only at the intrinsic properties of genes—by considering only the preservation of the base sequences. Genes' relational properties are also of great causal importance, and these are not nearly so reliably copied to the next generation. Crossing-over in meiosis is a major source of evolutionary change, as are deletions, insertions, and translocations and inversions of the DNA sequence.

The search for facts about genes that distinguish them from all other systems of heredity and developmental causes continues. John Maynard Smith and Eos Szathmary have distinguished between "limited" and "unlimited" heredity systems (Maynard Smith and Szathmary 1995). They claim that only genes and memes (human ideas) display "unlimited" heredity: the possibility of limitless, open-ended evolution. Developmental systems theorists are unimpressed, citing pheromonal "cultural transmission" in eusocial insects as an inheritance system comparable to memes (and much better understood).

5.6 Developmental Systems and Extended Replicators

The developmental systems critique of gene selectionism concludes that nothing singles out genes as being sufficiently unique to justify the replicator/interactor distinction. Genes do not form a special class of "master molecules" different in kind from any other developmental factor. Hence genes are not the replicators. If anything, whole developmental systems are the replicators, but then the distinction between replicator and interactor is at best unclear. This argument is a most interesting and serious challenge to gene selectionism, and one of us (Griffiths) accepts it.

The positive proposal of the developmental systems theorists is that the fundamental unit of evolution is the life cycle. A life cycle is a developmental process that is able to put together a whole range of resources in such a way that the cycle is reconstructed. The matrix of resources that create a life cycle is the "developmental system" from which the theory takes its name. Life cycles form a hierarchy of evolutionary units similar to that described by more conventional hierarchical views of evolution (2.3). A "selfish gene" like a transposon has its own life cycle, and variants on this life cycle compete with one another. Organisms have life cycles, and so do groups like ant colonies. Variants on these life cycles also compete with one another. In this respect, developmental systems theory offers a vision of evolution similar to the hierarchical views of Elliott Sober and David Sloan Wilson, which we will encounter in chapter 8.

A developmental system is a very complex entity, raising the question of how a biologist could actually study such an object. Opponents of the developmental systems view see it as unmanageably holistic. If no element within the developmental system is more important than any other, then perhaps to understand the role of *any* element we have to understand the role of *every* element. But that seems to undercut the standard methodological strategy in science of understanding a system one element at a time. Defenders of the developmental systems view point out that in actual research, a biologist usually chooses to assume that many elements of a developmental system stay constant over time and studies the change over time in a few chosen elements. This approach simplifies a reality in which change over time in any one element is coupled to change over time in many others. Such research strategies are familiar from traditional evolutionary studies, in which biologists try to study change over time in a phenotypic trait without considering how all the other traits on which its fitness depends are changing. The success of such "atomistic" approaches depends on the actual degree to which the fitness of alternative forms is constant across contexts.

One proposed advantage of the developmental systems approach is that it allows the biologist to study change over time in elements of the developmental matrix or the life cycle that are not parts of the traditional phenotype—for which there is no gene. She can model, for example, the evolution of competing alternative pheromonal nest cultures or competing alternative methylation patterns. Another proposed advantage is that as a theoretical framework, the developmental systems approach continually draws attention to the interdependence of elements of the system, whereas gene selectionism deliberately thrusts it into the background. This, of course, is the flip side of the heuristic argument for gene selectionism: that it draws attention to the fact that the integration of biological organizations may break down due to competition between their parts (3.4).

There is at least one possible response to the developmental systems challenge, which is endorsed by one of us (Sterelny), but it involves a considerable revision of the gene selectionist idea. This response is the so-called *extended replicator* theory. The idea is to rescue the notion of genetic information in something like the way outlined at the end of section 5.4. The genome really can be said to *represent* developmental outcomes because representation depends not on correlation, but on function. The plans of a building are not the primary cause of a building. The relationship between plan and building is indirect. But plans do play a distinctive functional role in the construction of a building. The role of the plan is to make sure that the building comes out as planned. That is not the function of a bag of cement. Similarly, replicators are designed mechanisms: their biofunction is to contribute to the process through which phenotypes and genotypes reproduce themselves. Replicators play a privileged role in the developmental matrix because they are designed copying mechanisms. Some parent/offspring similarities result from elements of the developmental matrix that have been selected to produce those similarities: those elements are replicators. Replicators exist because they produce those similarities; that is what they are for (Agar 1996; Sterelny, Smith, and Dickison 1996). That is why they have the function of producing that phenotype, and hence why they represent that phenotype. So an informational idea of a replicator can be preserved. A consequence of this argument is an extension of the class of replicators. In this view, the full suite of developmental adaptations emerge as replicators. The genes are paradigmatic replicators, but not the only ones. Most of the extragenetic copying mechanisms that we have mentioned in this chapter are also replicators.

5.7   One True Story?

The debates over gene selection and its alternatives raise a difficult overarching problem. Most of the participants agree that each of these views can give some account of almost every feature of evolutionary history. There is no very marked empirical difference among them, as there was, for example, between Darwin's theory and its predecessors, or between the "modern synthesis" of Darwin and Mendel and older non-Mendelian versions of Darwinism. Heterozygote superiority does not refute gene selection in favor of the organism as the unit of selection (3.3), and extended phenotype examples do not turn the tables on the received view (3.4). We have just seen that nongenetic replication does not straightforwardly refute gene selection, but rather forces it to take more seriously its own formal definition of a replicator. If this is true, then what is the status of these disagreements? At times, gene selectionists seem to be claiming that their view is the only right view of evolution. But many of the arguments for gene selection (and other rivals of the received view) are heuristic. They allow us to see certain similarities more easily, help us to avoid errors we could easily make, and make us less likely to overlook important phenomena. These arguments suggest an alternative conception of gene selectionism. There are a number of more or less adequate descriptions of evolution, but the gene's eye view offers methodological advantages over its rivals, at least for some evolutionary questions. This question of whether disputes are factual or heuristic will arise as well about other rivals of the received view.

Further Reading

5.1 Developmental systems theory grew, not surprisingly, out of developmental biology and developmental psychology, perhaps beginning with Daniel S. Lehrman's critique of the ethological notion of instinct (1953) and continuing in, for example, Lehrman 1970, Gottlieb 1981, and Stent 1981. The work of Patrick Bateson (1976, 1983, 1991) has been important in this tradition. Lickliter and Berry (1990) have written a useful paper explaining why developmental biologists have always been frustrated with the genetic program concept. Susan Oyama's *The Ontogeny of Information* (1985) is regarded by many as *the* book of the developmental systems tradition. A new edition has just been published by Duke University Press, along with a volume of Oyama's collected papers. Griffiths and Gray (1994) attempt to state systematically the implications of the developmental systems approach for evolutionary theory. Their paper is reprinted in Hull and Ruse 1998, which has a good selection on developmental biology as well as units of selection. Gray 1992 is an excellent general introduction to the developmental systems approach, and Gray 1997 discusses further implications of these ideas. Schaffner's paper (in press) is an important attempt to assess the validity of the developmental systems critique of "gene-centered biology." It is accompanied by a number of useful peer commentaries.

5.2 The mechanisms of epigenetic inheritance are reviewed by Jablonka and Lamb (1995) and Jablonka and Szathmary (1995). These papers put a radical spin on these discoveries, while Maynard Smith and Szathmary (1995) play down their radical implications.

5.3 Lewontin (1974) makes a classic presentation of the pitfalls of partitioning traits into genetic and environmental components. A similar view is presented by Sober (1988a). Lewontin's more radical views can be found in

Lewontin 1982b, 1983, and 1991. Kitcher (in press) has written an important paper rejecting Lewontin's later views and defending the interactionist consensus. The defense of the "gene for a trait" locution by Sterelny and Kitcher (1988) is relevant here, and is attacked from a developmental systems perspective by Gray (1992).

5.4 In addition to Oyama, Griffiths, and Gray in the works cited above, Johnston (1987) rejects the notion of genetic information, as does Sarkar in two very substantial and important papers (1996, 1997). Maclaurin (1998) mounts a defense of genetic information. Nijhout (1990) has written a useful paper on the lack of fit between the program metaphor and actual molecular processes. Fox-Keller (1995) provides an extended but very readable discussion of the same topic. Moss (1992) also focuses on whether the idea of a genetic program has any basis in molecular reality. Chadarevian (1998) traces the growing disillusionment with the program concept in one field of molecular biology.

The literature on naturalizing intentional content is enormous. A quick introduction to the various alternatives is chapter 6 of Sterelny 1990. The first attempt to analyze intentional content in terms of causal information was made by Dretske (1981). Dretske 1983 is a useful summary of his theory together with peer commentary. The problems facing Dretske's theory are surveyed by Godfrey-Smith (1989, 1992). Millikan 1989a is a brief introduction to "teleosemantics"; its problems are surveyed in Godfrey-Smith 1994a, Neander 1995, and Godfrey-Smith 1996.

5.5, 5.6 For a good introduction to the complexity of the gene, and the indirectness of genetic copying, see Fogle 1990. The idea that genes are copied more "directly" is critiqued by Griesemer (1992b). Sterelny, Smith, and Dickison (1996) accept much of the critical case made by developmental systems theorists, but argue for the retention of the replicator concept in a revised and more general form. Griffiths and Gray (1997) reply to this paper; the ideas in it are developed further by Godfrey-Smith (in press-a).

5.7 Dawkins appears to change his mind, quite frequently, on whether gene selectionism is first among equals, or the only right view. In Dawkins 1982 he is pluralist, but both Dawkins 1976 and Dawkins 1989b seem less concessive. The pluralist position is defended by Sterelny and Kitcher (1988), Dugatkin and Reeve (1994), and Waters (1994a). Pluralism in philosophy of biology in general is defended with gusto by Dupré (1993) and attacked with equal vigor by Hull (1997; in press).

6

Mendel and Molecules

6.1 How Theories Relate: Displacement, Incorporation, and Integration

One problem in philosophy of science concerns the relationship between apparently different theories of the same domain. For example, in psychology, we have three apparently different ways of explaining human behavior. Cognitive psychology explains human behavior by seeing it as the result of information processing. Its program is to explain, say, our ability to predict others' behavior by characterizing the information about others we possess, the form in which that information is stored, and the techniques we use to process and deploy that information. But the neurosciences are also in the business of explaining human behavior. Those disciplines are gradually developing an account of the physiological mechanisms on which our behavioral abilities depend. Furthermore, we were not wholly incapable of explaining human behavior before the scientific developments of the twentieth century. For thousands of years we have had at our disposal a "folk psychology" through which we have explained the behavior of others. These explanations are couched in terms of beliefs, goals, emotions, moods, and the like. How do the explanations of folk psychology relate to those developed in the natural sciences? How do the two scientific programs relate to each other?

This general problem arises in biology as well. As we saw in section 2.2, heredity—parent/offspring similarity—is central to evolution. Unless offspring tend to resemble their parents more than they resemble some randomly chosen member of their parents' generation, natural selection is powerless to change the character of a population over time. But there seem to be two different theoretical programs through which this central phenomenon can be studied. The first of these dates back to Gregor Mendel's work in the mid-nineteenth century; the second began when the rediscovery of his work at the beginning of the twentieth century prompted a search for its cellular and molecular basis. What follows is a cartoon version of these programs; we go into more detail in section 6.2.

It was Mendel who hit on the idea of genes as discrete units of inheritance while studying the results of pea breeding experiments in the 1860s. When he focused on two states of a single character, round versus wrinkled seeds in true-breeding pea lineages, he noted that first-generation hybrids were all round, but that second-generation hybrids were not. Some—about ¾—were round, but ¼ were not. When he considered not just one, but two traits, seed texture and flower color, once again the first-generation hybrids were uniformly round-seeded, yellow-flowered peas. But the second-generation hybrids were not. Roughly ⁹⁄₁₆ of the second generation were like the first-generation hybrids. But about ³⁄₁₆ were yellow-flowered, wrinkled-seeded peas; about ³⁄₁₆ were green-flowered, round-seeded peas; and about ¹⁄₁₆ were both green-flowered and wrinkle-seeded.

Mendel realized that these results fell into place with the following assumptions:

1. Phenotypic traits such as color and texture are determined by a unitary hereditary factor. These factors can exist in alternative forms, or *alleles*.

2. The gametes of an organism (the pollen or the ova) carry just one of the alternate character states of these traits (one of the factors for yellow or green; round or wrinkled).

3. When an organism is formed from two gametes that carry rival factors for one trait, one dominates the other. In this case, the factor for round is *dominant* over the factor for wrinkled. In other words, the factor for wrinkled is *recessive*.

4. When a first-generation hybrid organism (the *first filial* or F_1 generation) forms gametes, about 50% of the gametes carry one factor, and about 50% carry the other.

5. The factors for traits that are not alternatives to one another—in this case, flower color and seed shape—are inherited independently of one another. From the fact that a gamete carries the *wrinkled* factor, we can tell nothing about whether it carries the *yellow* factor, and vice versa.

As we shall see in section 6.2, after the rediscovery of Mendel's work around 1900, much was added to this picture, and it was altered in important ways. But biologists have continued to investigate heredity by studying the

Box 6.1 What Is an Allele?

Mendelian genetics defines genes, and hence variants of the same gene, through their effects on phenotypes rather than by appeal to their intrinsic physical structures. So when do we have two genes, each of which may exist in a variety of forms? When do we have different alleles of one gene? Since genes can affect more than one trait, we cannot assume that a gene that affects, say, antenna structure in fruit flies is distinct from one that affects their wing length.

Genetic complementation was a central technique in answering this question. Suppose we have two mutant flies: one with short wings, and another with wrinkled antennae. We wish to know whether we have two different mutated alleles of the same gene or mutant forms of two different genes. Mutant forms of different genes (typically) *complement* one another. That is, if we cross the short-winged fly with the wrinkled-antenna fly, and the result is phenotypically normal offspring, we can infer that the mutations are of distinct genes at different loci. We have discovered that the genes are *complementary*. The phenotypically normal offspring result because the gametes from the parent with wrinkled antennae have an unmutated, *wild-type* allele for wing length, and the gametes from the short-winged parent have an unmutated, wild-type allele for antenna form. So the offspring get one unmutated allele for each gene, and are hence phenotypically normal. The offspring are heterozygotes at both loci, with the normal (wild-type) allele dominant over the mutant allele. Clearly, this explanation of why the offspring are normal assumes that the mutations were of separate genes, hence the inference from complementation to alleles of distinct genes. On the other hand, if the hybrid generation is phenotypically unusual, we can infer that we have two mutations of the same gene, and hence two different alleles of the one gene.

patterns of parent/offspring similarity manifested in an organism's phenotype. This program is sometimes known as *transmission genetics*. The debate about human intelligence is one particularly controversial example of such studies.

Shortly after the rediscovery of Mendel's work, a second closely related program developed: an investigation into first the cellular and then the molecular basis of heredity. While the molecular basis of hereditary factors—protein versus nucleic acid—remained in dispute until the mid-twentieth century, their cellular basis in chromosomes was soon discovered. As early as 1903, Walter Sutton showed that meiosis explains our second principle, Mendel's *law of segregation*. For meiosis results in each gamete receiving just one of a homologous pair of chromosomes. Somewhat later, in T. H. Morgan's famous fly lab, the discovery of the physical location of genes on chromosomes undercut principle 5, the *law of independent assortment*. When genes are located on the same chromosome, the inheritance of one is not independent of the inheritance of the other. Further on down the track it was discovered that nucleic acids were the critical molecules making up the genes. Then, in 1953, James D. Watson and Francis Crick developed the famous double helix model of the structure of DNA. Since then, discoveries have come thick and fast.

How, then, might these theoretical programs be related? One possibility is the *displacement* of one program by another—that is, one program can show that another is simply mistaken. The geological program of plate tectonics displaced the conception of earth history in which the position of the continents was taken to be fixed. Much more controversially, Paul and Patti Churchland argue that folk psychology is being displaced by the neurosciences. It was once expected that folk psychological explanations of behavior could be "reduced" to neurophysiological explanations. The idea was to define the concepts of folk psychology—moods, emotions, and cognitive states—in neurophysiological terms. Fear, for example, might turn out to be a specific form of arousal of the autonomic nervous system. Most philosophers of mind are physicalists and think that there is nothing to the mind except the physical brain and the wider physical context it inhabits. However, it is now generally accepted that though the emotions do depend on the physiology of the nervous system, they do so in complex ways that vary from individual to individual and over time. So there is wide agreement that psychological concepts like belief and desire cannot be defined in neuroscientific terms. The Churchlands take this to be a symptom that there is something wrong with folk psychology. In their view, the failure of reduction suggests that the neurosciences should displace folk psychology (P. Churchland 1986; P. M. Churchland 1989).

A second possibility is that one program *incorporates* or absorbs the other—that the first is shown to be just a special case of the second. Planetary motions in the solar system are well described by Kepler's three laws of planetary motion:

1. The orbits of the planets are ellipses with the sun at a common focus.

2. The line joining a planet to the sun sweeps out equal areas in equal periods of time.

3. The squares of the periods of any two planets' orbits are proportional to the cubes of their mean distance from the sun.

Reduction takes place when such laws are shown to be a special case of a more general system of laws. Thus Kepler's laws were shown (with minor corrections) to be a special case of Newton's laws of motion. They can be deduced from, and hence are reduced to, those more general laws. As we shall see, "reduction" is an ambiguous notion, but construed this way, it explains why nothing is lost in the move from the old theoretical framework to the new one. The first theoretical framework is shown to have limited validity by its successor framework; it is incorporated within its successor.

Displacement and incorporation should probably be seen as two ends of a continuum rather than two sharply distinct fates. The fate of Newton's theory is often seen as intermediate between incorporation and displacement. Newton's theory correctly predicts how objects move in space and time at low speeds. At these speeds, the predictions of a theory in which an object has an absolute location in space and time are almost exactly the same as those of a theory in which an object's location is relative to the observer's frame of reference. Relativistic physics is both more accurate and covers a wider array of cases than Newtonian mechanics, but Newton's framework is shown to have some partial validity by its successor.

A third possibility is that two programs can be *integrated*. The classic theory of gases describes the lawlike relationships between observable quantities such as pressure, volume, and temperature. The kinetic theory of gases explains these relationships as the effect of random movements of large ensembles of molecules, each with a quantity of kinetic energy, which it can transfer by impact to other molecules. The explanation of the laws in terms of molecular motion supports the claim that gases are "nothing but" ensembles of molecules in motion. The ontology of the first theory—gases, heat, and pressure—is reduced to the ontology of the second theory—molecules and kinetic energy. We have here a second concept of "reduction": the objects described by one theory are "reduced to" the apparently very different entities postulated by another theory. The classic theory of gases relating pressure, volume, and temperature is sometimes called the *phenomenological* theory of gases because the properties it deals with are observable phenomena. A reduction in this second sense explains the regularities among these observable properties by appeal to the properties of their unobservable constituents.

The distinction between incorporation and integration is not sharp. If the ontological reduction is simple—if there are definitions or bridge laws link-ing the concepts of a reduced theory to the concepts of a reducing theory—then integration can turn into incorporation. The chemical property of va-lency, which measures the capacity of an element to form compounds with other elements, turns out to have a straightforward physical basis in an atom's configuration of electrons. Valency is definable in physical terms. So some chemical generalizations about the combinatory power of atoms will turn out to be special cases of physical principles about electron bonds. They can be deduced from physical generalizations via these bridge laws or defini-tions. Usually, however, it is at least practically necessary to continue to use phenomenological theories. Trying to calculate the efficiency of a heat pump in a freezer by tracking individual molecules would be a thankless task. And, as we shall see, there can be more fundamental reasons that block incorporation.

Prima facie, the relationship between molecular and Mendelian genetics includes elements of both incorporation and integration. Molecular mecha-nisms, we might suppose, explain the regularities in parent/offspring simi-larity revealed in Mendelian genetics. Molecular genetics seems to be a superior and more general successor to Mendelian genetics. Mendel's origi-nal laws are reasonably accurate in a limited range of cases because some of the DNA segments described by modern molecular biology are passed on from one generation to another in roughly the way Mendel postulated. When Mendel's laws are not honored, the new theory can explain what is happening instead. These considerations suggest partial incorporation. Mo-lecular genetics also seems to reduce earlier genetic theories ontologically. Surely there is nothing more to genes than the DNA studied by molecular biologists? Classic Mendelian genetics is a phenomenological theory, for it involves observable patterns in the inheritance of phenotypic characteris-tics. Just as the phenomenological theory of gases, relating the observable quantities of heat, pressure, and volume, is explained by features of their microscopic constituents, so too are the generalizations of classic Mendelian genetics explained by microscopic constituents of genes. Yet for the same reasons that the phenomenological theory of gases remains useful in practice, transmission genetics retains some practical value.

No one doubts that there is something right about this picture of the relationship between Mendelian and molecular genetics. Everyone agrees that the genetic material is made up of DNA and associated molecular struc-tures, and that the behavior of these molecular structures underlies the regu-larities observed by earlier geneticists. However, there is an influential group of philosophers of biology, starting with Hull (1974), who think that the relationship between classic genetics and molecular biology is vastly more complicated than the parallels with heat, valency, or planetary motion suggest. Over this chapter and the next we shall focus on the relations between molecular and Mendelian genetics. In this discussion, the following themes will all be prominent:

1. To what extent does molecular biology vindicate the central ideas of Mendelian genetics, explaining the molecular mechanisms that underlie the patterns of similarity and difference among relatives? To what extent does molecular biology require a revision of these ideas?

2. To what extent can transmission genetics and molecular genetics be developed independently of each other? The chemical property of valency is linked via a bridge law or definition to the configuration of an atom's electrons. According to the antireductionists, the concepts of transmission genetics are not definable in any comparable way. Molecular biology illuminates many aspects of earlier genetic theory, but in complex and indirect ways. Mendelian genetics contains theoretical concepts, such as the idea that one allele is "dominant" to another, whose explanation in molecular biology varies case by case. The idea of dominance has no single, natural correlate at the molecular level. Furthermore, molecular biological explanations often refer to the wider cellular context in which molecular events occur. This seems to run counter to the idea that the behavior of larger entities is being explained in terms of their smaller constituents. So, although the transmission of similarity from parent to offspring depends on molecular mechanisms and their context, these patterns can be studied in relative independence from molecular biology. The two theories are linked by the fact that in any given case, we can explain the observable similarity between parent and offspring in molecular terms, but since these explanations vary from case to case, their integration is not tight.

3. Entwined with these specifically biological themes are more general ones about the right way to conceive of the relationship between scientific programs. Here the general issue of *reduction* looms large. As we have already noted, "reduction" is a many ways ambiguous notion. Three ideas, at least, are in play:

a. An idea that historically has been very prominent in the discussion of reduction is the idea of *theoretical unification*. According to this conception, the aim of science is to develop systems of laws or generalizations. Particular branches of science are characterized by the laws or generalizations that they discover. We have already seen an example in planetary science, Kepler's three laws of motion. Theoretical unification was achieved when these laws were shown to be, with minor corrections, a special case of Newton's laws of motion. More controversially, and with much more correction, Newton's laws are seen as a special case of relativistic laws. Many philosophers of science interpret the relations between the generalizations of chemistry and those of physics in the same way. The generalizations of chemistry are shown to be special cases of those of physics with the aid of various bridge laws defining chemical properties in physical terms. A definition of valency in terms of electron shells is an example of such a bridge law. So theoretical unification involves the incorporation of the laws of a reduced theory into those of the reducing theory, either directly or via the aid of bridge laws. Thus one aspect of scientific progress is the construction of an increasingly general, unified conception of nature's laws.

As we shall see, it is this sense of reduction that is most under the gun in the antireductionist consensus. Hull and the other antireductionists have raised doubts about the existence of suitable bridge laws. But as we shall see in section 15.2, it is not at all clear that we should think of the branches of biology as being in the business of formulating laws or generalizations. This whole conception of reduction and the nature of science, based as it is on physics and chemistry, may not fit biology well.

b. An important "reductive" research strategy in contemporary science is explanation by *decomposition*. How do we work out what is going on in some domain? By taking it apart and studying the components in isolation. If the system cannot be decomposed physically, we can decompose it methodologically. We do this by keeping every component but one constant, and studying the behavior of the system when that one component changes. For instance, we can establish a *norm of reaction* for a genotype by studying how a clone of plants grows when we vary different aspects of the environment, one by one. Variation in the system as a whole is studied by controlling potential sources of variation and allowing only one focal component to vary.

Those who argue for the importance of *holistic* approaches to science and against reductionism often have this conception of reduction in mind. They oppose it by arguing for the importance of *emergent* phenomena. For example, it is common to suggest that ecosystems cannot be understood by decompositional methods because crucial ecological phenomena arise only out of the interaction of many components of a system. Whatever the merits of this idea, it is important to realize that it is quite different from the view that Hull and his allies put forward. There are, however, echoes of this idea in the view that the cellular context in which a gene acts is so important that the strategy of explanation by decomposition is undermined (7.3).

> c. A third sense of reduction is the idea that a scientific explanation must include an identifiable mechanism—it cannot depend on "miracles." One reason why the proponents of continental drift remained in the minority in the period between the two world wars was that it was impossible to see how the continents *could* shift. The mechanisms proposed were unworkable. So continental drift was unpopular as a scientific theory because it depended on a spooky mechanism, a process that could not be understood as a concatenation of ordinary physical and chemical processes. The objects, mechanisms, and processes of a scientific theory must involve nothing spooky: no additions to the standard mechanical processes of the world.

We take this third idea to be an uncontroversial version of reductionism. For instance, a standard puzzle about memory is posed by the fact that humans are very good at recognizing human faces in their normal orientation, but not if the face is inverted. Explaining this phenomenon by detailing the physical changes in the parts of the brain involved in memory is in this sense a reductive process, however complex the relation between a psychological description of what we can remember and a neuroscientific description of changes in neural connectivity might be, for an account of the neural substrate would show that memory involves nothing spooky or occult. In this sense, molecular explanations of dominance or of the independent assortment of traits are reductive explanations, however complex they are, for they show that nothing spooky is in play.

So one sense of reduction clearly involves the incorporation of the reduced theory into the reducing theory. But the two other senses may not: they are compatible with the two theories being integrated without one being incorporated within the other. Consider, for example, the fact that genes are often *pleiotropic;* that is, they have effects on more than one trait. Explanation by decomposition may be an effective strategy for studying this phenomenon even if the relationship between pleiotropy and the molecular mechanisms that explain it is too complex and varied for there to be a bridge law defining it in molecular terms.

We have no interest in haggling over which of these various ideas deserves to be called "reduction." The important point is to recognize their differences, and the fact that the relationship between real theories in science will rarely fit exactly one of these definitions cleanly. So the reader is warned: in this and the next chapter, a number of balls are in the air. We first sketch the empirical background of this controversy, and then proceed to the theoretical upshot.

6.2 What Is Mendelian Genetics?

Mendelian genetics is the theory that grew by elaboration and development of the laws of segregation and independent assortment after these were re-discovered at the beginning of the twentieth century. The first Mendelians realized that the pattern of inheritance of some biological traits could be explained by postulating a pair of factors underlying each trait—a pair of *alleles* occupying a *locus* on a chromosome. The law of segregation says that the two alleles are separated in the formation of the gametes (sex cells), with each gamete receiving only one allele. Although the alleles from two gametes are united in the *zygote* (the fertilized egg), they do not mix together, and they are separated again to form the next generation. The law of independent assortment says that the probability of a gamete receiving a particular allele at one locus is independent of which allele it receives at another locus. This second "law" was subsequently discovered to be widely violated. There are *linkages* of varying strength between loci: the stronger the linkage, the more likely the alleles are to be inherited together.

Both the original Mendelian "laws" and the exceptions to them were discovered through breeding experiments. In his seminal presentation of the antireductionist consensus, Hull followed the geneticist Theodosius Dobzhansky in using this methodological fact to distinguish the new molecular genetics. Molecular genetics is concerned with the intrinsic nature of the hereditary material; it proceeds by looking inside the cell. In contrast, "genetics is concerned with gene differences; the operation employed to discover a gene is hybridization: parents differing in some trait are crossed and the distribution of the trait in hybrid progeny is observed" (Dobzhansky 1970, 167; quoted in Hull 1974, 23).

The outcomes of breeding experiments, however, were very quickly related to *cytology*—the study of the structure and activity of cells. The discovery of chromosomes provided an explanation for the phenomenon of gene linkage. The genetic material in the cell nucleus consists of several chromosomes. If we assume that genes occur in a line along each chromosome, then genes on different chromosomes will assort independently, while those on the same chromosome will be linked together. A further cytological observation explains the fact that the links between genes can differ in strength. Chromosomes come in homologous pairs, and one of the pair is passed on

Figure 6.1 Mitosis, meiosis, and crossing-over. (a) Mitosis is the process by which cells multiply and organisms grow. It is represented here for one pair of homologous chromosomes (the two copies of the same chromosome contributed by the organism's two parents). During *interphase* the cell's DNA is replicated, so that when the chromosomes become condensed and visible in *prophase,* each consists of two *chromatids* connected by a *centromere*. During *metaphase* the nuclear membrane disintegrates, and microtubules from the centromeres join to those of the *spindle*. During *anaphase,* the chromatids are drawn apart by the spindle. During *telophase,* two new nuclear membranes form. The cell can then split into two. (b) Meiosis, or reduction division, forms four haploid sex cells by two successive divisions of one diploid cell. The process is represented here for two pairs of homologous chromosomes. The first division resembles mitosis, although there are important differences. Most importantly, *crossing over* occurs during prophase I, something that is very rare in mitosis. The second division is not preceded by DNA replication, and so produces haploid cells with half the diploid chromosome number. (c) (Adapted from Alberts et al. 1994, 100.) Crossing-over is a process in which pairs of homologous chromosomes line up with one another and exchange segments. Where the mother and father were not genetically identical, this can create new gene combinations.

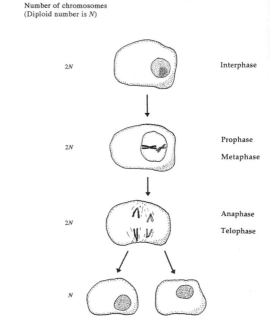

a)

Number of chromosomes
(Diploid number is *N*)

$2N$ — Interphase

$2N$ — Prophase / Metaphase

$2N$ — Anaphase / Telophase

N

to each gamete. During meiosis, homologous chromosomes cross over and recombine, so that a part of each chromosome is exchanged with the other (see figure 6.1c). The probability of two linked genes being separated by *crossing-over,* thus breaking the link between them, can be greater or smaller depending on how close together they are on a chromosome.

Two other important elements of Mendelian genetics are its account of the relations between genes and phenotypes and its account of the relations between the pairs of alleles that occupy a locus. It was natural for early Mendelians to adopt the hypothesis that there is a single gene for each phenotypic trait. It soon became clear, however, that this hypothesis could not be defended in the face of pleiotropic genes and polygenic traits. *Pleiotropy* refers to the phenomenon of one gene having many effects. Hull gives the nice example of an allele that affects both the eye color of *Drosophila* (fruit flies) and the shape of the spermatheca (an organ in females for storing

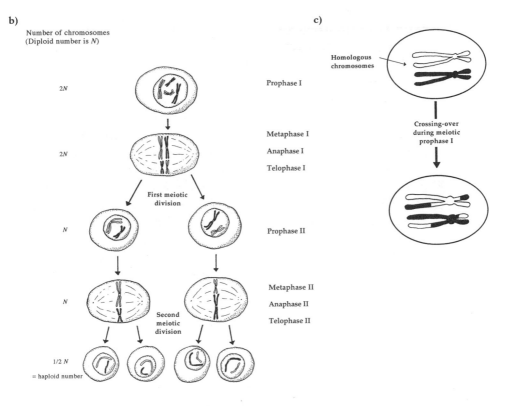

b)

Number of chromosomes
(Diploid number is N)

2N — Prophase I

2N — Metaphase I / Anaphase I / Telophase I

First meiotic division

N — Prophase II

N — Metaphase II / Anaphase II / Telophase II

Second meiotic division

1/2 N = haploid number

c)

Homologous chromosomes

Crossing-over during meiotic prophase I

sperm). *Polygenic* traits, such as human height, are affected by many different genes. Furthermore, some genes interact *epistatically:* the effect of an allelic substitution at one locus depends on which alleles are present at one or more other loci. The relation between genes and phenotypes is thus not one-to-one, but many-to-many.

The way in which the two alleles at a single locus interact to create their distinctive effect is similarly complex. An allele can be characterized as *dominant* or *recessive* relative to some other allele that can occupy the same locus. When two different alleles occur together, if the heterozygote, *Aa,* has a phenotype identical to that of an organism with two copies of one of the alleles—say, *AA*—then *A* is dominant and *a* is recessive. Numerous other categories of dominance were defined by classic geneticists. When the heterozygote expresses a trait more extremely than either homozygote, the alleles are said to be *overdominant.* When the heterozygote expresses the traits of both homozygotes, the alleles are said to be *codominant.* An allele of a pleiotropic gene may be dominant with respect to some of its effects and recessive with respect to others.

Box 6.2 Genetic Atomism

In the growth of theories of heredity and development, the gene has been pressed into service to play a number of distinct biological roles. One is transmission: the production of offspring/parent similarity. But another is mutation: the creation of an unheralded phenotypic form in offspring. Yet a third is recombination: the reshuffling of traits in the phenotype of the next generation that occurred separately in the last, and vice versa. Recombination thus defines the "grain" of inheritance. Finally, genes must somehow function in the development of the organisms that carry them.

The simplest hypothesis is that the gene is the fundamental unit of all four processes. This hypothesis was developed by Morgan and his school in the 1920s. One way of interpreting the further developments in both transmission genetics and molecular genetics since that time is that these roles have been separated. For example, the fundamental unit of mutation (the single base) is distinct from that of function (the codon, a three-base sequence), and that is different again from the unit of recombination (Portin 1993, 781).

Mendelian genetics discovers phenomena that are revealed through breeding experiments, so the explanation of dominance, overdominance, codominance, and similar effects lies outside its scope. Genes interact with one another to determine the norm of reaction of a genotype, and this interacts with environmental variables to determine a phenotype. Mendelian genetics can describe the differences made to this process when one allele is substituted for another at a particular locus on a chromosome, but it does not explain the mechanical bases of these differences. It is part of the role of molecular genetics to uncover these underlying mechanisms. The theorists who expected to reduce Mendelian genetics to molecular biology expected to find one or a few molecular mechanisms that would explain how gene substitutions cause phenotypic differences. This would have allowed them, for example, to identify the phenomenon of dominance with one or a few specific molecular mechanisms. The antireductionist consensus is generated by the fact that expectations of this sort have not been fulfilled.

6.3 Molecular Genetics: Transcription and Translation

The phrase *molecular genetics* refers to the study of the chemical nature of the hereditary material and its molecular surroundings. Chromosomes had long

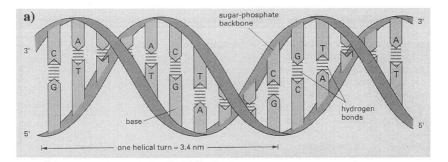

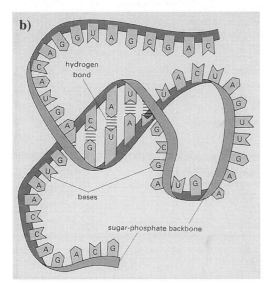

Figure 6.2   (a) The double-
stranded helical structure of DNA.
(b) The single-stranded structure
of RNA, which is the genetic material in viruses and some bacteria.
(Adapted from Alberts et al. 1994,
101.)

been known to contain nucleic acids, such as DNA, and proteins, such as
histones. It finally became clear at the beginning of the 1950s that DNA was
the critical ingredient of the genes. In 1953 Watson and Crick produced a
successful model of the molecular structure of DNA. Since then, much has
been discovered about its molecular machinery. In this context, these discoveries all contribute to a common theme: they highlight the critical role
of the cellular environment in structuring the effect of DNA sequences
on an organism's phenotype. The causal chain between DNA and phenotype
is indirect and complex not just in having many links; it also has many
branches. As we shall see, different cellular environments link identical DNA
sequences to quite different phenotypic outcomes.

   It was clear as soon as the structure of DNA was elucidated that this structure explains some of the phenomena observed by transmission geneticists.
DNA plays its central role in life because it can be both replicated and read.

Box 6.3 DNA as Code and Replicator

DNA can be reliably replicated because guanine and adenine form hydrogen bonds with cytosine and thymine, respectively, and only with them. When the double helix is split apart, each half specifies how to reconstruct the other by forming G-C and A-T bonds. Later research has revealed how DNA functions in the formation of the proteins that make up the structural and functional elements of cells. A single strand of messenger RNA (mRNA) is transcribed by *RNA polymerase enzymes* from one half of the double strand of DNA. The DNA sequence specifies the mRNA transcript by means of the same complementary pairing that allows DNA replication (except that in the mRNA transcript, the base uracil replaces thymine). Within the DNA sequence there is a region beginning with a start and ending with a stop signal. These signals form a *reading frame*. Within the reading frame, the bases divide into three-base sequences, counting from the start signal. Each of these triples is a codon. Hence *frameshift mutations* can cause transcription of the sequence to begin at a new point by redefining the reading frame. A sequence that had been segmented into the codons, say, __/AAG/AGG/GUU/__ can become re-divided into __A/AGA/GGG/UU__/.

The critical feature underlying its replicability is its *complementarity*—the fact that when the double helix splits into two single strands, each uniquely specifies the other. Each base in the sequence will pair with only one other base.

DNA reading depends on two main mechanisms, *transcription* and *translation*. First, DNA specifies *messenger RNA (mRNA)* by the same unique pairing mechanism involved in its replication. The resulting mRNA transcript, like its DNA template, is organized into three-base sequences called *codons*. This *primary transcript* plays a central role in protein synthesis, as the codons specify particular amino acids. These amino acids, in turn, are the constituents of proteins. However, it would be wrong to suppose that DNA specifies proteins in the sense of uniquely determining a particular protein. Different primary RNA transcripts can be transcribed from the same DNA sequences. It is also possible for sequences transcribed as different mRNAs to overlap one another (see box 6.3). So the relation between a given DNA sequence and the mRNA input to the protein-making system is one-to-many. When we consider the reading mechanisms of eukaryotic cells, this basic message gets further support.

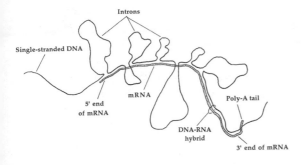

Figure 6.3 Introns can be located by artificially inducing an edited mRNA transcript to bind to a single strand of the DNA from which it was transcribed. Each section of the mRNA *hybridizes* with the section of the DNA from which it was transcribed. The leftover loops of DNA are the introns; the corresponding sections of the mRNA were spliced out during posttranscriptional processing. (Redrawn from Arms and Camp 1987, 205.)

In eukaryotic cells, such as those of plants, animals, and fungi, the primary transcript of mRNA is further processed by the enzymatic machinery of the cell. "Tails" and "caps" are added to the mRNA transcript, and extensive portions are cut out and discarded. These discarded segments are referred to as *introns*. The segments that are retained and spliced together to form the final mRNA are known as *exons*. Alternative splicing patterns, of which there are many examples, make it possible to produce several final mRNA transcripts from the same DNA sequence. Finally, it has recently been discovered that some primary mRNA transcripts may be edited in detail, one base at a time, before proceeding to the translation phase. Some mRNAs are edited (by converting a C into a U) so as to produce a *stop codon* in the middle of the transcript so that it codes for a different, shorter protein. Notice, already, the complex, indirect, and equivocal nature of the relationship between the DNA sequences in chromosomes and their phenotypic consequences. In what follows, this message gets yet more support.

Translation from mRNA to protein occurs with the help of devices called *ribosomes* and a second form of RNA, *transfer RNA (tRNA)*, which acts as a physical link between the amino acids that are the constituents of proteins and the final mRNA transcript. The ribosome moves along the mRNA, creating chains of amino acids that are then folded into proteins. The genetic code is *degenerate*—different codons specify the same amino acid—but it is never *ambiguous:* the same codon is never linked via its various intermediaries to more than one amino acid.

Even in the accompanying technical boxes we have barely scratched the surface of the complex machinery that mediates between DNA and protein construction. But the take-home message is simple: One DNA sequence can

Box 6.4 The Genetic Code

In a rather dubious metaphor, the genome of an organism is often regarded as a coded description of the organism as a whole. But there is a sense in which it really is a code for the proteins in the organism. Proteins are made from a stock of twenty different amino acids. So the basic function of the genetic code is to specify those amino acids in the right sequence. Each amino acid is specified by a three-base sequence drawn from the mRNA bases uracil, adenine, guanine, and cytosine. But since there are sixty-four (4 × 4 × 4) possible three-base sequences, there are sixty-four different codons, and hence there is *degeneracy* in the coding system. That is, more than one three-base sequence can code for the same amino acid. AUG codes for the amino acid methionine, and since all newly synthesized proteins start with methionine, AUG functions as the *start codon*. But there are three *stop codons* (UGA, UAA, and UAG), and sixty-one codons that code for amino acids. The degree of redundancy ranges from leucine, coded by six sequences (UUA, UUG, CUU, CUC, CUA, CUG) to tryptophan, coded only by UGG. An additional source of degeneracy is the differences between the coding mechanism of the genes in the cell nucleus and those in the mitochondria. UGA is not a stop codon for mitochondrial DNA. But though the code is degenerate, it is never *ambiguous:* one codon is always mapped onto one, and only one, amino acid.

be input to mechanisms that yield different protein sequences. So though the RNA codon/tRNA anticodon/amino acid system is not ambiguous in that anticodons always attach to the same codon and are always attached to the same amino acid, this is merely an unambiguous subsystem within a system fraught with ambiguity. It is a system that maps the same DNA sequences onto different proteins and, further, to different phenotypic outcomes. The one-to-many character of the DNA/phenotype relationship is even more apparent when we consider the regulation of genes—the mechanisms that turn them on and off.

6.4   Gene Regulation

A skin cell and a brain cell are very different from each other—and they and their descendants will probably remain that way. Tissue differentiation is often a one-way street. Once a cell lineage has become a lineage of one par-

Box 6.5 Reading the Code

Only one strand of the DNA double helix is read, since DNA can be read from only one end, the 5′ end. From this strand, an mRNA strand is constructed as each base in the 5′ strand is paired with its complementary base. The codons of the genetic code are sequences in this mRNA strand.

Actual protein synthesis takes place at structures called *ribosomes* in the cell cytoplasm. Transfer RNAs (tRNA) are chunks of RNA in the cell cytoplasm, each consisting of three bases. Each tRNA binds at one end to a specific amino acid and at the other, again via the base pairing mechanism in which each base has a unique partner, to the mRNA at the ribosomes. So each codon of the mRNA is recognized by a tRNA *anticodon* with an amino acid attached. As the amino acids are lined up and attached by tRNA to mRNA at the ribosome, they form bonds with their neighbors, and a sequence of amino acids is built. This sequential order, in the right molecular context, specifies the protein.

As we have noted, the genetic code is degenerate. Where it is degenerate, it is usually so at the third position in a codon. So mutations that affect the third position are often *silent:* they have no effect on the amino acid being made. But they can affect the rate at which it is made. For the rate at which the code is read depends on the stock of available reading chemicals. The building of a protein depends on the supply of tRNA in the cell cytoplasm. The range of tRNAs that a cell synthesizes helps to determine the assembly of amino acids into proteins.

ticular tissue type, it usually does not revert to some earlier, more plastic form. Early cell biologists took very seriously indeed the idea that the hereditary material was divided up between the different tissue types, so that the hereditary material for skin went to skin cells and the hereditary material for nerves went to nerve cells, and only the sex cells retained a full copy (the *mosaic theory*). But this hypothesis was disproved. In fact, most cells have the complete genome. The differences between them are due to mechanisms of gene regulation and cell line heredity. These mechanisms are being discovered at an impressive rate, and any attempt to summarize them here would be quickly out of date. Furthermore, even the mechanisms already known are far too varied and complex to describe in a text of this kind. So we offer here some very general observations about these mechanisms, which will play a role in the arguments over reductionism.

Figure 6.4 Transcription and translation. (a) Each base of DNA is transcribed into the corresponding RNA base, producing a strand of messenger RNA. (b) Each codon of the mRNA transcript matches the anticodon on one end of a transfer RNA. The other end of each tRNA carries a specific amino acid. Ribosomes (not illustrated) move along the mRNA, translating it into a chain of amino acids—one of the polypeptide chains of which proteins are composed. (Adapted from Alberts et al. 1994, 108.)

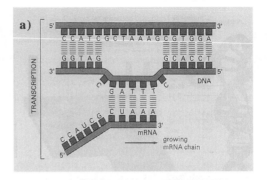

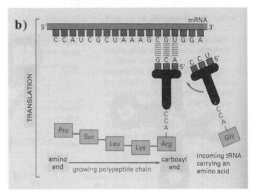

The expression of a DNA sequence can be controlled at almost every stage of the process between the sequence itself and the functional protein it produces. Various posttranscriptional mechanisms operate on the mRNA transcript, as we have already described. Each of these offers a point of intervention affecting the final protein. Splicing and editing affect the type of protein translated, and other processes affect the quantity translated. Since two forms of RNA play an essential role in this process, the rate of translation of mRNA to protein is affected by the availability of tRNAs (which are synthesized from other regions of the genome) and by the rate at which mRNAs are degraded so that they become unavailable for translation.

Gene regulation through control of transcription has been known for much longer than these posttranscriptional processes. The most intensively studied and best understood form of gene regulation involves *regulatory sequences,* short stretches of DNA that bind to certain characteristic classes of *regulatory proteins.* Transcription of DNA depends on an enzyme called *RNA polymerase,* which splits the double helix and begins the transcription process.

Regulatory proteins affect the ability of RNA polymerase to bind to the regulatory sequences and initiate transcription.

The DNA sequences that are transcribed into mRNA are preceded by *promoter* sequences, to which RNA polymerase attaches itself. In prokaryotic cells, such as the bacterium *E. coli,* regulation is relatively simple. Regulatory sequences lie adjacent to the promoters. Some of these bind *repressors,* negative regulatory proteins that interfere with RNA polymerase binding. Others bind *transcription factors,* positive regulatory proteins that facilitate RNA polymerase binding. In eukaryotic cells, such as those of plants and animals, things are much more complex. The RNA polymerases that transcribe eukaryotic genes typically require a whole complex of transcription factors to be present for them to initiate transcription. This complex machinery enables the overall rate of transcription to be influenced by many different factors, contributing to the ability of eukaryotic cells to create many different cell types from the differential activation of a single genome.

Transcription in eukaryotes is also affected by the organization of DNA into chromosomes. Chromosomes are composed of a material called *chromatin,* which consists mainly of DNA and structural molecules called *histones.* The long DNA molecule can be condensed in various ways in chromatin structures. The most compressed forms are known as *heterochromatin,* and DNA in these forms cannot usually be transcribed into mRNA. This form of gene regulation plays a well-known role in female mammals. Females have two X chromosomes, one of which is rendered inactive by being compressed into a dense, heterochromatic *Barr body.*

A cell's pattern of gene activity is frequently passed on to descendant cells that originate from it by mitosis. Some cells pass on to their descendants not only the genome, but a complex of extragenomic factors that they have acquired during the process of tissue differentiation and which cause them to express those genes, and only those genes, needed in that tissue. The inactivation of the second X chromosome just described is a case in point. One or the other X chromosome is randomly chosen to become a dense, inactive Barr body in the founding cells of certain cell lineages. All cells in the lineage inherit the same pattern of inactivation. So female organisms are genetic mosaics, with different sets of X chromosome genes acting in different tissues.

Another mechanism of cell line heredity is *DNA methylation,* in which parents attach methyl groups to the DNA of their sperm or eggs. In vertebrates and some invertebrates, additional methyl groups can be attached to the bases cytosine or guanine. Heavily methylated sequences are not transcribed. An enzyme called *DNA methyltransferase* copies the methylation pattern when DNA is replicated. A gene that was turned off by methylation in the parent cell is thus turned off in daughter cells.

Overall, then, the same lesson as before applies: the connection between DNA sequence and phenotype is not just indirect, it's many-to-many. The effect of DNA sequences on phenotype is modulated by mechanisms that turn genes on and off, mechanisms that affect the rate at which "on" genes are transcribed and translated, and mechanisms that determine which proteins are eventually built from a transcribed sequence. So the relationship between DNA sequence and phenotype is many-to-many with a vengeance.

6.5 Are Genes Protein Makers?

Just as early research in genetics was guided by the ultimately untenable "one gene–one trait" concept, early research in molecular genetics was guided by a "one gene–one protein" concept. The classic molecular gene concept is a stretch of DNA that codes for a single polypeptide chain. We have not tied any of the foregoing discussion to this important gene concept, referring instead simply to DNA sequences. That is because the classic molecular gene is a highly problematic unit in light of the very processes of transcription and translation that we have just described. The original intent of the classic molecular gene concept was to identify a gene with the DNA sequence from which a particular protein is transcribed, via mRNA. But even ignoring the fact that reading frames may overlap, the relationship between DNA sequences and protein chains is many-to-many, not one-to-one. To see this, consider the role of regulatory sequences. These sequences do not themselves code for a protein (so, if they are independent genes, the classic molecular gene concept is already in trouble). But unless at least some noncoding regulatory machinery is included along with the transcribed sequence, the presence of a gene does not explain the presence of the relevant protein. If all regulatory and promoter sequences were adjacent to the transcribed sequences they regulate, we could regard the whole sequence as a single gene. Bacterial genetics more or less works this way. The *operon* of bacterial genetics consists of one or more transcribed sequences and their immediately adjacent promoter and regulatory sequences (see figure 6.5a). In eukaryote gene regulation, however, regulatory sequences may be distant from the sequences they regulate and may be involved in regulating many sequences. Genes coding for transcription factors may be arbitrarily distant from the genes transcribed, perhaps because eukaryote DNA can loop around to bring transcription factors bound to distant regulatory sites close to a gene being transcribed (see figure 6.5b,c). Other problems for the classic molecular gene concept arise because of posttranscriptional processes. Alternative splicing and editing may make several different proteins from one primary transcript.

The upshot, then, is that molecular biologists do not seem to use the term *gene* as a name of a specific molecular structure. Rather, it's used as a floating label whose reference is fixed by the local context of use. Molecular biologists often seem to use *genes* to mean "sequences of the sort(s) that are of interest in the process I am working on." Their rich background of shared assumptions makes this usage perfectly satisfactory. However, it then follows that there is no straightforward translation of talk about genes in Mendelian genetics to talk about genes in contemporary molecular genetics. As we shall see, the antireductionist consensus makes the further point that the relationship between genes and the structures molecular biology has identified— exons, introns, reading frames, promoters, repressors, mRNA, tRNA—is so complex that there can be no clean mapping of Mendelian genes to *any* molecular kinds. We cannot identify Mendelian genes with molecular genes, for *molecular gene* is not the name of one specific molecular kind. But we cannot identify them with any other molecular structure, either.

One possibility at this point is to see these considerations as arguing for the displacement of Mendelian genetics by molecular biology. Contemporary geneticists have proposed, for example, that the dominant/recessive distinction be replaced by a gain of function/loss of function distinction. Recessive phenotypes, according to this idea, are typically the result of an organism being saddled with two copies of a defective gene. The recessive phenotype develops because something does *not* happen. Moreover, though genes can lose function for more than one reason, this would still be a more cohesive molecular-level explanation than the dominant/recessive one. One problem with this revisionary idea is that the gain of function/loss of function distinction depends on how wild-type gene functions are defined. *Oncogenes,* for example, are dominant and represent an inappropriate (from the organism's point of view) gain of function leading to cancer. However, it might be argued that the true "function" of an oncogene is to remain silent in certain cell types, and it is a *loss* of function in its control system that leads to its *gaining* the ability to be expressed at the wrong time (Chambers, personal communication). A more straightforward problem is that some loss of function mutations are dominant; for example, in cases in which the loss of one allele lowers protein production below a critical threshold level.

Classic accounts of reduction acknowledged that the old theory would often have to be "corrected" before it could be reduced. The old theory might contain elements unconnected with its explanatory successes (but

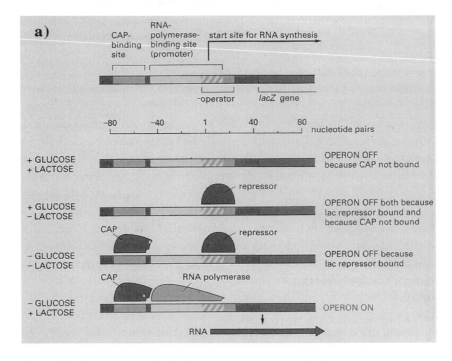

Figure 6.5. Gene regulation. (a) The *lac* operon in the bacterium *E.coli* was the first gene regulatory mechanism to be understood. The operon consists of a transcribed sequence plus one *promoter* site and one *repressor* site adjacent to the start site for mRNA transcription. The regulatory proteins bound at these sites respond to glucose and lactose concentrations. The regulatory factor CAP (catabolite activator protein) helps the enzyme RNA polymerase to open the double helix and initiate transcription of the DNA. The repressor protein stops this process from proceeding. This causes RNA polymerase to be bound and transcription to commence only when there is a low concentration of glucose and a high concentration of lactose. The resulting gene product metabolizes lactose into glucose. (b) Gene regulation in eukaryotes is much more complicated than in bacteria. The TATA box is a sequence of T-A and A-T base pairs close to the start site for mRNA transcription. This sequence binds a collection of general transcription factors (involved in the same process for many other genes). Regulatory regions specific to the particular gene may exist far upstream of the TATA box, or even downstream of the transcribed sequence. (c) The regulatory proteins bound to these distant regulatory regions are thought to be brought into contact with those bound to the TATA box by looping of the DNA. (Adapted from Alberts et al. 1994, 420, 424, 429.)

perhaps responsible for its explanatory failings) that could not be derived from the new theory and the bridge principles. However, if too much correction were required to effect a reduction, this process would no longer be one of theory reduction, but of theory replacement—that is, of displacement rather than incorporation or integration. No one would dream of "correcting" the phlogiston theory of combustion to say that phlogiston is taken up in combustion rather than lost in combustion and then claiming to reduce the phlo-

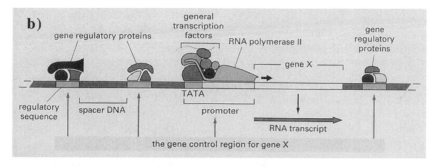

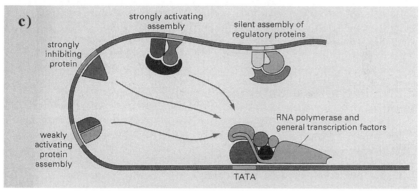

giston theory to the oxygen theory. The phlogiston theory was just wrong, and the oxygen theory displaced it. In one view, the "corrections" in Mendelian genetics that would be required in order to reduce it to molecular genetics are so large that this project resembles the frivolous proposal to "reduce" phlogiston to oxygen. So, just as the Churchlands take the irreducibility of psychological kinds to neural kinds to show that there really are no such things as beliefs, Rosenberg takes the irreducibility of classic genes to molecular genes to show that molecular genetics displaces Mendelian genetics:

> Molecular genetics reveals that there is no one single kind of thing that in fact does what Classical genetics tells us (classical) genes do. In this respect of course molecular genetics replaces classical Mendelian genetics. (Rosenberg 1997, 447)

One of the best current texts offers a summary review of "classical genetics," beginning with the claim that in classic genetics a gene is "a functional unit of inheritance usually corresponding to the segment of DNA coding for a single protein product" (Alberts et al. 1994, 1072). This, of course, is the classic *molecular* gene concept; Mendelian genes have disappeared from the map altogether.

The displacement view is not as widely accepted as either of the two alternatives. One alternative is the idea that Mendelian genetics is a viable science even though it does not reduce to molecular genetics: it can be integrated with, but not incorporated within, molecular genetics. The other alternative is that, despite appearances, reduction is possible after all. It is to these ideas that we now turn.

Further Reading

6.1   The classic account of theory reduction is given by Nagel (1961). See Boyd, Gasper, and Trout 1991, part III, for a selection of recent papers on reductionism from contemporary philosophy of science. Schaffner (1967) describes his more flexible "general reduction model." Chapter 9 of Schaffner 1993 contains a thorough survey of the literature on theory reduction since Nagel, including versions driven by the fashionable "semantic view of theories," which we have not discussed here.

As we note in the text, our picture of the history of genetics is very superficial. For serious treatments of this history, see (for the early days) Olby 1985, and for the development of Mendelian genetics in the fruit fly lab, Kohler 1994. For a very readable narrative of the molecular revolution, see Judson 1997. A more philosophically focused account of the history is given by Depew and Weber (1995). Mayr wears a historian's hat too: part III of Mayr 1982a is his account of the development of genetics. Dupré (1993) and Rosenberg (1994) present an interesting contrast. They essentially agree in thinking that the classic accounts of theoretical unification fail to fit biology. But whereas Dupré develops a case for thinking that the program of unification and the metaphysics that underlies it is wrong-headed, Rosenberg argues that unreduced biology cannot be regarded as an objective account of the way the world is. So their work is relevant throughout this and the next chapter.

6.2–6.5   The history of the gene concept is complex and controversial. Falk (1984, 1986) discusses its many transformations. Portin (1993) presents a good recent treatment. As usual, Keller and Lloyd (1992) provide a good entrée into the literature; Maienschein overviews the history of the concept, and Kitcher surveys its current uses. An authoritative source on modern molecular biology is Alberts et al. 1994. For accessible introductions to these difficult issues, see Moore 1993 or Mayr 1982a.

7

Reduction: For and Against

7.1 The Antireductionist Consensus

The classic account of theory reduction underpinning the incorporation of one theory into another is quite simple (Nagel 1961). The old and new theories are first made commensurable by providing translations from the vocabulary of one theory to that of the other. Then the old theory is shown to be deducible from the new theory, given these translations, and perhaps some restrictions on the range of systems for which the old theory is reasonably accurate. The translations from the vocabulary of the old theory to that of the new theory are known as *bridge principles* or *bridge laws*. In the case of classic and molecular genetics, the bridge principles would specify which molecular structures count as genes, how to recognize the dominance of one allele over another in molecular biology, and so forth. In the restricted range of cases in which classic genetics is accurate, it should be deducible from molecular genetics via these bridge principles.

The logical empiricist philosophers who originally developed this account of reduction supposed that bridge principles would always be available. They believed that the theoretical terms of a genuine scientific theory gained their meaning from the way the theory related them to observation and experiment. Hence it should always be possible to compare the vocabularies of two theories by translating them into a common vocabulary of observations. This view was challenged in the 1960s when Kuhn and Feyerabend argued that there is no theory-neutral observational vocabulary in which to state bridge principles (Feyerabend 1962; Kuhn 1970). Although this challenge has been extremely important in philosophy of science, it has never been one of the reasons for denying the reducibility of classic to molecular genetics, and so we will pass over it here. It has always been assumed by both sides in the debate that molecular biologists could determine how their theories would translate to the preferred vocabulary of genetics: the ratios of observable phenotypes produced by cross-breeding.

The antireductionists' most fundamental claim is that any number of different molecular arrangements could correspond to a single category in classic genetics. Bridge principles for the terms *gene, locus, allele, dominant,* and so forth would relate each of these Mendelian kinds to many different molecular kinds. Molecular genes coding for different mRNA transcripts function as alleles, but so do noncoding regulatory regions that affect transcription, and so do various forms of a sequence coding for a product involved in alternative splicings of some other gene product. A similar situation exists with the family of terms associated with the dominance relation between alleles. In Mendelian genetics, an allele is dominant if its characteristic effect is seen in the heterozygote. There are at least as many molecular ways to be dominant as there are ways for an allele to have a phenotypic effect, and as we have seen, there are many ways for an allele to have a phenotypic effect.

The fact that the bridge principles between Mendelian and molecular genetics have this one-to-many form means that the different instances of a single Mendelian kind may have no distinctive molecular property in common. Therefore the bridge principles are not lawlike. They do not connect a natural kind identified by hybridization and observation with a natural kind independently identified by molecular biology. What properties do the molecular structures that count as alleles all share? They have some effect on the phenotype, perhaps through their epistatic effect on the expression of alleles at other loci, and they occupy chromosomal locations that cause them to assort and recombine so that those phenotypic effects are expressed in Mendelian ratios. These properties are precisely those that Mendelian genetics ascribes to alleles. Molecular structures are recognized as alleles for no other reason than that they obey the principles of the old theory. The fact that alleles obey these principles cannot then be explained by the fact that molecular correlates of alleles obey them, since that is true by definition. The molecular ensembles that correspond to the Mendelian kinds do not emerge from molecular biology, but are constructed by grouping together diverse molecular events that look the same when viewed using the experimental techniques typical of classic genetics. The reduction relationship this generates is not one in which the new theory explains the old, but one in which the new and old theories represent complementary and mutually illuminating ways of viewing the same physical processes.

We do not see this as a troubling conclusion. There is nothing here to undercut the uncontroversial but important sense of reduction we identified in section 6.1, the ban on miracles. This idea of reduction has played, and continues to play, an important regulative role in scientific debate. As we noted in section 6.1, the evidence for continental drift was quite impressive even before World War II. But continental drift remained marginalized among geologists, in part because the mechanisms they proposed to shift continents were implausible. The "drifters" of the thirties conceived of continents as plowing through the ocean floor rather as a concrete slab might be pushed half through, half over the top of, a layer of earth. The proposed forces were too weak, and the stresses on the continental crusts would be far too great for them to survive the passage. Until drift could be backed up with a plausible physical mechanism, driftist explanations of continental movement were hard to accept (Le Grand 1988). Scientific theories cannot traffic in apparently miraculous mechanisms. There is a tree of explanatory dependence that links together all the different causal mechanisms postulated in science. That tree is rooted in fundamental physical processes. Through various different branchings, all scientific kinds depend on that root. The light molecular biology sheds on classic genetics is quite adequate show that inheritance in classic genetics is not mysterious or spooky in any way.

7.2   Reduction by Degrees?

When a macrolevel object, property, or process can be built in many different ways out of its microlevel constituents, we speak of that property (or object or process) being *multiply realized*. It is realized (made real) by different microlevel configurations. The claim that the theoretical entities of classic genetics are multiply realized at the molecular level is the core of the antireductionist consensus. Perhaps this argument looks so powerful only because we have supposed that the key kinds of Mendelian genetics are to be directly reduced to DNA sequences. But there is an important reductionist alternative. First, Mendelian kinds are reduced to gross features of cytology and development. Something is an allele just because it has a chromosomal location. Allele A dominates allele a because, for some complex developmental reason, the Aa heterozygote resembles the AA homozygote. Alleles on homologous chromosomes assort independently because there are two chromosomes that separate in meiosis, carrying one allele to each gametic cell. Second, these gross features of cytology—assortment, crossing-over, and the like—are explained by molecular biology. Molecular biology has shown that chromosomes are structures of histones and DNA, and is starting to explain how the cell moves these structures about in meiosis. The law of independent assortment is reduced to the molecular mechanisms of chromosome structure and meiosis. So Mendelian genetics is reduced to molecular biology in a two-stage rather than a one-stage process.

Kitcher has responded to this version of reductionism by arguing that the explanatory power of cytological features does not depend on their molecular implementation (Kitcher 1984). As Waters puts it in his critical response, according to Kitcher, the "gory details" of molecular mechanisms are irrelevant to the explanatory power of Mendelian principles (Waters 1994c). The gory details of the chromosome and its dance are not important in explaining the law of independent assortment. That law is fully explained when we know that there are two chromosomes and that one goes to each gamete. This "gory details" argument is a variant of the multiple realization thesis. Kitcher compares the situation to knowing why a round peg won't fit into a square hole, which is fully explained by the shapes of the two items, however they are physically realized. An explanation that specifies the molecular configurations of the peg and the hole is *too* detailed (Putnam 1978, 42). Any molecular configuration of the same shape would produce the same effect. Similarly, Kitcher argues, the generalization that explains independent assortment is an abstract statistical generalization about the effect of randomly dividing pairs of entities.

At this point we need to avoid becoming enmeshed in pointless squabbles about what counts as reduction and explanation. For, as we noted in discussing gene selection (4.2), there is a place for explanations that abstract away from the details of an event's causal history, but also for explanations that are rich in detail. A geometric explanation of why a round peg fails to fit into a square hole is a robust process explanation. Any variation in circumstances that preserves the gross geometry of peg and hole will yield the same outcome. An explanation of a particular peg's failing to fit a particular hole in terms of their precise physiochemical composition is an actual sequence explanation, for it gives the detailed, close-grained explanation of this particular event. These two explanatory strategies are compatible because they answer different questions. So the argument that Mendelian genetics can always ignore the gory details of its link with the molecular world is gratuitously strong. We might well want an actual sequence explanation of why, say, the sex chromosome and the sickle-cell gene sort independently. After all, some genetic diseases are sex-linked. However, there would remain a robust process explanation of independent segregation, one independent of the actual sequence explanation, just because there may be many molecular mechanisms that determine the placement of genes on particular chromosomes.

Robust process explanations are important when, and to the extent that, macroscopic processes are invariant over changes in the microscopic processes on which they depend in each particular case. If the antireductionist view that Mendelian categories and molecular ones are related in highly complex many-to-many ways is right, then Mendelian genetics is integrated with, and causally explained by, molecular biology, but it is methodologically and conceptually independent of that discipline. But that is not to deny the significance of individual, close-grained actual sequence explanations. Moreover, a sense of reduction is involved here too: the ban on miracles. The molecular explanation of meiosis shows that the law of independent assortment is mechanical; no spooky mechanism is involved.

7.3   Are Genes DNA Sequences Plus Contexts?

We have just considered the idea that the concepts of classic genetics can be reduced to molecular structures indirectly, via developmental biology. But perhaps the problem isn't with reduction as such, but with the proposal that classic genes reduce to DNA sequences alone. As we saw in sections 6.3 and 6.4, many molecular elements in addition to DNA play a role in the phenotypic effect of a DNA sequence. So perhaps classic fruit fly genes such as *wingless, white eye color,* and the like are DNA stretches *plus* the molecular machinery that uses them. In his seminal presentation of antireductionism, Hull remarked "The only plausible molecular correlate for a dominant gene is a highly specified molecular mechanism, not an isolated stretch of DNA" (Hull 1974, 24).

This idea gets off the ground because, as Hull notes, Mendelian geneticists have always been on the lookout for some concrete, structural object that they can identify as a gene. It was in this way, for example, that they made sense of the "position effects" first discovered in the 1920s, in which the same gene has a different effect on the phenotype when it is moved to a new location. If genes were actually *defined* by their position and function, the very idea of a position effect would have made no sense. Yet any structurally defined segment of DNA has the properties of an allele only because it is embedded in a much broader molecular context. So if we are to identify an allele with a specific molecular structure, the allele has to be the DNA plus this context.

The idea of identifying genes as DNA sequences in their context poses two problems. Hull himself thought there would be a serious problem because the bridge principles from the genes of classic genetics would have to specify unmanageably large chunks of the molecular context. Moreover, the relationship would still be one-to-many. Many different DNA sequences in different contexts would count as instances of the same allele. Second, this identification of a gene poses a problem for the decompositional strategy we identified in section 6.1 as one strand of reductionist thinking. The guiding assumption of this strategy is that the constituents' causal powers are relatively independent of their environment, so that the system can be taken apart and each part understood in isolation. If molecular biology had vindicated "bean-bag genetics," the idea that each trait of an organism is explained by the separate action of a particular gene for that trait, the decompositional strategy would have triumphed spectacularly (Kitcher 1984). But, on the contrary, this explanatory strategy may well be undercut by developments in cell biology. In addition to large-scale cytological events being explained by the action of their molecular constituents, molecular events are being explained in the context of the broader cellular milieu in which they occur. The large multinucleate cell that constitutes the early fruit fly embryo expresses different genes in different parts of its cytoplasm, and processes these gene products differently, because of the uneven distribution of chemicals in the cytoplasm. Hence large-scale developmental biology explains the action of individual genes just as much as the action of individual genes explains development.

So identifying genes with DNA sequences in their cellular milieu might not be a reductionist strategy at all, in one important sense of reduction. But reductionist or not, it is certainly a viable option. It has been recently defended by Neumann-Held (1998), whose takeoff point is the impossibility (in her view) of identifying genes with DNA sequences alone. We have already described how the effect of a gene depends on the broader molecular context of the cell (6.4). Neumann-Held argues that even whether a DNA sequence counts as a gene depends on the context in which it occurs. This context depends in turn on the processes by which cells differentiate and become part of larger units of biological organization. Neumann-Held suggests that a gene is a *process* that regularly results, at some stage in development, in the production of a particular protein. With rare exceptions, this process centrally involves a linear sequence of DNA, some parts of which correspond to the protein via the genetic code. But it also involves all the elements of the developmental matrix, inside and outside the cell, that regularly coincide at this stage in development to cause expression of the protein. Perhaps the most radical feature of Neumann-Held's proposal is that it makes genes themselves include environmental causes in development! Despite this feature, and the fact that she seems to "reduce" genetics to developmental biology rather than the other way around, Neumann-Held's radical proposal has some analogies with the reductionist views we are about to consider. The classic geneticists proposed that a gene was a unit underlying a given hereditary characteristic. Neumann-Held's proposal retains this property of genes at the expense of making genes simple sequences of DNA.

7.4 The Reductionist Anticonsensus

The antireductionist consensus depends on the complexity of the relationship between the genes identified by transmission genetics and molecular structures. As we have just seen, that consensus has been challenged. The inspiration of the reductionist anticonsensus can be summed up in a quote from the eminent geneticist Gunther Stent:

> What geneticist could take seriously any explication of "reductionism" which leads to the conclusion that molecular genetics does *not* amount to successful reduction of classical genetics? (Stent 1994, 501; italics in original)

Several philosophers of biology agree with Stent that if current philosophical accounts of reduction do not yield the desired conclusion, then it is the accounts of reduction that are at fault. Obviously, classic genetics was not just some horrible mistake, and we should not say about it what we say about phlogiston. So there must be some relation between the factors of inheritance identified by classic genetics and the molecular machinery discovered later on: a relation that explains the very considerable theoretical and predictive achievements of the classic tradition. In our view, the reductionist anticonsensus can be seen as raising, though not settling, the following questions:

1. Is the relationship between molecular and classic genetics strikingly different from, because it is more complex than, say, the relationship between heat, pressure, and volume and the kinetic properties of particle aggregates? If so, is it misleading to label both "reductions"?

2. As we have noted, the idea of reduction comes with a load of theoretical and ideological baggage. It is partly a legacy of sophisticated versions of positivist philosophy of science, a view of science that is dominated by models from physics and chemistry. So, if the relationship between more and less fundamental domains is typically complex, should classic theories of reduction, together with their attachments, be abandoned and replaced rather than updated?

3. Even if in some interesting sense classic genetics does reduce to molecular genetics, there may be an important sense in which macroscopic explanations remain independent of reducing explanations. As we have noted, actual sequence explanations do not exclude robust process explanations. So might not Mendelian genetics remain an independent theory more or less integrated with molecular genetics, rather than being incorporated by it?

Two of the most consistent advocates of reduction have been Kenneth Waters and Kenneth Schaffner. Schaffner was the philosopher who first suggested that classic genetics was being reduced to molecular biology (Schaffner 1967, 1969). The antireductionist consensus developed in response to this suggestion. In the thirty years since he first proposed this idea, Schaffner has developed a series of increasingly sophisticated and "data-driven" models of how theory reduction actually proceeds in the biological sciences (Schaffner 1993, 1996). Waters, too, has argued that the relation between Mendelian and molecular biology is at least in the spirit of the classic account of theory reduction, and has produced replies to many of the antireductionists' arguments (Waters 1994a,c).

Waters's reply to the multiple realization thesis has two elements. First, he argues that it depends on treating genes as causes of traits rather than causes of trait differences. Second, he doubts that there is an interesting, autonomous explanation of Mendelian facts through cell cytology. So if we are to explain Mendelian principles at all, the explanations will be molecular, and the "gory details" will matter. We start with multiple realizability. His reply begins with the observation that Mendelian genetics never claimed that there were genes for traits, only that there were genes for trait *differences*. A *red eye* allele in *Drosophila* does not really cause the production of red eyes. Instead, it makes the difference, in the presence of many other causes, between red eyes and eyes of some other color. If we think of genes as entities that code for phenotypic traits, we will reach the antireductionist conclusion that for any trait, many complex molecular arrangements can constitute the gene for that trait. However, if we concentrate on the idea of genes as difference makers, Waters claims that the entities that make such differences all turn out to have something in common: "The gene can be specified in molecular biology as a relatively short segment of DNA that functions as a biochemical unit" (Waters 1994c, 407). The phrase "functions as a biochemical unit" seems to nicely bring under one heading both coding regions and regulatory sequences of various kinds. This approach also fits well with the idea that genes are not self-sufficient causes, but difference makers in a larger causal process (a view developed most fully in Sterelny and Kitcher 1988). Clearly, alternative promoter or regulatory sequences that occupy allelic positions on the chromosomes could be difference makers in this sense, and thus genes in the sense of classic genetics.

In practice, though, Waters interprets his proposal as something like the classic molecular gene concept. This concept, he suggests, "is that of a gene for a linear sequence in a product at some stage of gene expression" (Waters 1994a, 178). So *gene* means "coding sequence," and regulatory sequences are really only parts of the coding-sequence genes that they regulate. There is an obvious problem with this proposal, which is that for some purposes, molecular biologists clearly do not regard some regulatory sequences as parts of the genes they identify. This is not surprising. In eukaryotes, transcription factors can bind to sites distant from the gene they regulate, and hence regulatory regions can assort independently of the gene they regulate. Waters suggests that the conversational context indicates to the molecular biologist which stage of gene expression is of interest, and hence whether to adopt a wider or narrower conception of the gene. While this explains how molecular biologists understand one another's usage, it hardly defends the view that *gene* names a single molecular unit.

Neumann-Held (1998) has expressed considerable skepticism about Waters's definition of *gene*. She suggests that it adds no more than a verbal unity to the diverse molecular units that geneticists refer to as genes. According to Waters's proposal, we must rely on the conversational context to determine whether to include introns, adjacent regulatory regions, distant regulatory regions, coding regions for transcription factors that bind to the regulatory regions, or coding sequences for factors involved in splicing or editing in "the gene." This suggests that *gene* does not really name a unit of molecular biology, but is shorthand for any of several different units. As we noted in section 6.5, *gene* is used in molecular biology as a shifting tag rather than as a name for a specific molecular kind. A few examples from the literature will illustrate the diversity of its actual use. First, *gene* means "coding region": "In this chapter we use *gene* to refer only to the DNA that is transcribed into RNA . . . , although the classic view of a gene would include the gene control region as well" (Alberts et al. 1994, 423). Second, Alberts et al. use *gene* to name the unit of function in transcription: "This definition [of gene] includes the entire functional unit, encompassing coding DNA sequences, noncoding regulatory DNA sequences, and introns" (Alberts et al. 1994, G-10). Biologists also use *gene* to mean a sequence of exons. There are further alternatives, which will become more important as posttranscriptional processing becomes better understood. So Waters's defense of reductionism against the multiple realization argument is at best problematic. The shifting use of *gene* undercuts the first stage of Waters's argument, which identifies genes as difference makers and hence suggests that they are identifiable molecular kinds.

Second, Waters doubts that there is any decent explanation of Mendelian principles in terms of cytology. This charge looks very plausible when leveled at some of Kitcher's examples. For instance, Kitcher suggests that the law of independent assortment is explained by the fact that pairs of alleles are situated on pairs of chromosomes, one of which goes to each daughter cell. Stripped of any details explaining *why* chromosomes separate—stripping that is necessary because these details vary case by case—this is a pretty thin explanation, and Waters's skepticism looks justified. However, other findings in developmental biology suggest that there really is a robust and interesting level of structure between Mendelian patterns and molecular structure. For example, one important and much reinterpreted concept in developmental biology is that of the *morphogenetic field* (for the most recent interpretation, see Gilbert, Opitz, and Raff 1996). A morphogenetic field is a region of the developing embryo that acts as a unit. A developing embryo, in this view, is a mosaic of such three-dimensional regions. Within each region, cells interact strongly with one another; between regions, there are relatively weak interactions. The precursors of the segments in the body of an arthropod emerge in this fashion long before they develop the physical boundaries that mark segments in the adult. Morphogenetic fields are set up by the action of genes in combination with environmental influences and the existing cytoplasm. The action of the genes is significantly influenced by the different chemical milieu of each field. For example—and very roughly—the initial distribution of chemical traces in the arthropod egg determines the differential gene expression in various areas of the egg that sets up the first fields. Differences in the fields lead to further differential gene expression, which further differentiates the fields from one another and creates fields within fields, and so forth.

The morphogenetic field concept provides an example of the sort of large-scale explanation that might not be illuminated by a case-by-case reduction to molecular processes. Developmental biologists and geneticists have long considered it a real possibility that such large-scale patterns in development *canalize* development toward certain outcomes (Waddington 1959; Kauffman 1993; Goodwin 1994; Wagner 1996; Wagner, Booth, and Homayoun 1997). Development compensates for minor changes in the genome in just the same way that it compensates for minor changes in environmental inputs, protecting important developmental outcomes against interference from such variation. Across a wide range of parameters, developmental outcomes will be invariant, and a robust process explanation identifies the space within which development is or is not canalized. An example of this phenomenon may be provided by the reversion of the bithorax mutation in *Drosophila*. This mutation in the important homeobox regulatory genes converts a segment of the fly into a copy of the segment carrying the wings, yielding a four-winged mutant fly in place of the usual two-winged wild type. However, bithorax strains need continuing selection to maintain the mutant form. Left to itself, the lineage will revert to the wild type (H. F. Nijhout, personal communication).

The idea that developmental outcomes can be stable in the face of underlying genetic variation has two implications. First, it makes the generalizations of developmental biology multiply realizable at the molecular level, creating the kind of theoretical independence for these generalizations that the antireductionists claimed to identify. Second, it creates the possibility of explaining gene action in terms of biological processes at a larger scale, and thus undermines the basic intuitions concerning the direction of explanation in science that motivate reductionism in the decompositional sense. There will, of course, be a close-grained actual sequence explanation for any particular developmental outcome. But the fact that that developmental sequence gives rise to, say, a two-winged rather than a four-winged fruit fly is itself explained by biological processes at a broader scale. Hence the existence and importance of actual sequence explanations does not undercut the explanatory importance of larger-scale explanations.

Schaffner has given the contrast between general robust process explanations and case-by-case explanations his own distinctive twist. He argues that the role of actual sequence explanations is filled in a special way through work on model organisms. Their role is both to exhibit and hence demystify the actual causal mechanisms involved in particular biological processes—say, the expression of segmentation genes in a fruit fly embryo's development—and to serve as a rough, partial template for similar explanations of the same process in other organisms. Thus theories in the biomedical sciences are characterized by a mixture of broad and narrow causal generalizations. The broad generalizations resemble traditional scientific laws, but the narrow ones elucidate the workings of particular model systems and are not expected to be applicable in any unmodified form to other cases. Model systems, which may be individual gene systems such as the *lac* operon in the bacterium *E. coli* or whole organisms such as the nematode *C. elegans,* act as exemplars (roughly, inspirational case studies) for work on less well studied systems of a similar type.

Our conclusion after reviewing both the "antireductionist consensus" and the "reductionist anticonsensus" is that nobody wins. Rather, considerable progress in understanding the relationship between molecular biology and classic genetics has been made under both headings. It has become clear that the reducing theory is not really independent from the theory it is supposed to reduce. Molecular genetics did not emerge cleanly as a new discipline with categories and laws that explained the successes of its predecessor. Instead, molecular biology has subsumed and enriched classic genetics, turning it into the modern transmission genetics that still plays a crucial role in determining the actual functions of stretches of DNA. In part this is because molecular biology, or molecular genetics, is misnamed. It is not a simple extension of biochemistry, but rather the study of how biochemical and other physical laws operate in the complex and varied cellular contexts that evolution has produced. The concepts of classic genetics, most notably *gene* itself, continue to play a role in molecular biology, although perhaps as little more than shorthand for the various DNA sequences and collections of interacting DNA sequences used in molecular biological explanations of organisms and their traits.

Further Reading

7.1–7.3   An extended treatment of the debate over genetics and reductionism has recently been given by Sahotra Sarkar (1998). Hull's original presentation of antireductionism in his *Philosophy of Biological Science* (1974) is still an excellent introduction to this debate. Other important presentations are those of William C. Wimsatt (1974, 1976, 1994, in press); Phillip Kitcher (1984); Alexander Rosenberg (1985), and Dupré (1993). Kitcher's account is developed further in Kitcher 1989 and Culp and Kitcher 1989. The parallels between these issues in biology and psychology are especially evident in Jerry Fodor's presentation (1974, 1975). Neumann-Held (1998) outlines her concept of the contextualized or constructionist gene. There is a brief note in *Nature* objecting to ideas along these lines as conflating the distinction between what a gene is and how it is used (Epp 1997). Griffiths and Neumann-Held (in press) reply to this objection.

7.4   Waters defends reductionism in a number of papers (1994a,c). Schaffner's views are developed in Schaffner 1969, 1993, 1996, with a good concise summary in Schaffner 1993. Gilbert, Opitz, and Raff 1996 is a good, not overly technical account of the morphogenetic field concept and its importance within developmental biology. The relationship between developmental and molecular biology, with special reference to that concept, is explored in a recent paper by Richard Burian (1997). Rosenberg (1997) argues against the idea that developmental biology has any macrolevel explanatory generalizations; in his view, the morphogenetic field concept and its ilk are descriptive, but not explanatory. So his paper is a critique of antireductionism from the perspective of developmental biology. But he interprets the antireductionist position as having a commitment to "top-down" causal explanations, so the antireductionism under Rosenberg's gun is a stronger position than the "antireductionist consensus" that we outline here. It is, however, relevant to Neumann-Held's identification of a gene. Some of Rosenberg's ideas take off from an interesting paper on these issues by Wolpert (1995).

III

Organisms, Groups, and Species

8

Organisms, Groups, and Superorganisms

8.1 Interactors

Many wasps are parasites whose attacks are fatal to their hosts—they are *parasitoids*. Among these is a small wasp, *Copidosoma floridanum* (small enough so that a thousand can develop in a single caterpillar). A female wasp lays one or two eggs in the larva of a moth. After a female egg hatches, it divides into a clone of identical siblings. Most of these stay together in a larval mass until the final developmental stage of the caterpillar, when they develop, eating the caterpillar from within, mature into adult wasps, and disperse in search of new hosts. But a small fraction of these wasp embryos (fewer than 50 of a total of over 1,000) develop early in the caterpillar's life into large and well-armed larvae that travel through the body of the host, seeking out and eating any *C. floridanum* larvae that are not their siblings as well as competing parasites of other species. These warrior morphs are doomed to die without themselves reproducing. For when the normal larvae—their unprecocious sisters—complete their development and consume the host, these precocious larvae die with the host (Grbic, Ode, and Strand 1992; McMenamin and McMenamin 1994; Hardy 1995, 13).

This snapshot of wasp natural history poses the central puzzle of this chapter: Should we think of this clone of wasps as a single biological individual—a single interactor—despite its being physically scattered throughout the body of the host? The death of parts of organisms is normal. We are always shedding bits of ourselves for the greater good. So if the clone of wasps is a single organism, then the termination of the warrior morphs is no more surprising than our shedding skin. But if we think of the warrior morphs as separate individuals, their behavior is surprising indeed. What could explain their self-sacrifice? Organisms are often forced to take risks. Life is risky, and selection can at best design organisms to choose the lesser danger. But how could selection build in a developmental pathway that is not just risky, but is certain to lead to death without reproducing?

This chapter takes up the problem of identifying the individuals whose competitive struggles determine the success of replicator lineages—that is, the problem of identifying interactors. Our wasp example illustrates the themes to come: the relative importance of physical integration, genetic identity, and coadapted, especially altruistic, behavior.

In section 2.3 we distinguished between two fundamental challenges to the received view. One is based on the interactor/replicator conception of evolutionary history. The other is based on a hierarchical conception of evolution, and it is this conception that now moves to center stage. The organism is the most striking feature of the living world, and the evolution of organisms is a central feature of evolutionary history. However, according to the hierarchical conception of evolution, organisms are but one among several levels of organization in the living world. Hives, roosts, herds, troops, families, and other groups appear to form composite individuals: individuals that are composed of organisms. So socially organized groups of organisms may also be a level of biological organization of evolutionary significance.

Hierarchical views of evolution vary, but their central focus is on interactors. What are the interactors whose differential ecological success results in the differential growth of replicator lineages? How do we recognize interactors? Evolutionary theories differ in their answers to these questions. In this chapter we consider one *version of* hierarchical conceptions of evolution. We consider the idea that baboon troops, lion prides, beehives, termite mounds, and the like are collective individuals *(superorganisms); that is,* they are themselves interactors, rather than being just populations of interactors.

The complexity, integration, and coordination of eusocial insect societies and colonial marine invertebrates is one reason for thinking of them as superorganisms. But the main motivation for thinking that groups of organisms are themselves interactors has been the problem posed by altruism. Many group-living animals—for example, meerkats and prairie dogs—warn their conspecifics about predators. Other animals signal food finds, engage in collective defense against predators, care for offspring not their own, and even forgo reproduction entirely to care for their nestmates. Altruism reaches its most extreme form in those eusocial insect societies whose complexity and coordination are also so impressive. Since organisms are in competition with one another, such aid to others is surprising. We would expect ruthlessly selfish animals to be fitter than ones that sacrifice their own interests to help others. Hence altruistic behavior is an apparent paradox for evolutionary theory. But as we shall see, such behavior can be explained by *group selection,* a model of selection that recognizes groups of organisms as interactors in their own right.

In the following section we explain the challenge posed by altruism. In section 8.3 we consider the rise and fall of traditional group selection; in section 8.4, its contemporary replacement. In section 8.5, we focus on a skeptical response to these ideas. In section 8.6, we return to the evolution of the organism, and the light that it throws on collective individuals.

8.2 The Challenge of Altruism

Altruism is a puzzle. Imagine, for example, that you are a male vervet monkey in a tree, and that you notice an eagle. Do you give an alarm call, warning all the monkeys around you, or do you quietly hide? Selection should favor quiet hiding. For then you certainly will not be attacked by the eagle, and one of your rivals in the group may well be. This is especially so if calling would attract the predator's attention to *you* (Hauser 1996, 427–28). Over time, we would expect selection to weed out the trait of warning others about predators, as well as signaling the presence of food, contributing to collective defense (as defenders would lose out to cowards and skulkers), reproductive restraint, and caring for others' young. "Look out for Number One" should be Mother Nature's first and only rule.

What, then, could explain altruism? There seem to be three possibilities. First, altruism could be error. Organisms are not perfectly adapted to their environments. If they were, no predator would ever make a capture; no prey would ever escape; no parasite would ever penetrate a host's defenses; no potential host would escape parasitism. Perhaps altruism is just a manifestation of the inevitable frailties of organic design. The robin feeding cuckoo chicks is sacrificing its breeding season for the cuckoo, paying for its imperfect capacity to recognize its own young. Naked mole rats live in colonies in which only a single dominant female breeds. It seems that she suppresses the reproductive cycles of other females through hormones in her urine. Perhaps their restraint is merely a consequence of her manipulation of their endocrine systems. Perhaps the lioness who suckles another's cubs is being parasitized, her maternal mechanisms infiltrated and subverted. No recognition system is ever perfect; some propensity for error is inevitable. But not every type of error is equally costly. Rejecting your own true cub—who will then starve—is a much more costly mistake than tolerating an occasional interloper. So a lioness may feed others' cubs rather than risking rejection of her own as a side effect of making herself invulnerable to freeloaders.

No doubt some apparent altruism flows from imperfect design. It is not surprising that cuckoos can parasitize other birds, for the asymmetry in selection pressure is marked. There is no living cuckoo who has not succeeded in tricking a host into feeding it. In contrast, there are many living hosts whose parents lost a brood to a cuckoo. Independent evidence also exists for the view that cuckoo hosts are suboptimal cuckoo detectors. Some species that are rarely parasitized by cuckoos show strong antiparasite adaptations. They detect and expel cuckoo eggs and mob cuckoos near their nests (Moksnes et al. 1990). These species may have "caught up" with the cuckoo! The increased vigilance of these former hosts has forced the cuckoo to switch to new targets. However, it's hard to see how induced error could explain collective defense, alarm calling, or food signaling. So an "error hypothesis" will not explain all instances of altruism.

A second possibility is that altruism is an illusion. Some acts appear to sacrifice the agent's interests in favor of others, but do not actually do so once we have a full understanding of the costs and benefits involved. This possibility is central to contemporary debates on altruism. For example, ravens that give loud yells when they find large carcasses turn out not to be acting altruistically after all. These ravens are young birds with no territories of their own. Though their calling recruits others with whom they must share their food bonanza, if they did not call, they would be expelled by the territory owners. Recruiting other ravens swamps the territory owners' defenses (Heinrich 1990). Ostriches have an odd breeding system in which a male both broods egg clutches and supervises creches of chicks that include many young that are not his. This behavior apparently dilutes the effect of predation. The extra eggs and chicks give his own offspring a better chance of escape. Fish that leave their schools to "inspect" predators may be advertising their alertness to the predator rather than collecting information for their school. Mark Hauser has shown that monkeys that keep silent when they find choice food items are punished if detected in their silent gobbling. So perhaps the prudent monkey calls when he finds food because honesty is *for him* the best policy (Hauser 1996, 583). We must, however, be careful in depending on the effect of punishment on utility. Punishment typically imposes a cost on those that inflict it, and where it does, the appeal to punishment simply moves the bump under the rug. Punishment poses exactly the same problem as collective defense: Why not leave it to others? An analogous problem is well known from social and political theory: We cannot explain collective action by appeal to community sanctions, for those sanctions are themselves instances of collective action.

So-called *reciprocal altruism* is an important element of this strategy of explaining away altruism. If two or more animals can secure some resource by cooperating that neither could secure individually, individual selection could promote joint action. Social predators such as African wild dogs take and then share prey that no individual could kill by itself. It is in each dog's interest to act with the others, so long as the individual's share of the joint carcass is more valuable than any prey it could catch by itself. Reciprocal altruism takes this apparently unproblematic form of cooperation as its model and extends it to cases in which the partners do not reap their individual benefits simultaneously. Primate social life provides apparent examples: Franz de Waal (1982) describes chimpanzee coalitions in which one partner benefits by achieving alpha male status, and his supporter benefits by being allowed access to females in estrus. But the best-known example of reciprocal altruism is found in vampire bats, which share blood. These bats starve unless they feed every couple of days, and hunting failure is quite common. So reciprocation is an essential element in vampire bat life. Successful bats share with those who fail, but bats that give are bats that receive (Wilkinson 1990).

So apparently altruistic behavior might be explained as mere error, or it might not really be altruism at all. A third alternative is to explain the altruistic behavior of individuals by appealing to selection on the collectives of which they are members. There is no difficulty in seeing how a baboon troop whose adult males cooperate effectively in defense against carnivores would do better than one saddled with males each of which tries to hide behind the others. An ant nest defended by guards prepared to sacrifice themselves for their colony is more likely to survive attack than one less zealously defended. In this view, the collectives themselves are interactors. The cooperative baboon troop characterized by collective defense is an interactor, and one with importantly different traits than a selfish troop. It is more likely to survive and to found new troops: troops like itself. So the population of baboon troops is a salient level of biological organization, a level at which interactors—baboon troops—vary and compete with differential success. In doing so, they promote the replication of some gene lineages (genes with extended phenotypic effects on troop character) and suppress the replication of others.

One of the most difficult issues within contemporary evolutionary theory revolves around this option, and it has provoked two quite different critical responses. First, evolutionary hypotheses about collective interactors have been challenged empirically: perhaps there is an unnoticed benefit to the individual organism. Perhaps the vervet monkey's alarm call means "I see you, eagle, so it's pointless trying to catch me." The monkey, for his own benefit, advertises his alertness to the potential predator; any benefit to his fellows is a side effect. The male baboon that charges a leopard menacing his troop advertises the fact that he is far more dangerous than any of the leopard's other potential meals (Zahavi and Zahavi 1997). As we shall see in chapter 10, testing competing evolutionary hypotheses is not easy. Nonetheless, though it may be hard to find out whether these hypotheses of individual advantage are correct, it's reasonably clear what they are claiming.

A second challenge to group selection involves redescribing selective episodes in other terms, so that selection on collective interactors seems to disappear. Selection on groups has often been reinterpreted as individual selection tied to the individual's social environment. The pukeko—Australasia's purple swamphen—often lives in hierarchically structured groups that breed communally, defend territory jointly, and in which the dominant birds allow subordinate males and females to breed (Jamieson and Craig 1993). We might interpret the relatively egalitarian distribution of breeding opportunities, and the fairly harmonious life of the group, as adaptive consequences of pukeko family group selection. Harmonious and egalitarian groups defend territory better than others, for subordinates are not put in a position in which they are better off defecting. An alternative is to see them as consequences of selection on individual pukekos. In an environment in which others are tolerant, it pays an individual to tolerate the breeding attempts of others in the extended family. In a tolerant group there is no obvious individual gain in "defecting" to intolerance (unless a single defector changes the character of the whole group). So selection favors tolerance, but only because tolerant birds are usually in the company of others with similar dispositions.

Thus our assessment of hierarchical conceptions of evolution turns on two factors. One is empirical: How good are the counterhypotheses? Though we shall occasionally offer our own hunches, assessing these counterhypotheses will not be our main focus. The other is conceptual: When there are alternative descriptions of the same evolutionary episode, which should we prefer? For the most part, we shall argue that the alternative accounts are equivalent. But there are cases in which, in our view, the individual selectionist redescriptions are implausible. Evolution sometimes does involve selection between competing collectives.

8.3 Group Selection: Take 1

In the fifties and sixties an influential group of evolutionary biologists explained much social behavior by appeal to its effect in promoting the good of the social group. Pecking orders and other dominance hierarchies, for example, minimized wasteful conflict within the group. Male courtship displays ensured that only the best and the fittest had mates. In the culmination of this tradition, Vero Wynne-Edwards argued that many species have mechanisms to ensure that groups do not overexploit their resource base (Wynne-Edwards 1962, 1986). His central example of altruism was reproductive restraint. The "central function" of territoriality in birds and other higher animals, he argued, is "limiting the numbers of occupants per unit area of habitat." Dominance hierarchies and communal breeding systems also limit populations. These social mechanisms have population regulation as their "underlying primary function" (1986, 9). Wynne-Edwards argued that these mechanisms evolve through group selection. Populations without such mechanisms are apt to go extinct by eroding their own resource base.

As we noted in section 8.1, one response to apparently altruistic phenomena is to argue that the appearance of altruism is misleading, and that the behavior is really to the advantage of the agent in question. Such *individual selectionist* counterhypotheses were often quite persuasive. Moreover, in contrast to hypotheses invoking group selection, they were empirically testable. Take reproductive restraint. It is true that many birds lay a clutch of eggs below their physiological capacity, for many species can re-lay if they lose all or part of their clutch. Moreover, experimental manipulations have shown that birds can often fledge more chicks than they hatch. But David Lack showed that such reproductive restraint can benefit the individual (Lack 1966). The winter casualty rates among small songbirds are very high, and a bird fledging fewer but larger young may well see more of its nestlings survive to their first breeding season. Moreover, there is a trade-off between reproductive effort in any one breeding season and a bird's chance of itself surviving the winter. So a less than maximal effort this season may be an investment in the bird's prospects of having another.

What of the more spectacular examples of altruism? Perhaps a robin is being canny in choosing not to lay all the eggs she can, but a bee that stings an intruder at the certain cost of her own life cannot be saving anything for a rainy day. The idea of *kin selection* has played an important role in explaining away the appearance of altruism even in these spectacular cases. The intuitive idea of kin selection is simple. We can think of an organism's *fitness* as its expected contribution to the next generation's gene pool. The reproductively triumphant organism increases the proportion of its own distinctive genes in the gene pool of the next generation. But an organism shares many of its genes, including its distinctive genes, with its relatives, especially its close relatives. Having children is just one way to be causally responsible for making copies of your own genes. Another is by helping your relatives to reproduce. There have been hints of this idea in evolutionary theory back to

Box 8.1 Inclusive Fitness

Inclusive fitness is an important theoretical tool, but, as Grafen (1982) has pointed out, it is very hard to measure. Suppose that Alex has one child, and helps his brother to have three children. To calculate Alex's inclusive fitness, we need to know:

1. how many children the brother would have had if Alex had not helped him

2. how many children Alex would have had if he had not helped his brother

3. to what extent Alex's relatives helped him to have his one actual child

Only with all this data can we really measure Alex's causal contribution to the number of copies of his genes in the next generation. If we forget to "strip out" everyone's personal fitness, we will count each child more than once when we measure inclusive fitness! This mistake makes kin selection appear much more powerful than it really is. If Grafen is right, this mistake is common among biologists.

Philosophers of biology make their own mistake, which is to think that "classical fitness" (how many children you have) is a less accurate measure than inclusive fitness. But in fact, classical and inclusive fitness models of selection are equivalent. If we calculate the average classical fitness of a kin group whose members help one another, we will detect the fitness increase that drives selection for mutual aid.

Darwin, but it was first made explicit by William D. Hamilton (Hamilton 1964a,b, 1996). Hamilton defined a new notion called *inclusive fitness*, designed to measure the number of copies of its genes for which an individual is causally responsible. If Alex has a "personal fitness" of one child, but also helps his brother to have three children rather than one, those two extra relatives contribute to Alex's inclusive fitness.

Most of the eusocial insects (termites are the exception) have a peculiar genetic system by virtue of which sisters are very closely related. In organisms with standard genetic systems, each individual has one chance in two of sharing any particular gene with a full sibling. Ants, bees, wasps, and the like have *haplodiploid genetic systems*. Females develop from fertilized eggs and have genes from both parents. They are diploid organisms like us, with two copies of each chromosome. Males, however, develop from unfertilized eggs. Males have no father and no sons. They have a random selection of one from each of their mother's pairs of chromosomes (perhaps modified by crossing-over), and they transmit all their genes to their daughters. They are haploid, with one copy of each chromosome. In diploid species, mothers are as closely related to their daughters as to their sisters. But haplodiploid full sisters are more closely related to one another than to their daughters, for they share *all* their paternally derived genes. Two full sisters thus have three chances in four of sharing any given gene; one in two if it is a maternal gene, and one in one if it is a paternal gene. A mother and daughter have only one chance in two of sharing any given gene, since half the daughter's genes come from the father. Because sisters are so closely related, they can project their genes into the future particularly effectively by aiding sisters, and the queen is sister to the workers. This was Hamilton's kin-selective explanation of why eusocial insect workers help the queen rather than having offspring of their own. It should not be forgotten that eusocial insect nests often have more than one queen, and that queens often store sperm from more than one male. So many of the workers are not one another's full sisters. Even so, it's widely accepted that Hamilton's kin selection mechanism has been central to the evolution of the spectacularly altruistic behaviors of eusocial insects.

When the theory of kin selection was first developed, it was taken to show that kin-selected behavior was only apparently altruistic. The self-sacrificial bee was acting in her own reproductive interests, albeit her indirect reproductive interests. In protecting the hive, she did her best to project her own genes into the future. So kin selectionist explanations were seen as superior individual selectionist alternatives to those involving group selection. They explained the most striking examples of altruistic animal behavior and complex cooperation without recourse to group selection. We shall revisit this issue in section 8.4.

The critics of group selection developed alternative explanations for many of the phenomena group selection was supposed to explain. But they also pointed to a deep problem with the idea of altruistic groups outcompeting selfish ones. Altruistic groups seem very vulnerable to subversion from within. Imagine a rabbit population varying in levels of reproductive restraint. Rabbits in restrained warrens, let's suppose, delay first breeding. Restrained warrens do much better in harsh winters, for most rabbits in profligate warrens starve. Even so, if there is migration between warrens, or if rabbits that lack restraint arise by mutation in restrained warrens, the unrestrained rabbits will gain the benefits of living in a restrained warren without paying the costs. Hence the unrestrained rabbits will undermine restraint. If group selection favors reproductive restraint and individual selection selects against it, what would we expect to happen? The power of selection depends on generation time and the amount of variation in the population. Remember, effective selection is cumulative selection, so the shorter the generation time, the more selective episodes, and the more effective selection will be. And selection is powerless without variation, so (within limits) the more variation, the more powerfully selection acts. Hence the shorter generation time of individual selection (since groups last longer than their members) and the greater variety of individuals (since there must be many more individuals than groups) suggests that in a race between the two selection processes, individual selection usually wins. Moreover, once the individually advantageous trait of fecundity appeared in a warren, we would expect the proportion of fecund rabbits within the warren to rise inexorably. It would seem that for selection for restraint on rabbit groups to be effective, either fecund warrens would have to go extinct very rapidly and be replaced by colonists from restrained ones, or there would have to be some mechanism that tended to block the establishment of fecund invaders of restrained warrens.

No one thinks that group selection is inherently impossible. In a series of experiments on flour beetles, Michael Wade showed that group properties do respond to selection on groups (Wade 1976; discussed in Sober 1984b, 264–66). Moreover, it is quite widely thought that group selection explains some unusual sex ratios. In section 3.3, we explained why the normal equilibrium sex ratio is roughly 50/50: if the ratio is unbalanced, the rarer sex becomes the most valuable. But this within-group process can be countered by group selection. Imagine an insect that feeds on abundant but widely scattered carcasses. These carcasses are typically found by one or a few mated females, which lay eggs in them. The progeny consume the resource, then mate and disperse. Almost all die, but a few find and colonize new carcasses. The more that disperse, the more likely it is that there will be some successful searchers. A female-biased sex ratio allows more successful colonizations, for a lone mated female can colonize successfully, but a male cannot. Under these circumstances, despite continued within-group selection for a balanced sex ratio, a female-biased sex ratio can, and apparently has, evolved. But though there can be selection on groups, until very recently it was received wisdom that group selection is effective only in rather specialized circumstances.

8.4 Group Selection: Take 2

Contemporary defenders of group selection, most notably David Sloan Wilson and lately Elliot Sober, defend a cut-down, austere concept of an interactor. What features of an organism enable it to function as an individual in competition with other individuals? What makes an organism an interactor? Wilson and Sober define an interactor as an entity whose parts share a *common fate,* rather than one whose parts form a complex organization. An organism is more than a population of cells, not because those cells form a complex system of organs, but because those cells share a common fate. Their reproductive fate is locked together on a single causal trajectory. Similarly, a group of beavers in a lodge is an interactor if their fitness is linked together on a common causal trajectory. Beaver traits that affect that trajectory for better or worse can be visible to selection through the fate of that beaver collective. Obviously, common fate comes in degrees. No one suggests that the fate of the individual beavers is as interconnected as the fate of the cells of one beaver. Beavers in a lodge, unlike the cells in an individual beaver, vary in fitness. Nonetheless, some particular traits of the beavers in the lodge affect all those beavers, and in the same way. Wilson and Sober propose that common fate is defined on a trait-by-trait basis. If the beavers cooperate in the construction and maintenance of their dam and lodge, then that characteristic will have a common effect on all the beavers in the collective. Those beavers will be affected, and in the same way, by this feature of group's members, and hence with respect to this trait—the dam-building trait—the beavers share a common fate.

Thus D. S. Wilson introduced and defended the idea of trait groups. *Trait groups* are groups of organisms, each of which feels the influence of the others with respect to some trait. If the trait is dam building, the trait group is the group of beavers that live and shelter behind the dam. Different traits will pick out different groups. In the most obvious examples, these groups are homogenous with respect to the trait in question. A group of beavers, all of which live behind their dam and maintain it jointly, would constitute such a group. However, trait groups need not be homogeneous. All the beavers that live behind a dam are a trait group even if a few are freeloaders. For all those beavers, and only those beavers, are in "the sphere of influence" of those that build the dam.

Wilson argued that the trait group is a unit of selection. Trait groups composed of altruistic animals can outcompete trait groups of selfish animals. This can be true even if *within every trait group* the selfish individuals are outcompeting the altruistic individuals. Sober showed that Wilson's idea that selection can work one way in every group and the opposite way in the ensemble of those groups is an instance of *Simpson's paradox* (Sober 1993, 98–102). This paradox relies on the assumption that although altruist groups contain some selfish individuals, they contain more altruists (and vice versa). Whenever there is a correlation between having a trait and interacting with

Table 8.1   Simpson's Paradox in Action: Before and After Selection

	Selfish individuals before selection	Selfish individuals after selection	Who is fitter than whom?	Altruistic individuals before selection	Altruistic individuals after selection
Selfish group	40	20	>	5	0
Altruistic group	5	8	>	40	40
Combined population	45	28	<	45	40

others that have the same trait—that is, whenever there are trait groups—Simpson's paradox can appear. In tables 8.1 and 8.2 we can see this paradox in action. Both tables represent the fate of the same organisms before and after selection in a single breeding season. In the first table we give absolute numbers; in the second, the same facts are represented by fitness values; thus, for example, the selfish individuals in the selfish group had a 50/50 chance of breeding. In each group, the altruists are less fit than the selfish. Successful breeding is tough in the selfish group, but it's especially tough for the altruists in that group: none managed to succeed. Life is easier in the altruistic group—helping helps—but freeloading clearly pays off. But more altruists in the population as a whole breed. The second table represents the same facts in terms of fitness rather than actual numbers, but the pattern is the same. The engine of the paradox is that *everyone* is fitter in the altruist group, and that most altruists are in that group. Conversely, most selfish individuals are in the less fit selfish group.

As we saw in section 8.3, critics of group selection have argued that we do not need selection on groups to explain animal cooperation, and have presented alternative explanations for apparently altruistic behaviors. Wilson's response has been to argue that these alleged alternatives to group selection are in fact instances of it. For example, he argued that kin selection is a special case of trait group selection. If members of a family treat one another in ways that contrast with their behavior to those outside the family, then kin groups are trait groups. If, say, a chimpanzee will share food only with her offspring, then she and her offspring form a trait group: the chimps within the sphere of influence of the food-sharing trait. If all chimp mothers behave this way, then with respect to food sharing the local population will be divided into mother-focused family groups. This sort of division of populations can lead to the evolution of altruism. Within the kin group, of course, a selfish individual that "defects" from the cooperative norm will do better than its

Table 8.2   Simpson's Paradox in Action: Fitness Calculations

	Selfish individuals' fitness	Who is fitter than whom?	Altruistic individuals' fitness
Selfish group	0.5 ($\times$ 40 = 20)	>	0 ($\times$ 5 = 0)
Altruistic group	1.6 ($\times$ 5 = 8)	>	1 ($\times$ 40 = 40)
Combined population	28	<	40

altruistic relatives, for the defector will enjoy the benefits of aid without bearing the costs of giving it. A female lioness that did not allow her sisters' cubs to suckle would improve the prospects of her own cubs reaching maturity. But a pride of altruist lions will raise more cubs than a mixed pride, which in turn will do better than a wholly selfish pride. So if the benefit of altruism is great enough, and the cost is small enough, the average fitness of the altruists can be greater than the average fitness of the defectors. Hence when the kin groups dissolve back into the general population before the next round of breeding, the proportion of altruists can rise despite freeloading in mixed groups. So genetic relatedness is important, but only because it can generate the correlation between *having a trait* and *interacting with others with the same trait* that is needed to drive Simpson's paradox. Since kin tend to resemble one another, kin groups containing one altruist are likely to contain others; similarly with selfishness. So where kin form trait groups, kin selection can have important evolutionary consequences. Where kin do not form trait groups because kin do not interact with one another in any distinctive way, organisms will still have inclusive fitnesses, but nothing will come of it. If none of an individual's traits differentially boosts the fitness of his relatives, then none of those traits boosts the replication of the gene(s) associated with them.

In Wilson's account of kin selection, kin groups are the interactors, so kin selection is a variant of group selection, not an alternative to it. Wilson and Sober treat reciprocal altruism in the same way. The exchange of favors has been seen as an explanation of cooperation that appeals to individual rather than group benefit. Each participant benefits from the exchange, though not necessarily at the same time. According to Wilson and Sober's alternative analysis, reciprocation evolves when selection favors reciprocating groups over other groups. A pair of cooperating bats sharing blood share a common fate. Wilson and Sober press their reanalysis of reciprocation through an amusing thought experiment (Wilson and Sober 1994). Imagine a cricket population that feeds on lilies scattered across a pond. The problem for the crickets is to get from one lily pad to the next. Wilson and Sober imagine the evolution of cooperative navigation across the pond, as pairs of crickets evolve the capacity to row between lily pads on dead leaves. Individual crickets cannot row effectively without a partner. The required coordination evolves by selection on pairs. Crickets better able to coordinate with their partners are fitter than their clumsier colleagues. But this adaptive advantage is visible to selection only through the increase in efficiency with which *a pair* reaches a lily pad. With respect to each trip, the partners share a common fate, and hence coordination evolves by group selection. The pair is an interactor, even if these are the only cooperative interactions between the crickets; even if a cricket rarely has the same partner twice; even if the great bulk of the cricket life cycle is between trips.

The relentless, though unplanned, march of evolution continues. A selfish mutant cricket arises that casts its partner adrift at the end of the trip. It does well when its partner is naive, but poorly when paired with another selfish morph, for each has a tendency to drown the other. Within-pair selection favors the mutant and causes the selfish behavior to spread. But evolution marches on: eventually a suppressor morph arises that prevents the selfish morph's behavior by clasping it when the two arrive at their destination. The clasping morph spreads through the population by pair selection alone, for whatever the nature of its partner, the two crickets benefit equally from every trip; there are no within-pair fitness differences. Throughout this whole evolutionary dynamic, Wilson and Sober think of the pair as a unit. The pair is the beneficiary of the joint behavior whose benefit is distributed over its members.

If Wilson and his allies are right, the two most central tools for explaining the evolution of social behavior—the kin selection hypothesis and reciprocal altruism—are versions of trait group selection. In their picture of evolution, the division of organisms into trait-defined groups plays a significant role in the evolution of a wide range of behaviors in many different lineages. These theorists do not expect trait group selection to be the only force acting in the evolution of some trait. Often a population of groups will include "mixed" groups, and the outcome will depend on the combination of selection between groups and selection within groups. For this reason, trait group models are sometimes called *intrademic* models of group selection, in contrast with the earlier, *interdemic* models. (A *deme* is a group of freely interbreeding individuals.) Although they acknowledge the power of "subversion" by individual selection, Wilson and Sober strongly deny that we should automatically expect within-group selection to swamp between-group

Figure 8.1 The older, interdemic model of group selection. S = selfish individuals; A = altruistic individuals. Groups with too few altruists go extinct, and their territory is recolonized by groups with a higher proportion of altruists. Notice, however, that even in altruistic groups, the proportion of selfish individuals creeps up. (Adapted from Dugatkin 1997, 20.)

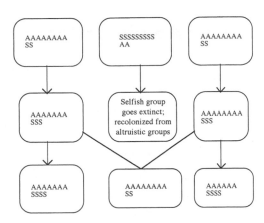

selection. This expectation rests on the assumptions that groups are few in number, long-lived by comparison with the individuals within them, and without effective defenses against subversion. If kin groups, water cricket pairs, chimpanzee coalitions, and the like are all trait groups, these assumptions are flawed.

Moreover, under traditional group selection models, the extinction of selfish groups, which are then replaced by altruist groups, was the only mechanism by which group selection for altruism could counteract individual selection for selfishness. Under Wilson's trait group model, group selection also exerts its effects through the greater productivity of altruist groups. Trait groups sooner or later merge into the general population, and the next generation of groups re-forms from that population. If groups that form from the general population tend to be differentiated, with altruists associating with altruists and freeloaders with freeloaders, then group selection acts through altruistic groups pumping more individuals into the general population from which new groups form, not just through the differential extinction of groups.

Putting all this together, first, Wilson argues that collective interactors need not be highly coadapted, physically integrated, and mutually coordinated groups like eusocial insects. A group whose members' fitness is linked together is a trait group, and trait groups are (potential) interactors. Second, he argues that the central alternatives to group selection as explanations of cooperation are in fact versions of group selection. Third, he argues that the assumptions underlying the view that group selection must be weak in comparison to individual selection are flawed. Figures 8.1 and 8.2 illustrate some of these themes.

Figure 8.2 The new trait group or intrademic selection model. S = selfish individuals; A = altruistic individuals. (a) Reproduction occurs in trait groups. (b) Within each trait group, S does better than A, but the greater productivity of A-dominated groups means that the number of A individuals is slightly greater in the population overall, while the unproductiveness of S-dominated groups causes a slight fall in the number of S individuals. (c) Trait groups blend back into the general population. (d) Trait groups form again in the next generation, and the proportion of A individuals is slightly greater than it was in the previous generation. Notice that this model depends on the productivity, not the extinction resistance, of altruistic groups. (Adapted from Dugatkin 1997, 20)

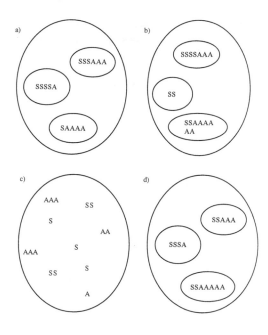

8.5 Population-Structured Evolution

We think that this new version of group selection is interesting and important. We think that the idea of trait groups identifies a very important element in many evolutionary histories. But are trait groups really interactors? Are they really collective individuals in competition in a population of other such collective individuals? An alternative view is to think of trait groups as a critical part of the environment that determines the fate of individual organisms. This alternative, in turn, raises an overarching question: Is there a single best way of describing evolutionary episodes? This question leads us to a third important point. We think that two conceptions of group selection coexist in the literature, and that trait group evolution characterizes only one of those conceptions.

Trait Groups: Interactors or Environments?

The division of a population into groups surely is important in evolution. But it does not necessarily follow that these groups function in ways that parallel the role of organisms. There is an alternative view of the evolutionary processes on which Wilson and his allies focus. According to this alternative view—*broad individualism*—trait groups are aspects of the environment in which selection occurs. Consider first an example for which the group selectionist argument is very persuasive. Eusocial insect groups really do seem to be interactors. They are cohesive, coadapted, and share a common fate. Ant colonies and beehives seem just as "visible" to selection as individual ants and bees. Many species of ants have elaborate warning and defense mechanisms. Often individual workers' defensive chemicals serve a second function of recruiting aid and alerting the colony to danger. Colonies with better warning systems last longer and found more new colonies, all else being equal, than less efficient ones. The colonies they found are likely to resemble the parent colonies in the vigor of their defensive responses. So those colonies become an increasingly dominant proportion of the colony population. Something like this seems to have occurred in the spread of aggressive "Africanized" bees in North America.

The idea that ant colonies are interactors is very plausible. But even when we consider evolution among these eusocial arthropods, there is an alternative version of events in which selection acts on individual insects. Ants with the disposition to respond to danger by defending against it while broadcasting warning odors are fitter in environments in which other ants have similar dispositions. In those environments, their signals will evoke the appropriate response. Their own responses, too, will be appropriate, for they will not be alone in responding. The close genetic relationships within the colony make it likely that an ant with these dispositions will be in similar company. So, instead of seeing warning behavior evolving via selection on ant colonies, we should see it as the result of selection on individual ants. Ants that warn and defend are fitter, on average, than ants that do not. This fitness will, of course, be manifested in their inclusive fitness, the effectiveness of their aid to their relatives.

As always, this fitness advantage depends on the environment in which the evolutionary change is taking place. In this case, a key feature of the environment is the *population structure* of the ant population itself. Ants that warn and defend are fitter only because the ant population is subdivided into colonies, each of which consists of close genetic relatives. Colonies turn out to be a key feature of the *selective environment*.

This reformulation extends to cases of reciprocal altruism. Instead of thinking of Wilson and Sober's hypothetical case of water cricket evolution being driven by group selection, we can think of it as being driven by frequency-dependent selection on individual crickets—selection driven by the relative frequencies of different types of crickets. Thus the clasping behavior is adaptive only when defectors are common. Apparent selection for cooperative groups is really selection for cooperative individuals, provided there are enough other cooperators in their environment. So group selection is converted into frequency-dependent selection on individual organisms.

One True Story?

In the face of these alternative accounts of real ant and hypothetical water cricket evolution, we must clearly consider the possibility that broad individualism and trait group selection are equivalent (Dugatkin and Reeve 1994). If they are, then Wilson's analysis is correct in the sense of being one of the adequate accounts of the evolution of altruism, but it need not be the uniquely correct account of altruism's evolution. Wilson and Sober are themselves pluralists in an important sense. In their view, there is no single evolutionary mechanism that plays the dominant role in every evolutionary episode. So there will be cases in which the organism is the unit of selection, cases in which the gene is the unit of selection, cases in which trait groups are the unit of selection, and cases in which selection operates simultaneously on a number of levels. But they are skeptical of responses to trait group selection that describe *the same episode* in more than one way. They claim that broad individualist redescriptions are apt to involve an "averaging fallacy" of the same kind that gene selectionists have been accused of perpetrating (4.1).

Wilson and Sober argue that broad individualism predicts the outcome of selection, but gives an inadequate account of the process. Suppose that cooperative dam building is evolving in a beaver population because the boost to beaver productivity behind well-maintained dams outweighs the cost of freeloading in mixed groups. A broad individualist account just averages the fitnesses of cooperators in all the groups in which they exist and compares this figure to the average for all defectors. But the average fitness figures for cooperation and defection sum the results of three different selective processes: selection in pure cooperator groups, in mixed groups, and in pure defector groups. This sum yields only the result of selection. The process of averaging bleaches out all the information about process, not to mention all the information about the mechanisms by virtue of which cooperating produces greater fitness than its rival. There is surely much justice in regarding this as a misleading picture of the evolution of dam building. Dam building is clearly a very different kind of adaptation than, say, good thermal insulation, and our account of its evolution should elucidate the difference in the route through which such a social adaptation evolves. Hence Wilson and Sober think that broad individualist redescriptions of selection trivialize the idea that the individual organism is the only interactor.

The proper account of an evolutionary change must retain information about the process of that change, not just report an average result. But there is a way of understanding broad individual selection so that it does not merely average the fitness of every type. Beaver fitness, in this view, has two components. One derives from the beaver's social environment. Beavers that live with other dam builders are fitter by virtue of that fact, for each and every beaver is the beneficiary of beaver contributions to dam maintenance, whether it contributes or not. The other component derives from the beaver's role in its social environment. Defecting beavers that do not contribute to dam building escape from its costs and dangers. So the very fittest beavers are defectors. Even so, the average builder can be fitter than the average defector, because most builders live with other builders. This fact boosts their average fitness, despite the fact that their fitness gets no extra boost from their role within their social environment. Most defectors live with equally defecting neighbors in bad neighborhoods. If they live with no other dam builders at all, they will be forced to depend on natural ponds, with all the attendant risks of natural variation in water levels. This depresses their fitness, despite the fact that these beavers are not penalized by the role they play within their group.

So in one view, cooperative dam building evolves because dam-building groups are more productive—sufficiently so to outweigh individual selection for freeloading. In the other, the dam-building beaver is fitter than the freeloader because it is likely to live in a building group, despite the fact that the beaver's behavior in that group gives it no relative advantage over the others. Freeloaders' fitnesses are not depressed by the role they play in their group. In mixed groups they have a relative advantage over some of their neighbors. But since most live with others that are equally uninvolved, their average fitness is depressed by the character of their social environment. In this alternative picture, the fitness of groups drops out. Both views recognize the importance of the division of the population into groups, and both recognize that an organism's fitness depends both on the character of the group it inhabits and its own character. We think these two pictures are equivalent. So Wilson and Sober's reason for rejecting pluralism are not convincing. A version of broad individualism can capture the process of selection, not just its outcome. Though the trait group conception is often a good heuristic for thinking about social evolution, it is not the only correct view of these evolutionary episodes.

Trait Groups and Superorganisms

In part 3 of this book, we have shifted from a focus on replication to a focus on interaction—on identifying the interactors whose competitive success or failure determines the fate of replicator lineages. But the "pluralism option" that we have just discussed raises the question of whether there can be an objective count of interactors. A parallel from philosophy of mind might illuminate our problem. For reasons both moral and scientific, there is much interest in the problem of identifying *persons* or *intentional agents*—complex systems with beliefs, desires, and purposes. We generally assume that most adult humans are intentional agents. But when in human development do we come to be agents? Are any other animals agents, and hence deserving of the respect and protection of the law? Is it possible in principle to build agents? Would a suitably well designed robot be an agent? If not, why not?

Dennett has argued that there is no objective count of agents. In his view, the identification of agents is relative to observers, and to their knowledge and purposes. For example, it may be useful to treat a chess-playing computer as an agent with chess-directed beliefs and purposes in order to predict its behavior and win the match. In Dennett's terminology, we may take such an *intentional stance* toward a well-designed complex system because it is useful to do so. But in principle, and often in practice, it is not *compulsory* to do so (Dennett 1987). It's often possible to play against a chess-playing computer by exploiting its computational design, by taking advantage of limitations on its ability to evaluate quiet positions.

Dennett is thoroughgoing in his pluralist conception of intentional agents. For some systems—most obviously other human beings—the intentional stance is usually inescapable. We are so complex that in most circumstances the only way we can predict one another's behavior is by treating one another as intentional agents. In other cases—such as the chess-playing computer—we really do have options, and both the intentional stance and other ways of predicting, explaining, and manipulating behavior are available. In explaining the behavior of thermostats, automatic door openers, and simple organisms, the intentional stance seems gratuitous and hence seems like unwarranted anthropomorphizing. Despite these differences, Dennett does not think there is a qualitative distinction between systems that we can usefully treat as if they were intentional agents and real, objective intentional systems. There is only a pragmatic gradient, from systems that are so simple that there is hardly any point in predicting their behavior by crediting them with beliefs and desires, through organisms that are so complex that we rarely have a practical alternative. Hence there can be no objective count of intentional agents in our world. Different observers with different needs and abilities would give a different but equally legitimate count.

Pluralism about high-level interactors raises a similar possibility. According to this line of thought, it can often be useful to take the "interactor stance" toward beaver families, baboon troops, chimp coalitions, and the like. We can treat these as interactors, but we need not. It is equally legitimate to treat them merely as being composed of interactors. But in thinking about whether some biological system is an interactor, is pluralism always a possibility? Or is there an objective distinction between biological systems that are only as-if interactors—for example, temporary coalitions—and real, objective interactors? A Dennett-like view would suggest that any biological system is an interactor only to the extent that it is predictively or heuristically useful to take an "interactor stance" toward it. Perhaps with some biological systems—complex organisms like us—that stance would be pragmatically inescapable. With others—such as Wilson and Sober's pair of cooperating and defecting water crickets—it would be somewhat forced. But the analogy to Dennett's take on philosophy of mind suggests that there is no fundamental or qualitative difference between the water cricket pair and an individual water cricket. Given our perspective and our epistemic limitations, it is harder for us to avoid treating water crickets as interactors than it is to avoid treating water cricket *pairs* as interactors. So we tend naturally to think that water crickets are real interactors and that water cricket pairs are only as-if interactors. But in this conception, that is as much a fact about us and our limitations as it is about biological reality.

We are inclined to reject this radical version of pluralism. Though of course there will be borderline cases, we think there is a real difference in kind between as-if interactors and organisms. This version of pluralism would allow two apparently different but equivalent accounts of evolutionary change in a population of organisms. One would be couched in terms of selection on individual organisms by virtue of their relative fitness. The other would identify the individual *cells* as interactors, and recognize two vectors in their fitness: a fitness component due to their role within their local population of cells, and the fitness they derive from the character of that population vis-à-vis other populations. Hence, radical pluralism fails to recognize the importance of the organism in evolution.

One difference between paradigmatic organisms such as ourselves and looser interactors such as water cricket pairs is that we are structured, integrated, and cohesive. Yet termite nests, ant colonies, and the like are also structured, integrated, and cohesive. If there is a qualitative distinction between organisms and the groups that we can sometimes usefully treat as interactors, should termite mounds and the like fall on the same side of the line as multicellular organisms? There is nothing inherently implausible in the idea that termite mounds are genuine superorganisms. Cells evolved, and then cell assemblages—organisms—evolved. More complex, layered interactors evolved from earlier and simpler ones. So if we reject the idea that there are superorganisms, we need to explain why has it been impossible for yet more complex interactors built from organisms to evolve. On the other hand, if superorganisms exist at all, they are clearly relatively rare. Hence we need to explain what is special about the organism as a level of biological organization in the history of life.

This contrast between water cricket pairs and termite mounds suggests that there is an ambiguity in the group selection debate. One strand of this debate consists of attempts to characterize population-structured selection. These attempts form the central pool of examples of those who defend trait group selection. We suggest that trait group selection and broad individualism give equivalent formulations of population-structured selection. Population-structured selection is a precondition for the evolution of real composite interactors. We shall refer to this stronger sense of high-level selection as *superorganism selection*. In the final section, we attempt to characterize these composite interactors and to distinguish superorganism selection from mere population-structured selection.

8.6 Organisms and Superorganisms

In section 8.5, we made three linked suggestions. First, we argued that some groups of organisms—temporary coalitions, foraging groups, cooperating bat pairs, and the like—can be seen as collective interactors, but need not be. Second, we rejected radical pluralism. Organisms are interactors, and as such are important and objective features of the biological landscape. Third, we suggested, rather more tentatively, that some collective individuals seem to be genuinely organism-like. It is this suggestion we develop in this final section.

There is an intuitive contrast between wild dog hunting coalitions, baboon troops, lionesses sharing cub care, and vervet warning coalitions on the one hand and eusocial insect communities and colonial invertebrates on the other. Recall from section 2.1 the Siphonophora, the colonial marine invertebrates that combine to build jellyfish-like creatures, of which the Portuguese man-of-war is a typical example. In these colonies, the various participating individuals *(zooids)* are specialized for particular morphological roles. Some become flotation cells, others become little jet propulsion devices, and still others are specialized for prey capture and defense, digestion, and even reproduction. Yet each zooid develops from a single fertilized egg, and in many species each zooid reproduces separately. These collective individuals, called *medusae,* seem very different from a lodge full of squabbling beavers.

Other cases are just puzzling. We sometimes think of symbiotic alliances as single organisms. We regard lichens as single organisms, although each lichen is a symbiotic association between a fungus and an alga. Leafcutter ants live in equally obligatory symbiosis with a particular species of fungus that they cultivate and feed on. When virgin queens leave on their nuptial flights, they carry with them a sample of the fungus in a special pouch in their mouths. Despite the similarity between these two cases, we tend to think of the leafcutter ant and its associated fungus as two separate, associated lineages. Yet why should we think this of them, but not lichens?

In section 3.4 we discussed the organism and various attempts to define it. Our main point there was that there is no single definition. Instead, "the organism" turns out to be a highly contestable notion. Let's briefly recapitulate the main points we made there, with an eye on what they might tell us about superorganisms. If there is a common-sense view of the organism, it is the idea that organisms are complex, coadapted, and physically integrated. They have differentiated parts. They are physically cohesive, with an inside and an outside. Since many metabolic processes depend on the existence of this inside/outside distinction, organisms are often equipped with homeostatic mechanisms to ensure that the inside remains stable despite variation outside. One major problem with this definition is that it fits plants very badly.

One alternative to physical cohesion as a defining property is genetic identity. Hence Daniel Janzen argued that, although a field of dandelions may consist of thousands of distinct physiological units, they are parts of a single genetic individual, for they are all parts of a clone (Janzen 1977). The same would be true of the clone of larval wasps with which this chapter begins. In Janzen's view, the multiplication of genetically identical dandelions in the field is growth, not reproduction. The individual dandelion plants are not in competition with one another in any sense relevant to evolution. The success of one over another has no distinctive evolutionary consequences. Only competition between the different clones has any evolutionary upshot.

Dawkins responded to Janzen by arguing that the evolution of adaptive complexity in multicellular organisms depends on a developmental cycle that passes through a single-cell bottleneck. A genetic change can make important differences to the whole organism when development is funneled through this bottleneck. In an intriguing thought experiment, Dawkins compared two padlike growths floating on water, one of which reproduces through single-celled offspring, the other through having chunks break off and grow. The second form might exhibit cellular evolution, but the new pads will never have a structure different from that of their parent, for a genetic change will never reconfigure the whole plant. We think Dawkins's idea is important, yet, as we noted in section 3.4, this potential for overall change is less restricted than he suggests. Many invertebrates go through dramatic changes in form during their life history. So do some vertebrates: amphibians change form quite dramatically, and many fishes change sex. Plants often change very considerably over their life histories. The juvenile form and foliage of many New Zealand trees is strikingly different from that of the adult. At any point in the life history in which a global reorganization takes place, a change affecting that developmental cascade could have global consequences. So, though developmental integration is important, its nature is difficult to define.

As we have seen, Wilson and Sober think of organisms as populations of cells with a common fate. We read Leo Buss's (1987) work on the evolution of individuality as a defense of a somewhat stronger version of this idea. According to Buss, organisms are, among much else, assemblages of cells built by clonal replication. The integration of the organism depends on the accuracy of the processes through which the cells' genomes are cloned. If replication were inaccurate, we would get not an organism, but a mosaic of tribes of cells containing different replicators, and hence with different evolutionary interests. As we have seen, genetic replication is complex and indirect (5.3; 5.5; 6.3; 6.4). Buss argues that these complexities are adaptations to suppress genetic diversity within an organism, and to control it when it does arise. Not only are there proofreading and repair mechanisms that help suppress mutation, but also other mechanisms that localize mutation within one part of the organism, or ensure that it has no access to the germ line (Buss 1985, 93; Buss 1987, 33). *Apoptosis*—programmed cell suicide—also functions to suppress competition between cell lineages (Legrand 1997). We can see Dawkins's developmental bottleneck as simply one example of an adaptation ensuring that no DNA in a given organism can replicate except by aiding the replication of other DNA. If there were no such developmental bottleneck, the way would be open for distinct cell lineages to control different avenues of reproduction. Seeing development from a single cell as just one instance of a more general phenomenon explains how it can be so important for understanding the evolution of some organisms while being absent in, for example, many plants.

In the face of these different ways of conceiving of organisms, we have a number of options. First, we could try to show that one of these ideas gets it right, that it isolates *the* essential feature of being an organism. Second, we could argue for a "package deal" conception of the organism. We could agree that these different ideas all latch onto something important in the evolutionary invention of the organism, but that no one of them is sufficient.

This might be an attractive option to those who think that organisms play a unique role in evolutionary history, which explains why the received view is so plausible. The uniqueness of the organism might be due to the *combination* of these features. Third, we could argue that there is no single notion of the organism. Perhaps each of these criteria define an important biological natural kind. The memberships of these kinds overlap, but we have already seen many examples—symbiotic associations, colonial quasi-organisms, clones—in which they are not identical.

There is a particular version of this third line of thought that we think is quite important, for we think it explains an important polarity in evolutionary thinking about the organism. There is a tension between characterizing the organism as a unique and uniquely important feature of the biological landscape and using it to characterize a role that is exemplified particularly clearly by paradigmatic organisms. One approach attempts to characterize what is distinctive about organisms as an organizational level of nature. Dawkins exemplifies this approach: his notion of the *vehicle* is an attempt to characterize the distinctive evolutionary role of organisms. But a different approach focuses on what makes organisms exemplary instances of the more general category of interactor. Hull exemplifies this approach in his notion of the *interactor*.

We regard these issues as open and very difficult. But we claim that organisms are objectively interactors, and that some collective individuals are enough like organisms in their crucial respects to be real superorganisms. Hence they too are objective interactors. Our discussion of different "definitions of the organism" suggests two ways in which these ideas might be developed and defended.

One way is to emphasize the importance of physical cohesion and the existence of a physical boundary between the organic system and the rest of the world. Physical boundaries are important in two ways. First, a physical boundary gives us a clear and natural segmentation of an evolutionary process. We can easily distinguish between a baboon's phenotype and the environment that makes that phenotype adaptive. Where there are evolutionary episodes with no objective boundary between an adaptation and the environment that makes that trait adaptive, the pluralist option is a live one. In thinking of water cricket evolution, we can "boundary-shift." We can focus on one cricket type—the clasper—and regard the rest, including the clasper's partner, as the environment. Alternatively, we can focus on the rowing pair, and treat everything else as environment. Nothing seems to make one boundary right and the other wrong. Often, in thinking of adaptive change, relative stability can establish this boundary. When we think of adaptations to aridity by small mammals, we can take the climate as fixed and the phenotype as changeable. But this is not true of the evolution of social traits: the evolution of clasping changes the environment of the water cricket. Since in these cases we lack any way of making an objective distinction between the organic system and the environment, there is an important sense in which baboon troops, kin groups, and temporary coalitions do not have determinate phenotypes. We cannot draw an objective boundary between their design and the environment for which they are designed. That is why we can reinterpret trait group selection as selection on individuals in a particular population structure. But the same is not true of all collective individuals. Termite mounds, beehives, ant colonies, and, even more obviously, the colonial marine invertebrates have boundaries. They have an inside and an outside. There is no greater difficulty in segmenting termite mound from termite mound environment than in doing the same with the individual termites. Termite mounds are integrated, cohesive, and have a physical boundary.

There is a second reason why the existence of a physical boundary between organic systems and the rest of the world is important: As a physical boundary develops, the units within the boundary become increasingly important to one another. They become the dominant element of one another's environment. As cells cease to be under the direct control of the external environment, their selective environment becomes the community of which they are a part. Cells in one organism do not interact with cells outside that organism except indirectly, via those cells' effects on the organism they are in. In contrast, the members of baboon troops, bat pairs, wolf packs, and the like continue to have direct interactions with many creatures outside their groups. Baboon troop members interact directly with many organisms, including members of other troops. Such permeable groups interact with the environment as a collective with respect to some selective agents and as individuals with respect to others. Botflies interact with a nest of fledglings as individuals; one may be attacked without the others being affected. But raccoons or snakes interact with the kin group as a whole: a successful attack on the nest will mean the loss of all the eggs. So groups become more like organisms as their members become the dominant features of one another's environment.

The existence of a physical boundary is one of the conditions that promote the evolution of adaptations that suppress competition within the group. This role of a physical boundary leads to our second strategy for distinguishing between objective and as-if interactors. We suspect that there is a connection between having a life of one's own—interacting directly with other biological individuals—and having a fate of one's own—having one's reproductive fate not irrevocably tied to that of others. Wilson and Sober define interactors through the idea of a common fate. In their view, super-organisms evolve as competition within the group is suppressed. When this suppression is permanent and relatively complete, a true biological individual has evolved (see also Buss 1987, 184). We noted above that different traits will divide a population into different trait groups. Dam building divides the beaver population into one set of groups; kinship and alarm calling may well divide it into different groups. But as competition is suppressed, more and more traits will pick out the same groups of organisms. Many different ant traits will divide the ant population into the same groups: their colonies. Complexity, integration, and collective adaptation thus gradually emerge as within-group competition is minimized, controlled, or eliminated.

So the problem is unsolved. The status of the inside/outside barrier in the evolution and identification of an organism-like level of organization is still very much open. Though we think there is a significant, theoretically motivated distinction between trait groups and superorganisms, the nature of superorganisms, and their status within evolutionary theory, are not well understood.

Further Reading

8.1, 8.2 For a discussion of the general idea of levels of organization in nature, their significance, and how to identify then, see Wimsatt 1994. The problem of altruism has been discussed extensively in recent work on the evolution of social behavior. For typical responses to the problem, see Wilson 1975, Krebs and Davies 1981, 1984, and Trivers 1985. These works all assume, rather than seriously argue for, the failure of group selective explanations of altruism. The same is true in spades of Cronin 1991. This bias in the literature is set straight by Sober and Wilson (1998). For a bold attempt to read virtually all altruistic behavior as advertisement of an individual's quality for individual benefit, see Zahavi and Zahavi 1997. For good surveys of the biological phenomena of altruism, see Dugatkin 1997; Brown (1987) looks specifically at birds. For a most unusual perspective, surveying bacteria, see the papers in Shapiro and Dworkin 1997. Most of these papers are technical and specialized, but the two introductory survey papers are accessible.

8.3 The classic defense of group selection is presented by Wynne-Edwards (1962); he returns to the fray in Wynne-Edwards 1986. Somewhat earlier, and also influential, was Allee's work (1951). The classic critique of these ideas is by Williams (1966), though Maynard Smith (1964, 1976) has also been influential. Wade (1978) argues that both Williams and Maynard Smith, as well as the other critics, make simplifying assumptions about the process that rig the game against group selection. Brandon and Burian 1984 remains a very useful work on this issue; it includes selections from Wynne-Edwards, Williams, Maynard Smith, Wade, and an early statement by D. S. Wilson of his line of thought. Sober 1984b remains an important discussion of these issues. Sober and Wilson (1994) review the whole units of selection problem from their perspective. Both Sober 1994 and Hull and Ruse 1998 contain good selections on this issue.

8.4, 8.5 The view of D. S. Wilson (not to be confused with E. O. Wilson, the founder of sociobiology) on these issues has been developing over some years: see Wilson 1983, 1989, 1992, 1997. The idea that kin selection is an instance of group selection has been independently developed by Colwell (1981). Sober has been a powerful ally of Wilson's in their recent collaborations; see Wilson and Sober 1994 and Sober and Wilson 1998. Their 1994 paper was published in *Behavioral and Brain Sciences* with a wide range of critical responses. Wilson also edited a special issue of *American Naturalist* (vol. 150, supplement, July 1997), which includes papers on the evolutionary transition from cell to individual, on bees as superorganisms, on symbiotic associations as superorganisms, and on group selection in human evolution. These papers are technical but important.

Three responses to trait group selection, all defending some version of pluralism, are Maynard Smith 1987, Dugatkin and Reeve 1994, and Sterelny 1996b. Two recent empirical works apply Wilson-like group selection ideas. Herbers and Stuart (1996) consider a species of ant in which queen number varies between nests, and argue that the nest is the functional unit of selection. Aviles (1986) discusses a spider species with a sex ratio strongly biased in favor of females, and argues that the spider colony is the unit of selection, with female-biased ratios being favored, because colonies split to found new ones only once they reach the right size, and they often go extinct before doing so.

8.6 Janzen's views on the organism are given in Janzen 1977; Buss sketches his views in Buss 1985, 1987. The symbiotic origin of the eukaryotic cell is taken up from Buss's perspective by Blackstone (1995). Maynard Smith and Szathmary (1995) take up the whole issue of transitions between levels of biological organization in their very important but difficult book. The evolution and survival of the developmental cycle has been discussed in two recent papers by Grosberg and Strathmann (1998) and by Fagerstrom, Briscoe, and Sunnucks (1998). Dawkins (1982) discusses his views at length in the final chapter of *The Extended Phenotype,* and more recently in Dawkins 1990, 1994. His take on group selection is interesting, and opens up an extra option, one we have not discussed explicitly in this chapter. Dawkins's general picture of evolution, and his distinction between replication and interaction, is neutral on the existence of high-level interactors. But as a matter of fact, he is very skeptical about group selection. This is not because he thinks that all adaptation is properly seen as the adaptation of individual organisms. He does not defend broad individualist accounts of phenomena dear to the hearts of group selectionists. Rather, he assimilates these cases into his category of extended phenotypic effects: these are genes whose route to the next generation does not go via building a vehicle. In Dawkins's view, not all phenotypic gene action is congealed into a single vehicle.

The superorganism concept has a long history in ecology and evolutionary theory. An early version of the view is given by Wheeler (1923). Wilson and Sober (1989) revive it, and Mitchell and Page (1992) respond. Seeley (1989, 1996) and Moritz and Southwick (1992) defend in detail a superorganism conception of honeybee colonies. E. O. Wilson and Hölldobler defend the idea that ant colonies are superorganisms in their superb works (Hölldobler and Wilson 1990, 1994), though not in great detail. Bourke and Franks (1995) are skeptical, but not dismissive.

9

Species

9.1 Are Species Real?

Not every distinction that seems real to us is real. We perceive others' speech as discontinuous, as divided into discrete sounds, words, and sentences, but the acoustic signal is usually continuous. So the fact that organisms seem to us to be parceled out into reasonably discrete groups does not by itself show that species are an objective feature of the living world. They could be just an artifact of our limited temporal perspective on the history of life. One view of evolutionary history suggests that species cannot be real. If a smooth continuum of change links us to the earlier primates from which we evolved, then there can be no fundamental difference between (say) *Homo sapiens* and *Homo erectus*. Our recognition of those species depends, the thought goes, on our temporal standpoint. We track our slowly changing lineage backward in time until we come to organisms that seem similar to one another, but different from today's humans. They are different enough that, had we discovered them alive on a remote island, we would have thought them a separate species. We call them members of the species *H. erectus*. If the human lineage continues to change, some future hominids, seeing themselves as typical, might see us as the intermediate gradation between two other hominid species, *Homo future sapiens* and *Homo post erectus*. Unless there are large evolutionary jumps, the lineage in which we find ourselves can be equally well segmented into species in many ways.

The idea, then, is that if phenotypic change does not proceed by large jumps *(saltations)*, then species are not objectively identifiable over time. Moreover, change is unlikely to proceed in this way. Plant species are occasionally created in a single generation by hybridization, but that is rarely true of animals. Major mutations that create a marked difference between parent and offspring are extremely unlikely to be viable. Organisms, depending on their size and kind, have thousands of genes composed of millions of bases. So the number of possible base sequences is huge. The great majority of these combinations cannot build an organism, still less an organism decently suited to its environment. If you took a viable genome and scrambled the DNA, randomly reordering the base sequences, your chances of coming up with a working organism would be vanishingly small. Your chances would be no better than those of writing a new novel by scrambling the letters of an old one. Yet that is what a major mutation does: it scrambles a fragment of an organism's DNA. Moreover, organisms are integrated wholes. If a minor mutation (a single change at a single locus) chanced to have a major phenotypic effect—say, doubling tooth size—the result would probably be catastrophe, for the necessary alterations elsewhere would not be made. So new kinds of animals rarely arise abruptly—on human time scales—from their ancestors.

Fortunately, evolutionary gradualism does not really imply that species distinctions are illusions. It is true that the differences between parent and viable offspring are likely to be small. Since viable offspring develop from coadapted developmental resources, any major change in those resources is likely to derail development, not generate significant change. However, there is no similar argument against rapid change in *population-level* properties. Species can quickly go extinct, change their range, change their role in an ecosystem, or change in genetic diversity. These changes occur on ecological rather than geological time scales. For example, a species hit by a new predator can be forced through a population bottleneck that strips it of much of its previous genetic diversity. Australian rabbits from before and after the myxomatosis epidemic look similar, behave in similar ways, eat the same things. No major phenotypic change has occurred in individual lineages of rabbits since they were attacked by the virus. But there probably has been considerable change in the rabbit population as a whole, since it probably contains a large proportion of rabbits that are resistant to the virus, and the population structure has changed through alterations in the density and distribution of rabbits through Australia. So if populations are species by virtue of population-level properties, speciation need not be smooth, gradual, and seamless. Furthermore, the most important contemporary theories identify species and speciation through population-level properties. One well-known approach, the *biological species concept,* identifies species by asking a question about populations: is this population reproductively isolated? In turn, reproductive isolation is a property of a population, and one it can acquire quickly. A change in the course of a river, a change in pigmentation pattern, or a change in daily activity cycles can cause reproductive isolation. This concept

Figure 9.1 An example of punctuated equilibrium. *Cope's rule* is a rule
of thumb stating that descendant species tend to be larger than the founding species of a lineage. In the evolutionary history depicted here, we
have an instance of Cope's rule: the
surviving species are all larger than
their common ancestor. Yet individual species phenotypes do not
change over time, even though there
has been phenotypic change in the
species lineage as a whole. For once
speciation takes place, the members of
daughter species have phenotypes distinct from those of the parent species.
In this case, differential species survival shifts the phenotype to the right
of the graph.

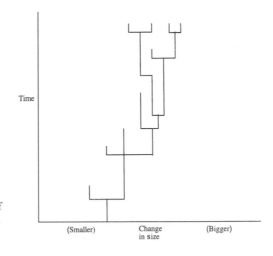

of species is controversial, but its rivals share with it the feature that matters
here: they identify species by features of populations and lineages, not of
individual phenotypes.

So evolutionary gradualism does not force us to be skeptical about the
reality of species. Recent developments within evolutionary theory, while
still controversial, underline this message. The prospects for realism about
species have improved in the last couple of decades through the development
of the theory of *punctuated equilibrium*. This hypothesis about evolution is
complex and controversial, but its essential element is the idea that the typical
life of a species involves a relatively sudden appearance followed by a comparatively long period of stasis, terminated either by extinction or by splitting
into daughter species. If the ideas behind punctuated equilibrium are right,
distinct types of organisms rarely arise by a gradual transformation of a parent
stock into a daughter stock. Instead, new forms typically arise relatively
rapidly (on geological time scales) when lineages divide. Species are born
through splits in a lineage and the subsequent reorganization of the fragments, not by the transformation of a whole lineage. If so, then the typical
life story of a species involves relatively well defined origins and termina-
tions. We expand on these population-level conceptions of species in section 9.2.

In our view, evolutionary theory lends no support to the idea that our
species classifications do not reflect objective features of the living world. The
division of organisms into species is an objective feature of the living world.

This is fortunate, for species have a very important role in biology as "score-keeping devices," as indices of the effects of evolutionary and ecological processes. For example, the *species-area effect* is the idea that the biodiversity of a region is linked to its area in a striking way. It suggests that diversity falls disproportionately. Two small national parks will not contain as much biodiversity as one large park of the same total size. This idea is very controversial, but if it is right, it has major consequences for environmental policies. All else being equal, one large protected area will be more valuable than a number of small ones. The species-area effect is an example of the way in which we use species counts as currency to measure stability and change in the living world. We could not use species in this way if our species categorizations were projections of our perceptual and temporal limitations onto seamless continuity in the organic world, finding boundaries where none exist in nature.

Let us grant, then, that species are real. How and why did they evolve, and what are the consequences of their evolution? In section 2.2, we sketched the received view of the origin of species. New species originate when isolated fragments of a population differentiate from the parental population as a result of selection and chance. As a particular fragment differentiates—and if it escapes extinction—sooner or later its members will cease to be potential mates for the parental population, and vice versa. It is widely accepted that there is something right about this view, but it cannot be the whole story about species and their importance.

The received view of species does not explain patterns in species formation. The title of one of ecology's most famous papers (Hutchinson 1959) asks "why are there so many kinds of animals?" Yet how many should we expect, if the received view is right? Moreover, there is structure to species diversity. There is geographic structure: tropical rainforests and coral reef communities are proverbially species-rich. There is phylogenetic structure, too. Some branches of the tree of life are much twiggier than others. There are appallingly many beetles and very few horseshoe crabs. Moreover, as we shall see in section 9.2, the received view fits some organisms better than others. Even if our classification of organisms into species reflects (no doubt imperfectly) objective differences in nature, it does not follow that there is any single species category. Plant species may be importantly unlike animal species, and bacterial species may be very unlike either of these. In thinking about the organism, we considered the possibility that we are using a single term for a number of kinds. The same possibility arises for species.

Moreover, the received view may understate the evolutionary importance of species. Eldredge argues that the received view treats species as an epiphenomenon of evolution, and in his view, this is its fundamental flaw. He has argued for a critical link between speciation and adaptive shifts. In his view, only speciation entrenches evolutionary change (Eldredge 1985b, 1989, 1995). The most ambitious claim of all is the idea that species themselves are units of selection. In chapter 8 we discussed the idea that local populations of organisms might be in competition with other local populations. Similar views have been defended for species. In this view, just as organisms have properties that make them more likely to survive and reproduce, species have properties that make them less likely to go extinct or more likely to speciate. The beetle lineage may be species-rich by virtue of lineage-level features, rather than traits of individual beetles or of particular beetle-building replicators.

Some of these questions are empirical, and hence are not ours to answer. It's not our job to explain beetle diversity. But most have conceptual and theoretical aspects, and it is on these that we shall concentrate. First we return to the identification of species. We then discuss the place of species in the overall tree of life. Finally, we go on to discuss species selection.

9.2   The Nature of Species

There have been three main views of the species category—three families of species definitions. So-called *phenetic species concepts* define species by appealing to some measure of overall morphological, genetic, or behavioral similarity. Species are seen as groups of similar organisms. This view has slid from favor. One problem it faces is the plethora of measures of similarity. Different methods of calculating similarity give different results, and the choice among them is arbitrary. If species are just collections of similar organisms, measured by one of the many different similarity measures available, then our species classifications are not the recognition of an objective distinction in nature, but instead result from a convention on how to define similarity (Ridley 1986).

Second, the organisms that make up a species are not always similar to one another. Most obviously, females, males, and juveniles can look very different. Moreover, there are many *polytypic* species: species whose members vary strikingly. There are butterfly species in which some individuals mimic one species and others, another (see Wickler 1968, chap. 2; Owen 1980, chap. 10). So different individuals of the same species can resemble members of another species more than other members of their own. There are species in which males have several different breeding strategies, and are hence quite unlike one another (a lovely example is an iguana, *Uta stansburiana,* with three different types of males; see Sinervo and Lively 1996). In social insect species, not only are there huge differences between members of different

Box 9.1 What is Genetic Similarity?

Much current taxonomic work seeks to compare the *genetic similarity* of two populations by sampling the DNA of both populations. It is important to remember that the similarity being measured here is similarity in base sequences. If the DNA of one organism is aligned with the homologous DNA of another, one can count the number (or percentage) of differences in bases: the number of times one has a T where the other has an A, and the like.

population 1	AAGGT CCTTA
population 2	AAGGC CCTAA
population 3	AAAGGT CCTTC
population 4	AGGT CCTTG

This notion of genetic similarity has the advantage of being able to sidestep the difficult issues of counting genes that we discussed in part 2. It is not, however, entirely innocent of theoretical assumptions. First, theorists choose the locations to compare based on their expectations about how long two lineages have been separated and how fast different chunks of the genome evolve. Moreover, populations 3 and 4 in our example will look very different from populations 1 and 2 unless we decide that a gene duplication added an extra A in population 3 and deleted one in population 4, so that the truly homologous sequences are

population 1	AAGGT CCTTA . . .
population 2	AAGGC CCTAA . . .
population 3	AA/GGT CCTTC . . .
population 4	A-GGT CCTTG . . .

Thus when we read that humans and chimps have 98% of their genes in common, what this means is that a randomly selected human and a random chimp are expected to match over 98% of their base sequences and vary at 2% of them. This finding is logically compatible with humans and chimps sharing no genes at all in the protein-coding sense of counting genes, since these genes are many bases long. But since the third position in a codon is often irrelevant to the amino acid coded, it is also compatible with humans and chimps being genetically identical in that same protein-coding sense.

castes; but members of the same caste sometimes vary strikingly between nests. In the fire ant *Solenopsis invicta,* some queens live in single-queen colonies and others in multiple-queen colonies: a difference with profound consequences for both queens and nests (Keller and Ross 1993). So uniformity within a species is by no means inevitable. More importantly, to the extent that species are uniform, that is part of what we want our account of species to explain. Similarity is not part of the definition of species, but part of the explanatory agenda.

If we reject phenetic approaches, we have two main alternatives. One of these (the second of our three styles of species definitions) is to identify the processes that create and sustain species and define species in terms of those processes. If species are created and sustained by barriers to gene flow, then we can define species by reproductive isolation, so that a species is a group within which genes can flow freely. This is the famous *biological species concept.* If the received view has a received species definition, it is the biological species concept. Finally, we can look to pattern rather than process. We can identify species with particular segments of the phylogenetic tree. In this view, species are lineages of ancestral/descendant populations. This is the approach adopted by the various *phylogenetic species concepts.*

Both process-based and pattern-based species concepts are historical in a broad sense. All the alternatives to phenetic definitions accept some version of the proposal of Michael Ghiselin (1974b) and David Hull (1978) that particular species are defined by their history. No intrinsic genotypic or phenotypic property is essential to being a member of a species (1.2). People born with the wrong number of chromosomes, eyes, or arms are still human beings. So the essential properties that make a particular organism a platypus, for example, are historical or relational. An animal is a platypus by virtue of its place in a pattern of ancestry and descent (its *phylogeny*). But to say that species are historical kinds is one thing; to say just which historical kinds is another. Why do we regard the domestic dog as a single species, rather than as a group of sister species? What do we need to find out to determine whether the Neanderthals were a separate species or a mere subspecies of *Homo sapiens?* Phylogeny is the correct *grouping criterion* for organisms, but it does not provide any obvious *ranking criteria* to determine which groups are species (Mishler and Brandon 1987). In other words, the facts of history and relatedness determine which organisms should be grouped together. But since groups of organisms are nested in successively larger ones—the domestic dog is nested in the dog/wolf/coyote group, which is nested along with cats, badgers, and stoats in the Carnivora, which is nested in the mammals— we need some way of telling *which* genealogically connected groups are species, which are subspecies, and which are superspecies. This "way of telling" is a ranking criterion.

The biological species concept takes reproductive community to be central to the role of species as evolutionary units. Adaptation and speciation require some isolating mechanism so that an incipient species, a small population in a new selective regime, can preserve the evolutionary innovations that develop within it. An unprotected population will be diluted by migration. Its distinctive gene complexes will disappear if there is substantial gene flow between it and the parent population. One migrant a generation is enough to prevent populations drifting apart through the accumulation of chance differences (Chambers, personal communication). If their divergence is driven by sustained selection, much more substantial migration is needed to homogenize the populations. Even so, Mayr and Eldredge have argued that there can be no special suite of adaptations without some form of isolation, and there can be no protection of that suite of adaptations without entrenching that isolation between parent and descendant populations. This is why Eldredge denies that species are just by-products of individual changes in individual populations. Imagine an isolated population of New Zealand rabbits that has acquired immunity to "1080," a standard poison. Traits typically depend on several genes, not just one. So the immunity will probably depend on a coadapted gene complex (and perhaps on other developmental resources), rather than a single gene. Immune rabbits will need the right set of genes, a set that becomes the common property of the rabbits in the isolated group. Now suppose that isolation breaks down, and rabbits migrate in and out of the population. Unless the isolated rabbits prefer to breed with their own kind, or unless the immunity genes are linked in inheritance in some way, interbreeding with the nonimmune parent population will break up the new coadapted gene complex, and the adaptive shift will be lost. Hybrids between the immune population and the nonimmune parental population will have a mix of both sets of genes. Very likely, this mixing will destroy the gene combination on which immunity depends. So until reproductive isolation has been established, adaptive change remains fragile. The biological species concept identifies a category of populations— reproductively isolated populations—that can evolve distinctively.

We conjecture that any solution to the species problem will incorporate substantial elements from the biological species concept. But in its raw form, it faces serious problems. First, the biological species concept has no good way of segmenting a lineage over time. Suppose, for example, that we stretched the notion of "potentially interbreeding" by supposing that an organism ceases to be conspecific with a member of a later generation if it

Figure 9.2 Tracking species over time. If we attempt to apply the interbreeding criterion over time, we lose the objectivity of our species distinctions. The left bar shows the species distinctions we might make if we choose A as our baseline individual, and hence define a species by including all and only A's potential mates. A would recognize B as a potential mate, but gradual change in the traits through which mates recognize one another means that C would have changed beyond A's recognition threshold. So the lineage is divided into two species, one including both A and B; the other, C and D. On the right, we see the species distinctions consequent on choosing B to be our baseline individual and defining as a species all and only B's potential mates. B would recognize both A and C as potential mates, since both are similar enough—about equally similar—to B. But D is beyond B's recognition threshold. So we get two species, one of which has A, B and C as members, and the other with D as a member.

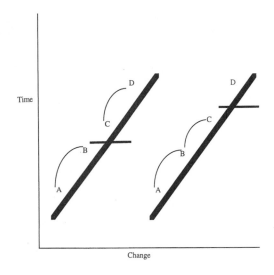

would not recognize that changed organism as a potential mate (McEvey 1993). Our segmentation of a gradually evolving lineage into species would then depend on our choice of baseline. There would be no objective speciation events (see figure 9.2). As Mayr himself realizes, this problem shows that the interbreeding criterion should not be applied to organisms at different times. If Abe and Adolf are members of different generations, they are in the same species if Adolf has descended from organisms conspecific with Abe by the interbreeding criterion and no speciation event has intervened in the genealogical tree. So to recognize species over time, the biological species concept needs supplementation by some definition of a speciation event.

A second kind of difficulty with the biological species concept is that the notion of a reproductively isolated community is an idealization, and we can legitimately choose different idealizations. We have to bear in mind two problems. First, there is the problem of real versus pseudo-division of a lineage. Groups that merely happen to breed only among themselves do not constitute a new lineage. The Queen's corgis are not a new species of dog, however scrupulously their pedigrees are preserved. A notion of reproductive isolation that disregards the Windsors' corgis' pedigrees is clearly appropriate to evolutionary biology. Other cases are more difficult, for they involve spatial separation. Impala are widespread across Africa. Spatially distant impala populations do not interbreed; indeed, they cannot interbreed. While there may be gene flow between some populations, others probably are genetically cut off from other populations. Even so, they are taken to be parts of the same species. *Ring species* pose particular problems, as they consist of chains of populations in which each link can breed with its neighbors, but populations separated by a number of links cannot, even if they come into contact. The literal ring of populations of black-backed gulls that circles the Arctic is a famous example. These are all cases of apparently divided populations, and our problem is to decide whether these divisions are real.

Second, lineages can be genuinely separate despite some gene flow. Hybridization occasionally takes place even between animals from paradigmatically distinct species. Major Mitchell cockatoos occasionally hybridize with galahs, but these two lineages are distinct. There are more problematic cases. In both Australia and New Zealand, the introduced mallard duck hybridizes freely with the native Pacific grey duck. Since mallard drakes are somewhat more aggressive than Pacific grey drakes, and because mallards adapt more readily to human-modified habitats, some are worried that the Pacific grey will disappear as a distinctive duck. Does this matter? Not if the two ducks are mere color variants of a single species. Since they freely hybridize, we might argue that they are not reproductively isolated, and hence they are members of a single species, though one with more morphological and genetic variety than most. In other views, their hybridization is an accident caused by human intervention. The two populations of ducks were on independent evolutionary trajectories before humans interfered, and are separate species. There are other examples. Human modification of New Zealand rivers together with the invasion of exotic predators has caused black stilts to hybridize with pied stilts often enough to threaten the survival of the black stilt lineage. Once more, it is not obvious whether this shows that the black stilt lineage had never really been reproductively isolated.

For these reasons, the reproductive criterion yields no unique segmentation of organisms into species. The notion of "potentially interbreeding" cannot be made fully precise (Kitcher 1989; O'Hara 1993). There is no objective count of protected gene pools. Gene flow really does come in degrees.

We think these problems are symptomatic of a third, deeper problem with the biological species concept: It fits multicellular animals much better than other forms of life. Occasional gene flow between separate animal species, and more regular gene flow between incipient species, is not a very severe problem for the biological species concept. We should expect there to be borderline cases of reproductive isolation in animals. However, lineage

Figure 9.3 When does speciation occur? (1) A species is distributed through its habitat. (2) Geographic change divides the species into two separate populations, A and B. (3) The geographic separation has become entrenched; no migration between A and B is possible. A and B are reproductively isolated from each other, but by extrinsic factors—factors external to the populations themselves. (4) Drift and selection have changed the phenotypes of both A and B; they are distinct both from each other and from their common ancestor. (5) Intrinsic isolating mechanisms have evolved. These in-

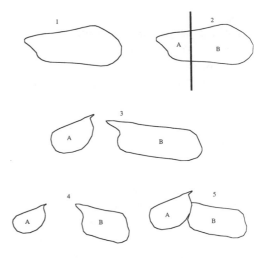

trinsic isolating mechanisms may be a side effect of other evolutionary changes, or they may be a result of selection against hybrids after the populations are back in contact. In any case, even though the populations are back in geographic contact, they remain distinct. Hence the separation at (2) was permanent. No one doubts that by (5) speciation has taken place. But should we regard speciation as incomplete or incipient until the establishment of intrinsic barriers to reproduction between the members of the two species, or is the establishment of permanent extrinsic barriers sufficient?

crossing is common among plants, among which gene flow across species boundaries is easier. Leigh Van Valen (1976) suggests that there may be oak species with easier access to genes from other species growing locally than to genes from the same species growing at a distance. Many single-celled organisms also pose a problem for the biological species concept, for in them gene exchange is decoupled from reproduction and is not limited to members of the same species. Bacteria that are radically different still nestle up to one another and exchange DNA plasmids. These examples suggest that the limitation of gene flow is just one of the factors that make a lineage "cohesive" (Templeton 1989). Conversely, in many species, gene flow between local populations is very limited (Ehrlich and Raven 1969). So we can have cohesiveness with little flow. The most extreme cases, of course, are species composed of obligatorily asexual organisms. These obviously escape the biological species concept, yet are not so rare or unimportant that they can be fudged away as a minor exception. So a protected gene pool is not all that matters in explaining the distinctness of a population, especially if genes are not all that is replicated in evolution. Phylogeny, shared environment, and exposure to a common selective regime must all be part of the cohesiveness of a species.

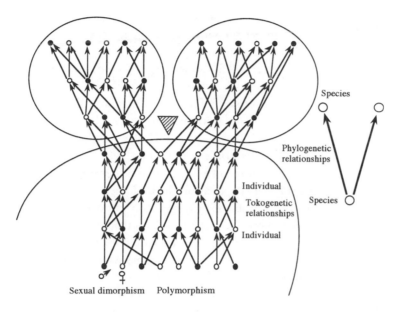

Figure 9.4 Phylogenetic species concepts base relationships between species, the *phylogenetic* relationships shown on the right of the diagram, on the actual family relationships between individual organisms, as shown on the left of the diagram. Here we see seven generations of sexual reproduction. During this period the descendants of the initial generation become separated into two streams that no longer interbreed with each other. If it becomes permanent, this splitting of the lineage into two parts will have been a speciation event. The single *stem species* will have been replaced by two *daughter species*. Cladists have yet another neologism for the individual relationships between one organism and another: *tokogenetic relationships*. (Adapted from Hennig 1966.)

These problems with the biological species concept lead us to the phylogenetic species concepts, which identify species through patterns in evolutionary history rather than through the causal process that generates those patterns. These concepts identify a species with a segment of a phylogenetic tree between two speciation events, or between speciation and extinction. Humans are not conspecific with protists, even though we descend from them, because the chain of descent that links us to them has often fractured. Phylogenetic species concepts are founded in the branching patterns of evolutionary history. Species come into existence as once cohesive lineages split. They cease to exist through true extinction—the death of all the individuals in a lineage—or by themselves splitting into daughter species. There may well be an advantage in identifying species by appealing to evolutionary patterns, for it's possible that no single process is responsible for cohesion in a lineage, or the breakup of that cohesion when lineages divide. So if we emphasize pattern rather than process, we can have a unitary account of what it is to be a species. We can get the same result—a lineage divides and a species is born—from different evolutionary mechanisms.

However, phylogenetic species definitions depend on the idea of a lineage dividing—on speciation. How do we count lineages? When do we have a single lineage rather than two or more ? There are 120 million feral brushtail possums in New Zealand, a far larger population than in their native Australia. Is this a second brushtail lineage a New Zealand endemic (since of course it would live nowhere else), or is it a part of a single lineage? Unless an alternative account of the division of a lineage can be developed, the phylogenetic species concept largely depends on the biological species concept, and thus inherits many of its problems. One obvious problem is posed by asexual species. A heroic response would be to deny that asexual organisms are parts of species. This is heroic indeed, for we would then have to explain just why "pseudo-populations" of, for example, some whiptail lizards seem to be species. Moreover, the distinction between asexuality and sexuality is not sharp. Rather, asexuality is the endpoint of a continuum of degrees of gene flow whose other endpoint is the promiscuity of plant hybridization (Templeton 1989).

An ideal solution to the problem of identifying dividing lineages would be to rework the cohesion concept as a theory of lineage splitting—to find a genuine equivalent in asexual organisms of reproductive isolation among sexual ones. Perhaps an idea from ecology can help at this point. Competition is most intense between members of the same species. Although asexual organisms are not in competition for a limited number of mating opportunities, they are in the competition to occupy a limited number of "living spaces" in their habitat. Alan Templeton uses this idea to supplement the criterion of reproductive isolation; Van Valen uses it instead of isolation. The *ecological species concept* attempts to define species in terms of their niches (Van Valen 1976). The problem with this approach is that it is unclear that ecological niches are the robust entities that these suggestions require. The controversies surrounding the niche concept are discussed at length in sections 11.4 and 11.5.

In sum, the most plausible account of species is that they are lineages between speciation events. The biological species concept, perhaps supplemented by the ecological species concept or by something else, reemerges as an account of speciation. Lineages split when their components become reproductively or demographically isolated from one another. Lineages converge when two formerly isolated lineages become a reproductive or ecological community through, for example, hybridization. However, both the

Box 9.2 A Flock of Species Concepts

Phenetic species concepts define species by appealing to the intrinsic similarities between organisms. The idea is to purge species identification of theoretical commitments. If our species identifications do not presuppose specific evolutionary theories, they can remain stable over change in our theoretical ideas, and can be used to test those ideas without circularity.

Biological species concepts define species by appealing to reproductive isolation. One version of the biological species concept is the recognition concept, which defines species as systems of mate recognition.

Cohesion species concepts generalize the biological species concept by recognizing that gene flow is not the only factor that holds one population together and makes it recognizably different from others. Alan Templeton, the first to formulate a species concept of this kind, argued that the members of a species play a distinctive role in an ecosystem, and that this role links the members of a species, making them different from other species. So he includes elements of both the biological and ecological species concepts when he defines a species as "the most inclusive group of organisms having the potential for genetic and/or demographic exchangeability" (Templeton 1989, 25).

Ecological species concepts define species by appealing to the fact that members of a species are in competition with one another, since they need the same resources. A species is a group of organisms whose members share an adaptive niche and can replace one another's descendants if they find more efficient ways to occupy that niche. Species are ecologically isolated by their distinctive niches.

Phylogenetic and evolutionary species concepts define species as segments of the tree of life. A species is a lineage of organisms, distinguished from other lineages by its distinctive evolutionary trajectory, and bounded in time by its origin in a speciation event and its disappearance by further speciation or extinction.

notion of reproductive isolation and the notion of an ecological niche are less satisfactory for these purposes than one could wish.

9.3 The One True Tree of Life

Our sympathy for some version of a phylogenetic species concept needs to be placed within an overall account of the tree of life and its best description. This is the task of *systematics*. We have from time to time used "scientific names" in our examples. We are members of the species *Homo sapiens*. *Sapiens* is the name of our particular species, but the name as a whole encodes the idea that *sapiens* is one of a closely related group of species, the genus of *Homo* species, of which we are the only survivor. Biological classification has traditionally recognized units larger than species. Until recently, our place in the big picture would have been sketched out by placing our *genus* in a larger group, the hominid *family,* which in turn is part of the primate *order.* Primates are mammals (a *class*); mammals are chordates (a *phylum*), and the chordate phylum is part of the *kingdom* of animals. Bells and whistles could be, and have been, added. Primates, for example, are placental mammals, so one could place an extra level, the Eutheria, between the primate group and the whole mammal group.

So a traditional view of species organizes them into a hierarchy of increasingly inclusive groups, or *taxonomic ranks:* species, genus, family, order, class, phylum, kingdom. The nesting is strict: each genus is a member of exactly one family; each family, of one order, and so on. But what is a genus? What is a family? The status of these larger categories is an important question. For, especially in paleontology, evolutionary patterns are often studied at the level of the genus or family, rather than by identifying individual species. The distribution, life span, and fate of particular species is often below the resolution the fossil record can give us. So when evolutionary theorists writing on the history of life compare the persistence of land versus shallow sea organisms through some mass extinction episode, or contrast the diversity of plants and animals, the information they extract will mostly be patterns of family extinction, survival, or spread. That makes it very important to ensure that when we compare, say, marine mollusk families with terrestrial arthropod families, we are comparing equivalent units.

Systematics has gone through a long period of controversy, some of it extraordinarily bitter (Hull 1988). But we think something like a consensus has emerged in favor of a *cladistic* conception of systematics. This consensus has been reflected in the shift in the name of the discipline from *taxonomy* to *systematics*. The controversy, we suspect, was largely generated by an attempt to have biological classification respect three goals at once:

1. A classification system should serve as a maximally efficient information store. In this view, we should choose a classification system—a way of grouping organisms—so that the full name of the organism encodes the greatest possible amount of information about the organism. Species names would thus group together maximally similar organisms. The classification of species would group together species so that clusters of the most similar species constituted a genus, clusters of the most similar genera a family, and so on. Identifying an organism's place in the classification scheme would then recover a rich array of typical features, at the different levels of generality indexed by different taxonomic ranks. The movement in taxonomy known as *phenetics* or *numerical taxonomy* is closely allied to this conception of the purpose of taxonomy.

2. A system of biological classification should reflect the disparity of organisms and the extent of their evolutionary change. This goal has been an important theme in traditional taxonomy. To take a very extreme example of this idea, the nineteenth-century anatomist Richard Owen elevated the human species to a subclass (Archencephala), equal in rank with all the other mammals combined, in recognition of our cognitive distance from them (Desmond 1982, 75). There are many less extreme cases. Standard biological classification puts the Cape Barron goose in a genus of its own in recognition of its divergence from all other gooselike birds. No one thinks that disparity is the only feature of life that classification should capture, but *evolutionary taxonomy* takes it to be an important aspect of taxonomy. Evolutionary taxonomists do not recognize birds as dinosaurs, despite the fact that birds originated as one branch of the dinosaur lineage, for birds have diverged profoundly from the rest of the dinosaurs.

3. A classification system should describe the branching pattern (and occasionally the fusing pattern) through which the tree of life has grown. Phylogenetic systematics, more often known as *cladistics* (*clade* = "branch"), incorporates this conception of systematics as history.

It is obvious that no classification system can fully satisfy all three criteria, because they make inconsistent recommendations. Consider a case in which two evolutionary lineages contain an identical number of species, but one lineage is conservative, with many similar species, and the other is not. In the second lineage, we find some species that are like the ancestor of the whole group, but many species that are very different from it, including some that are unlike any others. Since the branching pattern is the same, cladistics treats these lineages in exactly the same way, for cladists are concerned only with the pattern of speciation itself. Both evolutionary taxonomy and phenetic taxonomy may treat the two lineages differently. They are likely to group the nonconservative lineage into genera in ways that contrast with their treatment of the conservative lineage. They may well differ, one from another, in their treatment of single, very distinct species.

Moreover, since these criteria are so different, and have such different motivations, it's not easy to see how a "mixed criterion" classification system could combine them in any principled way. We accept a broadly cladistic conception of systematics, for we see conceptual problems facing both the information-storage and the disparity-capturing conceptions of classification. The information-storage conception relies on the idea of capturing patterns of similarity across groups of organisms. But—as cladists never tire of pointing out—similarity depends on the traits you measure. Are pigs and oysters similar by virtue of both being forbidden food to orthodox Jews? Phenetic taxonomists hope that if you measure and compare enough traits, this problem will be washed away. But if you place no restrictions on what counts as a trait, every two organisms are similar in infinite ways, and fail to be similar in infinite ways. So pigs and oysters are also similar in that neither has ten legs, neither eats spiders exclusively, and so on.

Phenetic taxonomists have often wanted to segregate taxonomy from theory. First, they wanted taxonomy to be stable across change in theory. More importantly, they thought that the pattern that classification reveals cannot be evidence for or against evolutionary theory if we appeal to evolutionary theory in constructing our system of classification. But the problem of deciding what to measure shows that the hope of theory-free classification is vain. Phenetic taxonomy needs a theoretically principled way of deciding what to measure. Moreover, even once we have decided on the traits to count, there are many different ways of calculating overall similarity between groups of organisms. In sum, it is just not clear that "biological similarity" is a well-defined notion.

Adaptive divergence may not be well defined either, as we discussed in section 1.6 and will argue at greater length in section 12.3. For example, it is very hard to see what could speak either for or against the traditional classification that places humans and chimps in different genera, despite their close evolutionary relationship. Although evolutionary taxonomists show a fair amount of intersubjective agreement in their judgments, both phenetic taxonomists and cladists complain that the idea of disparity rests on nothing but the educated intuition of the biologist constructing the classifications.

History has an objective structure. Suppose we share a more recent common ancestor with the chimp species than it shares with the gorilla species—that the ancestors of the gorillas diverged from a branch ancestral to both chimps and humans. If so, then historical phylogeny should put humans and chimps together in a more closely related group than any including both chimps and gorillas. If not, then it's just a mistake to think of humans and chimps together alone as a single group. It may not be clear whether a cladistic classification is right, for discovering species genealogies is not easy. But it's clear what those classifications claim. In contrast to other systematic ideas, there is nothing obscure about cladism's goals. Most importantly, as we shall see in section 10.7, the best kind of adaptationist thinking in evolutionary biology requires an amalgam of adaptive and historical hypotheses. So the phylogenetic information provided by cladistics is just the information we need to test adaptationist thinking in biology.

Let's sketch out cladistic ideas in a little more detail. As we see it, cladistics combines three central ideas. First and most important is the one we have touched on already: the point of systematics is the discovery and representation of evolutionary history. Systematics tells us who is more closely related to whom, where "more closely related" just means "shares a more recent common ancestor." So if the kiwis and emus shared an ancestor after the moas had gone their own way, the kiwis are more closely related to the emus than they are to the moas. The second element is a metaphysical claim. For cladists, real groups in nature are all, and only, *monophyletic* groups. Monophyletic groups are species groups that consist of a species and all, and only, its descendants. To the cladist true believer, there is no such thing as a reptile. "Reptile" does not name a real group, for there is no species that is ancestral to *all* the reptiles that is not also an ancestor of the birds (see, e.g., Archibald 1996, 22). Reptiles are not another real group in addition to the group that includes crocodiles, snakes, lizards, and birds, any more than a human family minus the eldest daughter is another real family.

The standard way in which cladists present their historical hypotheses is through a *cladogram,* a branching diagram that groups taxa by shared descent. The more recently an ancestor is shared, the more closely related the taxa. So, for example, a cladogram depicting species will show two species as most closely related *(sister species)* if they are hypothesized to share an ancestor that is ancestor to no other species.

While in general we are sympathetic to cladism, to the extent that cladists

Figure 9.5 A cladogram. If the history of descent represented here is correct, then the group birds/crocodiles/lizards/snakes is a natural group. It is *monophyletic*—that is, it consists of all, and only, the descendants of a particular ancestral species. In contrast, the snake/lizard/crocodile group is not a natural group. It is *paraphyletic,* since it contains only the descendants of a single ancestral species, but does not contain all of them; birds are left out. A group containing only birds and mammals would be even less natural. It would be a *polyphyletic group,* one containing species with no recent common ancestor. Cladists argue that only monophyletic groups are real.

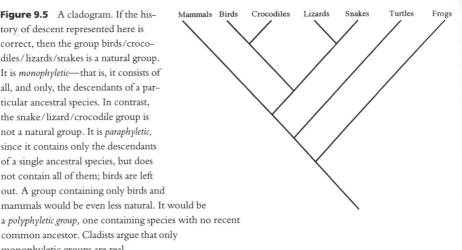

really do want to reject truncated monophyletic groups—groups that contain nothing but a single species' descendants, but not all of them—their views are too extreme. We think it quite likely that there can be good evolutionary hypotheses about such *paraphyletic* groups. For example, there may well be sensible evolutionary hypotheses about all the nonmarine mammals. That group is not a monophyletic clade, because there is no species ancestral to all the land-breeding mammals that is not also ancestral to the whales. Even so, it's easy to imagine events that affect all of, and only, that truncated group.

The third element of the cladist vision is methodological. The cladists have ideas on *how* to discover history. In particular, they have developed a theory about which traits are informative about evolutionary relationships within a group, as well as many techniques for using informative traits to construct a most probable evolutionary history. Let's first consider informative traits. First, *unique traits* are uninformative. A character that only the platypus bears (the male poison spur) tells us nothing about the relationship of the platypus to other mammals. Equally, *primitive traits* are irrelevant. The traits that all mammals inherited from the ancestral mammal tell us nothing about relationships within the class of mammals. A trait that arises before a branch of the tree of life emerges can tell us nothing about relationships within that branch. So internal fertilization, or having amniotic eggs, or having four limbs tells us nothing about the platypus's relationships to other mammals. Only *derived traits,* traits that vary within a group because of evolutionary change, are informative. The platypus's egg laying, electrolocation,

Box 9.3 Informative Sites

We have framed our discussion in terms of morphological and behavioral features of organisms, but DNA sequences can also be informative traits. In the following sequences, there are a good number of similarities and differences, but only one column of the matrix is informative; only one site carries information about the relationship between the taxa.

Taxon	1	A	G	G	T	C	C
Taxon	2	A	G	G	T	C	A
Taxon	3	A	T	G	T	C	T
Taxon	4	A	T	G	T	C	G

Clearly, the second site suggests grouping taxon 1 with taxon 2, and taxon 3 with taxon 4. The final site is unique; each taxon differs from the others there, so it tells us nothing about their relationships. The first, third, fourth, and fifth sites record a similarity, presumably inherited by all four from some common ancestor. So again, these similarities tell us nothing.

and much else groups the platypus with the two echidnas (their egg laying is a derived trait, not a primitive inheritance, for their eggs are large and the primitive mammal egg is tiny). These three share an ancestor with no other living mammal (Penny and Hasegawa 1997).

We shall spare our readers an account of the techniques used in reconstructing evolutionary history, for the details are complex and difficult. But the basic ideas are simple. If we could unambiguously identify derived traits, such reconstructions would be simple. If species A and species B are alone in sharing a derived trait, they are sister species. If A, B, and C share a derived trait (say, burrow nesting) and D does not, then A, B, and C are more closely related to one another than any are to D. The problem is that there are impostors: pseudo-present and pseudo-absent traits. D might have lost burrow nesting; C might have evolved it independently. Such independently evolved but qualitatively similar traits are *analogous* or *homoplastic* traits. Derived traits (and primitive traits) are *homologous*: taxa have them by *inheritance from a common ancestor.*

So the problem in reconstructing the past is to distinguish informative traits from fake presence and fake absence. Are humans and bonobos united by the shared derived trait of front-to-front copulation, or has each evolved this trait separately? Are mammals and birds sister groups by virtue of endothermic metabolism, or has the capacity to maintain body temperatures using

Figure 9.6   Resolving a phylogenetic tree by parsimony analysis. Under consideration are two hypotheses about the relationships among three taxa. One hypothesis groups A and B together: (AB)/C. The other groups B and C together: A/(BC). The primitive state of the character of interest (ascertained independently) is represented as 0, and the derived state

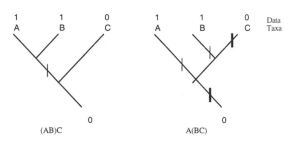

as 1. How do these two hypotheses compare as explanations of the character's distribution? The left-hand tree, representing the first hypothesis, is more parsimonious, as it requires only one change to explain the current distribution of the trait. The right-hand tree, and the second hypothesis, requires one or the other of two pairs of changes. The first possibility (heavy bars) is that 0 changed to 1 before the divergence of A, B, and C, and then changed back to O after C had diverged from B. The other possibility (light bars) is that 0 changed to 1 twice, first in the lineage leading to A after the (BC) lineage had diverged, and again in the lineage leading to B after this had diverged from C. (Redrawn from Sober 1988b, 246.)

internal energetic resources evolved independently? Occasionally a detailed inspection of the trait itself reveals independent evolution. No very deep investigation is required to establish that bat and bird flight are independent evolutionary achievements, for bat and bird wings are very different. But even if bat and bird wings had the same structure, there is another method that would reveal their independent evolution. This method of reconstructing evolutionary history depends on the idea of overall *parsimony*. If we construct an evolutionary tree in which bats and birds form one clade and all the other mammals another, we have to suppose that a host of bat and mammal traits evolved twice: once in the mammals and once in the bats. Alternatively, we might suppose that the shared traits of mammals and bats are primitive; that they are derived from an ancestor deep in the tree of life, one that lived before the evolution of bats, birds, and mammals. But then we have to assume a host of losses in the birds to explain the absence in birds of mammallike ears and the like. Either way, grouping bats with birds involves a very unparsimonious picture of past evolution. The most parsimonious hypothesis about an evolutionary tree is the one that requires the fewest possible evolutionary changes, for change is rare in comparison to non-change. Such a hypothesis is assumed to be most likely to capture the actual sequence of past changes (figure 9.6).

Time to sum up. While we think cladism presents the best view of systematics, biological classification nevertheless poses an unsolved problem. If we were to accept either evolutionary taxonomy, which builds disparity into its classification system, or phenetic taxonomy, which is based on the idea of nested levels of similarity, traditional taxonomic levels would be quite defensible. Within those taxonomic pictures, the idea of genus, family, order, and so on makes quite good sense. If cladism is the only defensible picture of systematics, the situation is more troubling. From that perspective, these taxonomic ranks make little sense. Cladists do not think there is a well-defined objective notion of the amount of evolutionary divergence. That, in part, is why they are cladists. Hence they do not think there will be any robust answer to the questions, when should we call a monophyletic group of species a genus? a family? an order? Only monophyletic groups should be called anything, for only they are well-defined chunks of the tree. But only silence greets the question, are the chimps plus humans a genus? It has long been received wisdom in taxonomy that there is something arbitrary about taxonomic classification above the species. These decisions are judgment calls. So cladists show only a somewhat more extreme version of a skepticism that has long existed. The problem of high taxonomic ranks would not matter except for the importance of the information expressed using them. Hence cladism reinforces the worry that when, for example, we consider divergent extinction and survival patterns, our data may not be robust, for our units may not be commensurable. Unfortunately, it does this without suggesting much of a cure.

9.4   Species Selection

Evolutionary trends have been a hot topic in evolutionary biology. Some evolutionary trends are shifts within a single species. For instance, the average size of red kangaroos has apparently shrunk significantly in the last 20,000 years, probably as a result of human hunting. Others involve tens of thousands of species and hundreds of millions of years. Vermeij documents and explains one such mega-trend, the growth of defensive structures in shelled marine invertebrates (Vermeij 1987). Intermediate between the mini-trend of shrinking kangaroos and the mega-trend of ever thicker, stronger, and gnarlier shells are some of the classic stories found in evolutionary textbooks. These include changes in the horse lineage (horses grew, as did their teeth, but their toe number shrank) and the growth in brain size in the *Homo* lineage.

What mechanisms might explain evolutionary trends? It turns out that some are the result of statistical artifacts and need no special explanation. Gould argues, for example, that the trend toward bigger horses is an accidental by-product of the near-extinction of the lineage. The diversity of the

Box 9.4 Terminological Terrorism

Cladistic theory is beset by a user-malign terminology, which we have mostly ignored in explaining its central ideas. But since this terminology is in standard use in cladist manifestos, we review here some of the most critical terms.

homology: a similarity between organisms that results from inheritance from a common ancestor. On the assumption that evolutionary change is the exception rather than the rule, homologies covary with one another and with phylogeny.

homoplasy: a similarity between organisms that has arisen independently in the lineages in question.

monophyletic group: a group of taxa consisting of a species and all, and only, its descendants.

paraphyletic group: a monophyletic group minus one or more of the ancestral species' descendant taxa.

polyphyletic group: a grouping of taxa whose common ancestor is deep in the tree (more than two speciation events deep) and that excludes the other descendants of that common ancestor.

plesiomorphic character: a trait that has been inherited unchanged from an ancestor.

apomorphic character: an evolutionarily novel trait. Clearly, *plesiomorphic* and *apomorphic* are relative terms, since all plesiomorphies begin as apomorphies, and should a taxon with an apomorphy have descendants, in them that character will be a plesiomorphy.

symplesiomorphy: a primitive trait; that is, a homology shared by a group that originated before that group came into existence and has been inherited as common property by all the members of the group. For this reason, a symplesiomorphy tells us nothing about relationships within a group.

synapomorphy: a shared derived trait; a homology shared by some, but not all, members of a group, and hence a trait relevant to determining the relationships within that group.

autapomorphy: a trait that is unique to a single taxon.

lineage declined rapidly, and the few survivors happened to be larger than the first horses (Gould 1996a, 57–73). In Dan McShea's useful terminology, such trends are *passive* rather than *driven* (McShea 1991, 1994, 1996b). But what might explain driven trends? One possibility is that a trend might be produced by correlated evolutionary change within a cluster of species as each responds in similar ways to the same evolutionary challenge. So we might suppose that as the Australian continent became hotter and drier, parallel evolutionary changes took place within many plant lineages. In species after species, leaves became smaller, tougher, with less porous surfaces. In short, many species evolved in similar ways, and for similar reasons, as they became adapted to aridity. Others became extinct, or confined to moist refuges.

There is, however, another possibility, opened up by the hypothesis of punctuated equilibrium. According to this hypothesis, species rarely change much from origin to extinction. If that is the case, then trends that result from correlated gradual transformation within species must be rather rare. Moreover, punctuated equilibrium suggests an alternative: evolutionary trends may be the result of the differential generation and extinction of species (Gould 1990). Australian trees, for example, are not just drought-proof; they are fireproof as well. A trend toward fire resistance in, say, Australian acacias might be generated by speciation and extinction of species, not change within species. Imagine a widespread and moderately fire-resistant acacia species in the early stages of the browning of Australia. As the environment changed, the range of the species would have been fragmented, creating the conditions for speciation (see figure 9.3). The resulting daughter species would have varied in fire resistance, with some being more resistant than the ancestor and some less so. But the less fire-resistant would have been likely to go extinct. So as further changes produced more fragmentation and speciation, the resulting new species would come from a pool of daughter species with an increased mean fire resistance. Once more, we would expect differential extinction of the less fire-resistant, followed perhaps by renewed speciation from the even more fire-resistant survivors, and so on. So differential extinction and speciation can build a trend toward fireproof acacias.

If this were the correct account of acacia evolution, we might conclude that fire-resistant acacias evolved by an mechanism analogous, though at a much grander scale, to the one that produced our camouflaged bittern of section 2.2. Variation, heritable differences in fitness, and cumulative selection build fire resistance, but the variant individuals are whole acacia species, not individual trees. But this inference would be too swift. As Elizabeth Vrba has made clear, a trend caused by differential extinction or speciation is a candidate for explanation by species selection. But it's only a candidate. *The species itself* must have characteristics that bias its chances of founding daughter species, going extinct, or both. Vrba makes an important distinction between *species selection*—the differential success of species by virtue of features of the species themselves—and *species sorting*—the differential success of species by virtue of features of their component organisms (Vrba 1984a,b,c, 1989, 1993). Her point is that extinction is often a side effect of evolution acting on individual organisms. If New Zealand's kakapo (an endangered large flightless nocturnal parrot) goes extinct as a result of predation by introduced stoats, that extinction will be the result of countless, sadly one-sided, kakapo-stoat interactions. Similarly, if lineage splits produced an array of hominid populations with different average brain sizes, and larger-brained hominids simply outcompeted their smaller-brained relatives, the disappearance of small-brained hominids would be a side effect of ecological processes at the level of individual organisms. Enough bad news for individuals in a population adds up to bad news for the population itself, without us needing to suppose there is any population-level process editing out the hominids of very little brain. So we have species selection only when an evolutionary trend depends on properties of the species itself.

Vrba is right to distinguish true species selection from species sorting. Her distinction forces us to consider what it is for a species itself to interact with its environment in ways that explain its success or failure. As we see it, a defense of species selection requires (1) an account of the distinction between a species trait and the traits of its component organisms; (2) a demonstration that species traits are causally salient; and (3) a case for thinking that species properties can be built or maintained by some type of feedback process, so that species traits result from cumulative selection. We will explore these ideas through two hypotheses. The first proposes that species selection can explain the range of a lineage. The second proposes that it can explain the heterogeneity of a lineage, and, more particularly, one mechanism for generating this heterogeneity: sexual reproduction. We think these ideas are quite plausible. Our aim, however, is not to defend species selection, but to explain what such hypotheses claim.

A necessary condition of species selection is that a species itself have properties; this is a consequence of Vrba's distinction. There is nothing mysterious about this idea. Ecologists sometimes distinguish between generalist and specialist species. Sometimes this is just a way of talking about the individual organisms of the species. Koalas, with a diet restricted to a few species of eucalypts, are obviously specialist individuals; the common brushtail possum is a species of generalists. But being a generalist is often a property of a species rather than of the organisms that constitute that species. Some eucalypts tolerate a wide range of climate and soil types. The immobility of trees and the relative stability of climate and soil structure implies that each individual tree has a very particular habitat, and hence each individual tree is a specialist, but these species as a whole are generalists. The common European cuckoo is a generalist nest parasite; it is not restricted to a particular host. Individual cuckoos do specialize, but they do not all attack the same species. Thompson (1994, 128–32) argues that parasitic species that attack a wide range of hosts, and hence are generalists *as species,* typically consist of organisms that specialize on particular hosts, and hence are specialists *as individuals.* In these cases, the property of being a generalist clearly depends on the properties of the component organisms in the species. But it is not identical to any of those properties.

So species do have properties. But do these properties enable a species to interact with its environment as a whole in ways that differentially affect the replication of its gene pool? Perhaps they do, and we can illustrate this possibility through some ideas about mass extinction. As we shall see in section 12.5, many paleontologists accept a distinction between *background* and *mass* extinction. Extinction is normal. Species go extinct all the time. But sometimes meteor strikes, massive volcanism, sudden climate change, and the like cause pulses of extinction. In mass extinction episodes, many species disappear suddenly and simultaneously. In one view, these episodes do not discriminate in favor of species well adapted to their pre-disaster environment. Nonetheless, the impact of these events is not random. In David Raup's terminology, the extinction rules are *wanton:* extinction is systematic but unrelated to how well the species is adapted (Raup 1991). Moreover, some of these rules seem to depend on species properties. Population size is a species characteristic, and one relevant to survival. Small populations are especially vulnerable to natural disaster and other bits of bad luck. Indeed, if a population is very small, one unlucky breeding season might finish it off. A small and widely dispersed population may be in real trouble if its numbers are further reduced, for the survivors will find it increasingly hard to find mates. A small but concentrated population is vulnerable to a merely local catastrophe. So population size may well be a causally salient population-level property. It is not, however, a property likely to play any role in evolution through species selection. For it is not likely to be heritable: a species with a large population is apt to give rise to small ones, and vice versa. So a species with a large population is unlikely to be large because it is a descendant of a species with a large population—a species that *survived because it is large.* So population size is a trait causally relevant to extinction and survival, but it is not heritable enough to be built or maintained by cumulative selection on a lineage. So it is not an adaptation of a species.

So far we have argued that species have traits, and that these traits may be causally salient: they may affect a species' prospects for extinction or speciation. But we have yet to find an example of a species trait that is both causally salient and passed on to daughter species. A candidate for such a trait is species distribution. The bat lineage is geographically widespread, but it is also ecologically widespread. Bats make their living in many different ways. Distribution, like population size, is causally relevant to surviving mass extinction events, and for rather similar reasons. Moreover, it may well be heritable. Suppose a lineage survives an extinction event—a meteor strike or a climate change—because it happens to be a little more widespread, in range and niche, than its rivals. The extinction event then enables it to extend further into vacated ranges and niches. The lineage radiates, and hence it becomes yet more widespread. So it survives the next extinction event, and the process repeats. In this case, range is a property of the lineage honed by cumulative selection, and the lineage has that property as a result of a selection process. David Jablonski (1987) defends a species selection theory of this kind for marine invertebrates.

One central example in the discussion of species selection is the evolution of sex. Sex is a puzzle for evolutionary theory, for it is expensive, both for individual organisms and the genes they replicate. Sex has obvious costs: the costs of sexual ornamentation and the time, trouble, and danger involved in finding a partner. It also has a more subtle, but more pervasive cost. Imagine an isolated stream that is home to ten platypuses, a population near the stream's capacity. Every year, each of the females gives birth to two offspring, a male and a female. About half of the offspring die, and about half replace adults that die. Then a female appears with a mutation that enables her to lay eggs asexually. She lays two eggs, which hatch into daughters that are her clones. Suppose she and one daughter survive to the next breeding season. In the population of ten, there are eight sexual and two asexual platypuses. The two asexuals give birth to four asexuals, so when the post-breeding population expands to twenty-two, there are now six asexuals. Suppose mortality then reduces the population to ten or eleven again. Unless mortality is preferential, there will still be three asexuals. So when the population swells to twenty-five on the next breeding round, there will be nine asexuals; when it shrinks back, there will probably be four, and so on. The invention of asexual reproduction means that the productivity of the population as a whole has gone up, but the extra young are asexual. Asexual reproduction increases the number of tickets in the survival lottery without increasing the number of winners. But it reserves the extra tickets for the asexuals. So unless their tickets are much worse, asexuality should sweep through the population.

Thus there is a very serious cost to sex. Yet sexual reproduction is typical among multicellular organisms. Asexual species do exist, but most seem short-lived (though not always: see Judson and Normark 1996). Asexuality has a cost too, though one that is paid over a much longer time frame. *Muller's ratchet* is the idea that asexuality is selected against in the long run by the accumulation of disadvantageous mutations. Imagine our clone of asexual platypuses in our stream. After a while, they will not be quite identical, because there will be copying errors when a genome is copied from mother to daughter. Many of these errors will be "silent," having no phenotypic consequences. But a few will make a difference, and the differences will probably be disadvantageous. Yet, because these organisms are asexual, the only way a mutation in a clone line can be lost, once it occurs, is if at the very same point on the genome where the copying error was made, another mutation returns the genome to the original state. The odds against that happening are very high. Otherwise, mutations are purged only by extinction of the clone lines that contain them. Mutations are added, but they are not taken away, hence the analogy of the ratchet. So the platypus clones will begin to increase their *genetic load* of mutations. Moreover, bad luck will from time to time eliminate one of the fit, undegraded clone lines. So the fitness of the fittest clone lines gradually goes down, as chance extinction picks off some and new mutations accumulate in others. Muller's ratchet keeps turning, and, depending on the frequency of mutation, the size of the genome, and the size of the population, eventually—though long after the end of the sexual platypus—in all the clone lines, disabling mutations will drive our population to extinction. The effect of Muller's ratchet can be masked by positive selection for mutations that mask the effects of others, and it may turn too slowly to matter for very large populations. But the ratchet seems likely to be important to small ones.

It is possible that sexual reproduction is maintained by species selection. Since sexual reproduction is a complex adaptation, its origin probably requires repeated selection over a large pool of variants, hence selection on individual organisms. But we might still explain its *persistence* by invoking the tendency of asexually reproducing clones to go extinct. If, for whatever reason, sexual reproduction arises, it persists, the idea goes, through the higher extinction probability of asexual daughter species, threatened in part by Muller's ratchet. For if sex were an option for our platypuses, all would have been well. The mutations in one clone line are unlikely to be identical to those in another. So if they could exchange genetic material, they could produce, among their range of genetically diverse offspring, platypuses without the mutated genes, and platypuses in which the effects of the mutation were masked by a dominant and functional gene. Moreover, the very same genetic recombination that resets Muller's ratchet to zero may also allow a lineage to evolve faster. For not all mutations are deleterious, and distinct advantageous changes can be joined by sexual reproduction. So sexually reproducing species may have an advantage over asexual ones, both in avoiding Muller's ratchet and in responding to environmental change through gene flow and recombination.

How well will this kite fly? The idea is certainly coherent. In discussing the distinction between species sorting and species selection, we argued that true species selection depends on species properties, and sex can be seen as a property of species. First, there are some species that are sexual, and to which the species-level benefits of sexuality accrue, despite the fact that most individual members of the species are asexual. A little bit of sex suffices to escape Muller's ratchet, and perhaps also to create the variability required for response to environmental change. Second, even in those species in which all the individuals are sexual, we can see sex as a species-level property by focusing on its consequences: the division of the population into two morphs, internal population structure, recombination, and a free flow of genetic material.

So the challenge to the species selection explanation of the persistence of sex is empirical, not conceptual. First, John Maynard Smith points out that the species selection hypothesis is vulnerable to a crucial empirical presupposition. It can work only if asexual defectors from sexual reproduction arise only rarely. For if such defection were frequent, most species would contain asexual variants. At that point selection for asexuality over sexuality within a species would subvert species-level selection against asexuality. It's the "subversion problem" for group selection (8.3) replayed at the level of species. This empirical presupposition is probably met in vertebrates, in which the derivation of asexual daughters from sexual parents requires a number of simultaneous changes. There are some asexually reproducing vertebrates, but the females of most all-female species require sperm from the males of allied species to initiate egg development, even though that sperm makes no genetic contribution. Only in a few lizard species are the females completely independent of all males. But this presupposition is not met in many plant and arthropod lineages, which nonetheless are predominantly sexual. So species selection seems unlikely to be the whole explanation for the predominance of sex (Maynard Smith 1989a, 165–80). Second, there are recent

alternatives to these species selection hypotheses that propose a direct individual benefit of sex. Since the offspring of a sexually reproducing individual
vary one from another, it is natural to suppose that sex pays off when the
environment is in some way unpredictable or variable. The most intriguing
suggestion along these lines is W. D. Hamilton's idea that the change generated by sex is a weapon in the war against parasites (Hamilton 1980).

Sex is one mechanism that generates a greater variety of genotypes in a
population. Perhaps, instead of species selection for sex, there is selection for
that variability itself. To see how this might work, consider the case of magnetotactic bacteria. These bacteria live in the sea near the boundary between
water and sediment. They come equipped with little compasses called mag-
netosomes, which they use to navigate away from oxygen-rich surface water
because oxygen is toxic to them. Put a Northern Hemisphere bacterium in
a southern ocean, and it will swim to the surface and die. Perhaps we should
expect these species to have short life spans, because from time to time the
earth's magnetic pole reverses, and that would set all the bacteria swimming
in the wrong direction toward their individual deaths and species extinction.
However, it turns out that, although the magnetite crystals in these organisms
are synthesized from certain genes, their polarity (their orientation toward
the North or South Pole) is not determined by these genes. This information
can be passed on only by transferring part of the magnet after cell division.
This acts as a seed for the compass in the new cell. But should this seeding
be incomplete, the polarity of the daughter's compass is randomized: there is
an equal probability of north- or south-seeking cells. Although under normal
circumstances the small percentage of the population with the reverse polarity would swim into a toxic zone, this heterogeneity within the population ensures the survival of the species should the magnetic field reverse
(Mann, Sparks, and Board 1990). There may have been bacteria with more
precise mechanisms for seeding their daughters with functional magneto-
somes, and if so, individual selection would have favored those bacteria.
Their adaptation would have saved some of their progeny from swimming
the wrong way. But any species in which this adaptation became universal
would go extinct at a reversal of the earth's magnetic field. So variability may
in itself be an important species-level property.

These bacteria are not just an elegant example of the importance of nongenetic replication. They also illustrate the potential evolutionary significance of variation within a species as a buffer against sudden environmental
change. Moreover, variety contributes to a species' evolutionary potential.
Lloyd and Gould argue that there is species selection in favor of this more
general property (developing an old idea of Lewontin's: see Lewontin 1957;

Godfrey-Smith 1996, 262–67). The greater the variation within a species, the better it can respond to selection as its conditions of life change (Lloyd and Gould 1993). So in this view, sex is just one case of a more general phenomenon of *phylogenetic plasticity:* the capacity of a lineage to change over evolutionary time. The difference between sexual and asexual species is obviously one key ingredient in plasticity, but not the only one. A species divided into many small populations explores more options (because of the greater importance of chance), and can fix adaptive changes more quickly, than an unfragmented species with the same population size and genetic variability. Thus a species' population structure is relevant to its capacity to change. So, as we saw in section 8.4, are mating systems and other behavioral characters that can divide the population into differing trait groups. No doubt genetic systems are relevant too.

Phylogenetic plasticity is clearly a property of lineages rather than their component organisms. It may well be causally relevant to extinction and speciation. It is likely to be inherited through speciation events. Population structure may not be transmitted across speciation events, but other elements of phylogenetic plasticity will usually be. Moreover, extinction events come in all sizes, so lineages are constantly being tested for a capacity to respond to change. The inheritance of their ancestors' extinction resistance by descendant species cannot be accidental. The capacity to change would be a consequence of cumulative selection. So Lloyd and Gould's basic idea has plenty of initial plausibility. We would not be surprised to find that phylogenetically plastic lineages are overrepresented in surviving lineages. But its empirical test we must, with relief, leave in others' hands.

Further Reading

9.1 Keller and Lloyd 1992, Sober 1994, and Hull and Ruse 1998 all have good sections on species. Depew and Weber (1995) nicely chart Darwinism's struggle with the problem of the reality of species; see especially chapters 11 and 12. For an introduction to the debate on Darwin's own views, see Beatty 1985. Eldredge (1985b, 1989, 1995) defends his views on the importance of species in evolution. His central argument derives from Mayr's theory that new species arise from small and isolated fragments of the original parent species. Mayr's theory of speciation is also ancestral to the ideas of punctuated equilibrium; see Mayr 1976a, part II, and Mayr 1988, part VII.

Punctuated equilibrium got off the ground with Eldredge and Gould's paper (1972). At the time, it seemed to be an application of Mayr's theory of speciation to our expectations about the fossil record: If speciation occurs in small, isolated populations, then intermediate fossils will be rare. Somewhat later, Gould, especially, gave it a more radical spin. His views were reported in the press as a refutation of Darwinism (see Gould 1980b,c 1983b; 1985)! Eldredge gives his version in Eldredge 1985a, which also reprints the original article. Perhaps provoked by Gould's radicalism, the final chapter of Dawkins's *The Blind Watchmaker* (1986) is a rather uncharitable interpretation of the significance of punctuated equilibrium. However, the same work has a fine discussion on the implausibility of evolution by jumps, as does Dawkins 1996. Dennett (1995) reviews these issues, from a stance close to Dawkins's, in section 10.3 of *Darwin's Dangerous Idea*. The defenders of punctuated equilibrium were at first read as questioning this element of received wisdom, but it is now clear that they do not. The first half of Somit and Peterson 1989 is devoted to debating these ideas. The inventors of punctuated equilibrium have distanced themselves from the most radical interpretation of their views (see Gould and Eldredge 1993). One of us (Sterelny) tries to sort out the ambiguities (Sterelny 1992b).

9.2 Species concepts have been seen as serving two functions. Seen one way, they specify the membership conditions for species: they tell us whether some arbitrarily chosen organism is (say) a member of *Canis familiaris*. This is sometimes known as the *species taxon* problem. Seen another way, they tell us what all species have in common—what all the populations we think of as species share. This is the sometimes called the *species category* problem. These distinctions are used in the readings that follow. Since we think an answer to the taxon problem should solve the category problem, and vice versa, we have not distinguished them in section 9.2.

The two classic defenses of the historical conception of species are Ghiselin 1974b and Hull 1978. Mayr is the great defender of the biological species concept: see Mayr 1982b, 1976a, part VI, and 1988, part VII. Problems for this concept are discussed by Ehrlich and Raven (1969) and O'Hara (1993, 1994; both fine papers) as well as by Kitcher (1989).

There are bafflingly many modern species concepts, perhaps more than twenty. Two recent attempts at an overview of the whole area are by Mayden (1997) and Hull (1997). Phylogenetic species concepts are defended by Wiley (1978), Ridley (1989), Mishler and Brandon (1987), and Kornet (1993). Van Valen (1976) introduces the ecological species concept. Templeton (1989) sets out and defends the cohesion concept of a species. A possible alternative to all these views is Paterson's "recognition" concept of the species, though we see this as a version of the biological species concept. Paterson's views are discussed with approval by Eldredge (1989), and are set out by McEvey (1993) and by Lambert and Spencer (1995). Many of the most important papers on species concepts are collected in Ereshefsky 1992. Much of the empirical literature on speciation is reviewed in Otte and Endler 1989, and more recently, in Lambert and Spencer 1995. Two new collections on species and species concepts are Claridge, Dawah, and Wilson 1997 and R. A. Wilson, in press, in which we both develop our further thoughts on species.

9.3 For a wonderfully readable, entertaining, and gossipy account of the extraordinary battles over systematics, see Hull 1988. For an overview of the main strands in systematic theory, though from a partisan cladist, see Ridley 1986. de Queiroz (1986) gives a short, punchy defense of cladism and its implications. He returns to the same theme in de Queiroz and Good 1997, though rather more technically. Panchen (1992) presents a thoughtful, philosophically informed, and interesting history of the development of systematic theory. It is sympathetic to, but by no means uncritical of, cladism. It includes a detailed discussion of cladistic techniques for discovering phylogeny. Brooks and McLennan (1991) give the cladistic prescription for world conquest ("our struggle"). Harvey and Pagel (1991) provide a careful, and for the most part readable (except for a brutal chapter 5), introduction to the problem of actually reconstructing history on cladistic principles. Sober (1988c) devotes himself to the same problem, treating it as a test bed for thinking about probabilistic reasoning in general. Minelli 1993 covers some of the same ground as Harvey and Pagel, but much more briefly, for it concentrates on giving an overview of the tree of life as we currently understand it. Sober 1994 has a section on systematics, including a chapter by Sokal defending phenetics and one by Mayr defending evolutionary taxonomy. Mayr defends his views further in Mayr 1976a, part V. Cronquist 1987 is an important critique of cladism in principle and in practice, arguing that its techniques cannot deliver on its claim to reconstruct phylogeny, and arguing that cladism cannot give an appropriate account of species known only from the fossil record. Donoghue and Cantino (1988) and Humphries and Chappill (1988) reply to Cronquist.

9.4 In section 9.4 we discuss the idea that species are interactors. It's also possible to defend a version of species selection in which species are replicators; see especially Williams 1992, though this idea is also briefly discussed in Dawkins 1982. An alternative way of distinguishing species sorting from species selection focuses on the relationship between a species' fitness and the fitness of its component organisms. This idea is defended by Damuth and

Heisler (1988). We think Lloyd and Gould (1993) have a similar view. Vrba (1984c, 1989) herself distinguishes species selection and species sorting in the same way we do, as does Sober in his discussion of species selection (1984b, 355–68). We defend our views in greater detail in Sterelny 1996a. For more on the idea of macroevolution as an explanation of lineage diversity, see Valentine 1990 and Jablonski 1987. For overviews, see Gilinsky 1986 and Grantham 1995. Damuth (1985) argues that species themselves do not have environments, and hence cannot be thought of as interacting as a whole with their environment. Species, he argues (and Eldredge agrees), are almost always divided across a number of distinct niches, so the real high-level units of selection are smaller units that occupy a single niche, local populations of species that he calls "avatars."

There is much good work on sex. Three classics are Williams 1975, Maynard Smith 1978, and Ghiselin 1974a. Maynard Smith summarizes his views in chapter 19 of Maynard Smith 1989a and in Maynard Smith 1988. Bell 1982 is long and in places difficult, but is a great synthesis of facts and ideas. Matt Ridley (1993) presents a very readable survey of recent work on the evolution of sex. Michod (1995) defends the view that the function of sex is gene repair, and argues against the alternative idea that sex is an adaptation to environmental unpredictability. It too is a breezy read. The most currently fashionable version of the "unpredictability" hypothesis is that of Hamilton, who argues that fast-evolving pathogens generate environmental unpredictability of the sort that selects for sex: the very success of the parental genotype encourages the evolution of pathogens well adapted to take advantage of it, and hence the success of that genotype in the next generation will be degraded. Hamilton 1988 is a good nontechnical introduction to this idea; the second volume of his collected papers *(Narrow Roads of Gene Land),* which is to appear shortly, will be largely devoted to this topic. Kondrashov 1993 is a good recent review article. The February 1996 issue of *Trends in Ecology and Evolution* is an up-to-date and mostly nontechnical special issue on sex and evolution. Nunney 1989 is a difficult technical paper defending group selection as an explanation of sex. On the more general issues of evolutionary plasticity, see Dawkins 1989a and Schull 1990.

IV

Evolutionary Explanations

10

Adaptation, Perfection, Function

10.1 Adaptation

As we noted in discussing theories of taxonomy (9.3), there are countless ways in which we can describe organisms. Egg laying, a poison spur, and an extraordinary bill are all striking and distinctive traits of the platypus. But the platypus has many other features less likely to be highlighted in natural history documentaries, such as the distance between the eyes divided by the inter-ear distance. That may seem a rather esoteric property, but anatomical descriptions of the platypus in texts on Australian fauna will include many that seem equally obscure: for instance, the length, shape, and weight of its various intestines. *Handbooks to the Birds of Anywhere* always specify the number of various types of wing and tail feathers. So organisms have many characteristics, some of which we routinely measure and describe, and others that languish unstudied. One pressing problem biologists face is making this choice: determining which aspects of an organism are important in its evolution, ecology, and development.

Among the traits biologists study, some are clearly special. As we discussed in section 2.2, some traits are favored by natural selection because they increase the relative fitness of their bearers. In other words, they are *adaptive*. A trait that exists because natural selection has favored it is called an *adaptation*. The eye-blink reflex exists because it protected the eyes of ancestral organisms and so increased their fitness. This reflex is an adaptation "for" protecting the eye. Each adaptation was selected for some effect or effects that influenced the fitness of its bearer.

However, despite the close links between these two concepts, adaptiveness is neither necessary nor sufficient for a trait to be an adaptation (Sober 1993, 84). The human appendix, for example, is an adaptation that is not adaptive. Humans no longer need to digest cellulose, and having this home for symbiotic bacteria that can break down that substance no longer increases our fitness. But the appendix is definitely an adaptation. It evolved through natural selection because it enhanced the fitness of our distant ancestors. So it's an adaptation without being adaptive. The appendix is a *vestigial* trait: a relic of previous selection. Conversely, the ability to read is adaptive without being an adaptation. Literacy is highly adaptive in most modern human societies, as the disadvantages suffered by dyslexic people testify. But the ability to read is probably a side effect of other, more ancient cognitive abilities. The invention of reading was probably much like the invention of computers. The use of computers did not originate in a few people with special new genes for programming. Computer use did not spread through the population because users had more children than nonusers. Our ability to read and use computers almost certainly depends on a set of more general cognitive capacities—capacities that have not changed in the few thousand years in which literacy has spread.

So some traits exist as a consequence of natural selection for one or more of their effects. These are *adaptations*. Some, but not all, of these traits continue to contribute to the fitness of organisms that have them. These traits are *adaptive*. Other traits are mere side effects of evolution, and these include some that happen now to be adaptive. A few Australian parrots have greatly increased in their range and numbers over the last century because they happen to have characteristics that suit them for the new habitat created by agriculture. The female spotted hyena has a hypertrophied clitoris that she uses in greeting ceremonies. But the clitoris is not large and penislike because it is used in such ceremonies. Rather, it is a side effect of selection for aggression and the hormones that drive it (we thank Richard Francis for this striking example). Other traits probably have no effect in themselves on fitness. We doubt that the ratio of inter-eye to inter-ear distance has ever in itself affected platypus life. That ratio is a mere epiphenomenon of the different evolutionary forces that built platypus eyes and ears.

Gould and Vrba have argued for a less obvious distinction among the traits of organisms. Very often a trait comes to play a role in an organism's life quite different from the one it played when it first evolved (Gould and Vrba 1982). The eighteenth-century French philosopher Voltaire accused his contemporaries of believing that the nose exists for holding spectacles in place. No evolutionist would make that mistake, but Gould and Vrba think that biologists are prone to subtler mistakes of the same sort. Feathers are very useful to birds in making wings. The superior efficiency of wings made of feathers may explain why birds rather than bats dominate the skies. But it is unlikely that feathers evolved from reptilian scales *because* they helped the ancestors of birds to fly better. It is thought that they evolved to assist in thermoregulation, and were later found to be useful for flight. Gould and Vrba call this process *exaptation*. A trait is an exaptation if it is an adaptation for one purpose but is now used—often in a modified form—for a different purpose. If the received story of feather evolution is right, feathers are adaptations for thermoregulation and exaptations for flight. Mammal ear bones are converted jaw bones; they are exapted for hearing. In older writings about evolution, this evolutionary pattern is often called *preadaptation:* feathers, for example, are preadaptations for flight. This older terminology is very misleading. The word *preadaptation* suggests that evolution is forward-looking—anticipating the future needs of the organism. Evolution by natural selection cannot look forward because it cannot incur costs in anticipation of later benefits: do not ask for credit, as extinction often offends!

Gould and Vrba think that a trait is an adaptation only for the purpose for which it was *first* selected. But what justifies this special status for the first of many selection pressures? The importance of the concept of adaptation in biology is that it explains the existence of many traits of the organisms we see around us. This explanation is not just a matter of how traits first arose, but of why they persisted and why they are still here today. If we want to understand why there are so many feathers in the world, their later use in flight is as relevant as their earlier use in thermoregulation. Adaptation is a process that happens in stages. Traits arise from new genetic structures. Some of them are adaptive, and hence are spread by natural selection. They become adaptations. They may spread so far that they become "fixed" in the population (possessed by every individual). Alternatively, they may spread to a certain frequency and no further. Later in evolutionary history, the lifestyle of the organism may change, and the trajectory of adaptation may change as well. In New Zealand, where, as far as we know, there were no native mammals except bats before human occupation, flying away from predators ceased to be part of the lifestyle of many birds. Flight ceased to be adaptive, and that had implications for the further evolution of those birds' wings. The wings of the New Zealand weka (a flightless rail) are vestiges of its old adaptations.

But a trait can be retained under changed ecological conditions if it does something else, something new, that is useful. Darwin gave some examples of this phenomenon when he discussed the evolution of emotions (Darwin 1965). He thought that many facial expressions were originally selected for some practical purpose, but were later selected because they had acquired a role in communication between members of the species. He suggested that the baring of the teeth by angry primates may originally have been selected as a preparation for attack or a demonstration of fighting ability. It then acquired a secondary use in signaling anger. That is why it still occurs in humans, who rarely fight with their teeth (see figure 14.1). Since these processes of *secondary adaptation* are probably very common, the adaptation/ exaptation distinction is not very useful except as an indication of the succession of evolutionary events. A trait is an adaptation for *all* the purposes it has served and which help to explain why it still exists. The important distinction is not between the first selection pressure and the others, but between all the selection processes and the processes that are happening today, but have played no role in past evolution. This is the distinction between "being adaptive" and "being an adaptation."

10.2 Function

The *function* of the heart is to pump blood. The heart also makes noises, but that is not part of its function. The function of the brow-raising response to surprise is to increase the visual field. This response also stretches the skin, but that is not part of its function. Distinctions like these are common in biology, but their equivalents in the physical sciences would seem bizarre. Physics does not tell us what the sun is "for." The sun has all sorts of effects, but there is no distinction between the effects it is "meant" to have and those that are accidental side effects. In an earlier phase of human thought we could have made such distinctions. The sun was created by God to warm the earth, and the fact that it warms Mars, where there are no creatures with souls, is a side effect. But the rise of modern science was marked by the expulsion of explanations in terms of purpose or function in favor of explanations in terms of natural laws. The sun came into existence because the expansion of matter from the Big Bang was not entirely regular, and all its effects, useful or useless, are equally unintended.

The conventional explanation of this difference between biology and physics is that biology studies the products of natural selection, while physics does not. Talking about functions is just a convenient way of talking about adaptations. If brow raising in surprise is an adaptation for increasing the visual field, then its function is to increase the visual field. Stretching the skin around the eyes has no known connection to reproductive fitness, so brow raising is probably not an adaptation for skin stretching, and skin stretching is not one of its functions. This view of function has been common among biologists for a long time. The architects of the received view even introduced a new name, *teleonomy,* to distance this biological understanding of functions and purposes from more traditional teleological ideas (Pittendrigh

1958). Konrad Lorenz, the co-founder of modern animal behavior studies *(ethology)*, describes this perspective very clearly:

> If we ask "What does a cat have sharp, curved claws for?" and answer simply "To catch mice with," this does not imply a profession of any mythical teleology, but the plain statement that catching mice is the function whose survival value, by the process of natural selection, has bred cats with this particular form of claw. Unless selection is at work, the question "What for?" cannot receive an answer with any real meaning. (Lorenz 1966, 9)

Philosophers call this the *etiological theory* of biological functions. An etiological theory explains something in terms of its origins, or *etiology*—in this case, its evolutionary origins. The functions of a biological trait are those effects for which it is an adaptation. A distinctive feature of the etiological theory is that a trait can have functions that it is unable to perform. The function of the white coat of a polar bear is to make the bear harder to see. There is no snow in most zoo polar bear enclosures, but it is still correct to point to the white bear on the gray concrete and say that it is white for the purpose of camouflage.

The etiological theory is the orthodox view in philosophy of biology, but it is not universally accepted. One influential criticism of this theory points out that people talked about biological functions long before evolutionary theory was invented. When William Harvey announced in the seventeenth century that the function of the heart is to pump blood, he didn't mean that it had evolved to pump blood—he thought that the heart was created by God. People used the concept of a biological function before having any idea of natural selection, so biological function cannot be about natural selection. This objection depends on the idea that the etiological theory is a *conceptual analysis* of function—that it is a theory about what people mean by the word *function*. Ruth Millikan (1989b) has argued that this is a mistake. The etiological theory of function is a scientific theory, not a conceptual analysis. No one objects to the theory that heat is molecular motion on the grounds that people understood the term *heat* long before anyone understood much about molecular motion. We are acquainted with heat, and develop various theories about what it is. One of those theories turns out to be the best. Millikan argues that the etiological theory—functions are effects promoted by natural selection—is the best theory of why organisms have functional traits. Indeed, apart from appeals to theology, it is our only such theory. It is without scientific rivals. Karen Neander makes a similar point (Neander 1991). She agrees that the etiological theory may not capture the definition of *function* ordinary people have in mind. But it may nonetheless capture the current biological conception of function.

The main rival of the etiological theory is the *propensity theory* (Bigelow and Pargetter 1987). According to the propensity theory, the functions of a trait are its *adaptive* effects, rather than the effects for which it is an *adaptation*. Functions are effects that increase an organism's propensity to reproduce. The etiological and propensity theories ask very different questions when trying to determine the function of a trait. Most people are able to learn to read fairly easily because they have typical human brain structures, rather than the slightly different structures found in people who are dyslexic. Is it the function of these structures to promote reading, or is this merely a side effect? The etiological theory asks why the structures evolved. Were they ever selected for producing reading? The propensity theory, on the other hand, asks whether people who can read typically have more offspring now than people who cannot read. If they do have more offspring, then it is the function of these brain structures to support reading.

Many people have been attracted to the propensity theory because it allows creatures with no evolutionary history to have biological functions. This point is often made using bizarre science fiction examples. Suppose a creature identical to you, atom for atom, were to arise through a random coming together of matter. Propensity theorists have a gut feeling that the heart of this creature would have the function of pumping blood and only the side effect of making heart noises. But according to the etiological theory, the creature would have no functions at all, because it would have no history of selection. It is unclear what significance to assign to gut feelings (often dignified with the name *intuitions*) about bizarre science fiction stories. Fortunately, the same point can be made using examples closer to the real world. Organisms can develop beneficial traits by mutation. If a bacterium incorporates a DNA plasmid from another bacterium that allows its new owner to synthesize a protein conferring resistance to a certain antibiotic for the very first time in the history of life, then according to the etiological theory, this protein has no function. Antibiotic resistance is a mere effect, for it does not explain the existence of the protein via the feedback loop of natural selection. Natural selection has not yet acted, as this variation has only just come into existence. Conferring resistance will *become* the protein's biological function only when bacteria with it have been favored by selection. According to the propensity theory, however, conferring resistance is the function of the protein from the moment it becomes useful.

Some biologists have also argued for an approach to function that concentrates on current adaptiveness rather than evolutionary history. They are not concerned with ordinary intuitions about when something has a function. They want to decouple claims about function from claims about evolutionary history because they have doubts about our ability to reconstruct evolutionary histories accurately. They think that if functional claims are implicitly claims about evolutionary histories, then functional analyses in biology will inherit all the uncertainties of these reconstructions (Reeve and Sherman 1993; Hauser 1996, 82–85). It is obviously desirable that notions like function and adaptation be defined in ways that make it possible to discover a trait's function. These authors' concern about our ability to confirm claims about function and hence adaptation reflects one of the most important debates in recent evolutionary theory—the debate over *adaptationism*—to which we turn in section 10.3.

We should not assume that biology traffics in only one type of function claim. Godfrey-Smith, Amundson, and Lauder all argue that there are two very different senses of *function* in biology (Godfrey-Smith 1993, 1994b; Amundson and Lauder 1994). Evolutionary biologists often use *function* in the sense defined by the etiological theory. Anatomists and physiologists, however, are not typically concerned with evolutionary history. They are interested in the activities an organism can perform: flying, digesting food, detecting viruses in its tissues, and the like. They explain how organisms perform these activities by *functional analysis*—by breaking down the overall task into parts that are performed by different parts of the organism. A biomechanical analysis of the knee joint explains how each part of the knee contributes to its ability to flex and bend. These functions of a biological trait are its *causal role functions*. Sometimes the causal role functions of a trait are the same as its etiological functions. The heart actually does pump blood, and that is what it was selected to do. In other cases the two kinds of function do not coincide. The redness of blood plays an essential causal role function in blushing, but our blood is not red because people who were able to blush had more children than other people.

We can think of the functions defined by the propensity theory as a special case of causal role functions. From a biological point of view, one of the most interesting properties of an organism is its capacity to survive and reproduce. Biological fitness is a measure of this capacity. Like any other capacity, an organism's fitness can be functionally analyzed. Each salient feature of the organism makes particular contributions to its ability to survive and reproduce; these contributions are *components of fitness*. In other words, these contributions are the causal role functions of those features relative to that capacity. They are the effects picked out by the propensity theory as the functions of those features. So if we concentrate on one particular capacity—the capacity to survive and reproduce—causal role functions and propensity functions coincide. The causal role conception of function is much wider than this, however, because it can be applied to any capacity whatsoever. The biomedical sciences, for example, functionally analyze the body's capacity to fail in various ways.

In this chapter, our primary focus is evolutionary biology. So we will be using the notion of function central in that domain, while recognizing that other branches of biology—biomedical science, physiology, and perhaps others—are often interested in the contribution of a part of a biological system to the activity of the system as a whole, without being concerned with historical questions.

10.3 The Attack on Adaptationism

Stephen Jay Gould and Richard Lewontin have compared the idea of biological adaptation to the ideas of Dr. Pangloss in Voltaire's eighteenth-century satire *Candide*. Dr. Pangloss believed that everything in the world was designed by a wise and loving God. Even sexually transmitted diseases like syphilis were really for the best in this best of all possible worlds. Gould and Lewontin accused modern evolutionists of the equally unrealistic belief that if an organism has a trait, then it must be, in evolutionary terms, the best trait the organism is capable of having (Gould and Lewontin 1978; see also Lewontin 1982b, 1985a, 1987). Gould and Lewontin's criticisms had three main components.

Confusing Adaptiveness with Adaptation

"Adaptationists" conclude that every useful trait exists *because* it is useful. If a bird flies south for the winter, the adaptationist concludes that this must be a behavioral adaptation for avoiding the cold. But what if the bird's ancestors lived in the south, and their habit of flying north each summer was favored by natural selection because of the abundant food resources of the brief northern summer? If all the bird's closest relatives live year-round in the south, then the evolutionary breakthrough is flying north for the summer boom. Then we might question whether flying back is an adaptation for avoiding the cold. Perhaps it is a side effect of flying north for the boom. A properly historical perspective on evolution is necessary in order to see where adaptive explanations are appropriate.

Overlooking Nonadaptationist Explanations

Gould and Lewontin's second criticism is that other kinds of biological explanations are unduly neglected in favor of adaptive explanations. Human arms have two bones rather than one in the forearm. Is this because it is adaptive to have two bones rather than one, or is it because humans are part of a large group of organisms that are designed that way? Within the group Tetrapoda—creatures with a characteristic four-limbed layout—organisms inherit the two-bone design, and they retain it unless there is powerful selection against it.

Part of the issue here is how to divide up an organism into "parts." What features of an organism are its traits? That is, what features of an organism have an evolutionary history to call their own? Mandrills are one of the larger Old World monkeys. Males have electric blue muzzles and a matching blue on their behind and genitals. Should we consider these colors to be part of a single evolving trait, the overall mandrill color scheme, or do the colors of these particularly salient parts of the male monkey have evolutionary histories to call their own? This is no simple question. One aspect of Gould and Lewontin's critique of adaptationism is the charge that adaptationists see organisms as a mosaic of separate parts, each of which has an independent evolutionary explanation. No one doubts that some traits can evolve independently of the rest of the organism. The beaks of the Galápagos finches change under selection without everything else changing. But Gould and Lewontin deny that the picture of the organism as a mosaic of traits is always or usually accurate. They argue, for instance, that the human chin is an inevitable effect of the way the jaw grows, but does not have any particular evolutionary purpose of its own. Seeking to explain the chin as a separate feature is bad biology.

The Unfalsifiability of the Adaptationist Program

Finally, Gould and Lewontin argue that adaptationism is unscientific because it cannot be disproved by experiment. In their view, adaptationists tell "just-so stories" about why a trait was selected in the evolutionary past and regard these stories as scientific explanations. In fact, these stories are only "how possibly explanations." They show that there is at least one way the trait *might have* evolved. This is a useful thing to do, because people are forever alleging that this or that unusual trait refutes the whole idea of natural selection. But it is not the same thing as a testable scientific explanation of how the trait *actually* evolved. Particular adaptive stories can be tested, as we discuss below, but Gould and Lewontin argue that this does not test the idea of adaptationism itself. Whenever a particular adaptive story is discredited, the adaptationist makes up a new story, or just promises to look for one. The possibility that the trait is not an adaptation is never considered.

10.4   What Is Adaptationism?

This critique of adaptationism has provoked a vigorous debate, one that is still very much in progress. But it has become clear that *adaptationism* does not name a single position. To the contrary: Godfrey-Smith (in press-c) argues that three distinct theses have been conflated in the controversies that followed in the wake of Gould and Lewontin's paper. He distinguishes between empirical adaptationism, explanatory adaptationism, and methodological adaptationism.

Empirical adaptationism was probably the main target of Gould and Lewontin. It is the idea that natural selection is by far the most powerful factor in evolutionary history, and that most of the biologically significant features of organisms are shaped almost entirely by natural selection. These features exist because of selection for one or more of their effects, and hence are adaptations.

This hypothesis is easily conflated with another, *explanatory adaptationism*. We suggested in section 2.1 that the explanatory agenda of evolutionary theory is dominated by the problems of diversity and adaptation. Explanatory adaptationism takes the existence of adaptation, especially complex adaptation, to be the central problem in evolutionary biology. Because natural selection is the only mechanism that produces complex adaptation, it is indeed the most important factor in evolutionary history. That is not necessarily because of its ubiquity or strength, but because it answers evolutionary biology's $64,000 question: What explains complex adaptation? Natural selection is the only satisfactory explanation of complex adaptation, even if it is highly constrained, and even if most features of organisms are not adaptations.

Dawkins is an explanatory adaptationist. The first chapter of *The Blind Watchmaker* (Dawkins 1986) is a perfect specimen of that view. But it is not at all obvious that he is an empirical adaptationist. The third chapter of *The Extended Phenotype* (Dawkins 1982) is a careful discussion of constraints on adaptation. Empirical and explanatory adaptationism are independent ideas. As Dawkins (at least in some moods) shows, we can certainly accept explanatory adaptationism without accepting empirical adaptationism. Equally, we can embrace empirical adaptationism without explanatory adaptationism, for as we shall see, explanatory adaptationism has certain presuppositions that the empirical adaptationist may deny. In particular, explanatory adaptationists are committed to an unorthodox definition of adaptation.

Though both are highly contestable, both empirical and explanatory adaptationism make important claims about the natural world. *Methodological adaptationism* makes no such claims. Rather, methodological adaptationists think that the best way to study biological systems is to look for good design. They look at adaptation as a good organizing concept in evolutionary theory.

There is something very plausible about explanatory adaptationism. The intricate, weird, and beautiful adaptations of the living word are genuinely striking. They scream out for explanation. Moreover, there is almost unanimous agreement that natural selection is indeed the only reasonable explanation of platypus electrolocation, bat facial anatomy, the fig tree/fig wasp symbiosis, and the like. Furthermore, this adaptationist idea is important to philosophy. The idea of natural selection has played an important part in refuting theistic arguments from design and in establishing a naturalistic conception of the universe. One of us (Sterelny) is at heart an explanatory adaptationist.

Nevertheless, explanatory adaptationism faces both empirical and conceptual challenges. Ronald Amundson (1998) makes the empirical challenge explicit. He distinguishes constraints on adaptation from constraints on morphology. The conservation of testicle number among the vertebrates, for example, may reflect no constraint on *adaptation*. The environment may not be asking a question that variation in testicle number would answer. Equally, the persistence of basic structural plans in large groups of related but ecologically diverse organisms might be adaptively neutral. Nonetheless, the conservation of these patterns requires explanation, which might be found in developmental and historical constraints on evolution. Constraints can be explanatorily important without being constraints on adaptation. Adaptation, in this view, is one great explanatory challenge that evolutionary biology faces, a challenge that the theory of natural selection meets. But, as Gould has often argued, the persistence of basic structural similarities across such vastly different lifestyles as those of the bats and the whales presents another challenge. The persistence of such similarities over hundreds of millions of years is as striking as the existence of complex adaptations, and it is *not* explained by natural selection. Natural selection explains adaptation and perhaps even diversity, but not this *persistence of type*. This challenge, we must mention, is itself controversial. Selection—so-called *stabilizing selection*—can act to prevent change, so perhaps it might explain the persistence of type after all. Moreover, persistence may be no more than chance. If change is rare compared with no change, we would expect, simply from chance, all the descendants of a species to manifest some of their ancestor's traits. Nevertheless, Amundson's challenge is clearly powerful.

In addition to this empirical challenge, explanatory adaptationism faces a conceptual one. In section 10.1 we fell in with the standard practice of defining adaptation by appealing to natural selection. Adaptations, by definition, are all, and only, the traits that exist by virtue of selection for their effects, current or past. But explanatory adaptationism does not make sense in this conception of adaptation. We cannot at the same time define adaptation as whatever natural selection causes and promote natural selection on the grounds that it is the explanation of a particularly puzzling phenomenon, namely, adaptation. If the theory of explanatory adaptationism is to mean something substantial, then adaptation, especially complex adaptation, must be characterized independently of its putative explanation, natural selection.

Empirical adaptationism is no less contested. It faces problems of both interpretation and testing. Let's look first at interpretation. Everyone agrees that all evolutionary trajectories depend on many factors. Tree kangaroos have a surprising array of adaptations to arboreal life, and all would agree that the evolution of these characters—for instance, the stiffened, counterweighted tail—depends on selection, history, and chance. Selection could not have made a counterweighted tail without the evolutionary possibilities the previous history of the lineage made available. What, then, does it mean to claim priority for one of these factors? If chance, selection, and history all play crucial roles, how can any one be more important than the others? Once we answer this challenge, we still face the empirical one: How can claims of relative importance be tested?

In the rest of this chapter we focus on empirical adaptationism. In section 10.5 we look in more detail at the biological explanations that are held up as alternatives to adaptationism. In sections 10.6 and 10.7, we return to the problem of formulating and testing adaptationist ideas.

10.5 Structuralism and the *Bauplan*

Gould and Lewontin have revived an old concept from continental European biology: the *bauplan*, or fundamental body plan, of an organism. A trait can be explained by pointing out its position in one of these fundamental body plans rather than by asking what adaptive purpose it serves. The existence of these two different varieties of biological explanation is endorsed by Darwin:

It is generally acknowledged that all organic beings have been formed on two great laws—Unity of Type and the Conditions of Existence. By unity of type is meant that fundamental agreement in structure, which we see in organic beings of the same class, and which is quite independent of their habits of life. On my theory, unity of type is explained by unity of descent. The expression of conditions of existence, so often insisted upon by the illustrious Cuvier, is fully embraced by the principle of natural selection. For natural selection acts by either now adapting the varying parts of each being to its organic conditions of life; or by having adapted them in long-past periods of time. (Darwin 1964, 206)

The first of these principles, the unity of type, was central to the advances of nineteenth-century biology that paved the way for *The Origin of Species*. Georges Cuvier, Richard Owen, and others made great strides in comparative anatomy—the structural comparison of the bodies of organisms of different species. It had been conventional since the eighteenth century to classify living creatures according to a hierarchical "system of nature"—species, genera, families, orders, classes, and phyla. Comparative anatomy demonstrated that the members of a genus or family share similarities that seem quite unrelated to the practical needs of their ecological lifestyles. A lobster shares its segmented body plan with the rest of the arthropods, and shares the distinctive fusing of the first few segments to form a head with the other crustaceans. Neither feature seems to have any particular connection with the lobster's lifestyle. It is as if each class of organisms was designed as a variation on a basic plan common to its order, and each family as a variation on a basic plan common to its class, and so forth down to individual species. Hence, many features of an individual species reflect its position in the system of nature. If we can find evidence that a species fits into a particular part of the system, we can predict that it will have not only the characteristic properties that caused us to place it in that part of the system, but other properties characteristic of the organisms in that part as well.

The law of the unity of type provides an alternative to explanation by adaptation (Darwin's "conditions of life"). We can explain by classifying. Lobsters have fused head segments because they are crustaceans. Pigeons find food by sight and dogs by smell because pigeons are birds and dogs are mammals, and those are the senses those groups typically use. *Explanation by classification* is familiar from the physical sciences. Like the system of nature, the periodic table of elements groups things in ways that predict their properties. We can infer that copper is ductile and conductive *because* it is a metal. Mendelev's discovery of the periodic table was hailed as a great scientific

achievement because it put a large quantity of information about the properties of different chemicals into a simple pattern, and because new discoveries fitted into roughly the same pattern.

One of Darwin's main achievements in the *Origin* was to turn contemporary comparative anatomy into an argument for his theory of evolution.
The apparently arbitrary resemblances between members of a family or genus
make perfect sense if all the species in the family or genus are descended from
a single ancestral species. All birds have a furcula, or wishbone, because their
common ancestor had one. All vertebrates have their spinal cords on the
dorsal side because that's where it was in their common ancestor. With a
single stroke, Darwin had turned the life's work of many scientists, including
many bitterly opposed to him, into support for his theory. Where does this
leave explanation by classification? In one sense, as Darwin says, the law of
the unity of type is subordinated to the law of the conditions of existence, or
adaptation—that is, to explanations that appeal to natural selection. The
"types" or "plans" are themselves the products of earlier evolution. One
could argue that explanation by classification simply begs the most interesting question, which is how the characteristics common to the whole
group evolved in the ancestral species. The only real explanation is one that
traces the origins of these characters by natural selection: "Hence, in fact,
the law of the Conditions of Existence is the higher law; as it includes,
through the inheritance of former adaptations, that of Unity of Type"
(Darwin 1964, 206).

This dismissal of unity of type may be too quick. It might still be true,
even after Darwin, that "all organic beings have been formed on two great
laws" (Darwin 1964, 206). Amundson and others point out that while the
special characters that mark out particular biological taxa may have had their
origins in natural selection, they have endured long after their adaptive
significance has disappeared. This is the basis of Amundson's challenge to
explanatory adaptationism. The independence of many of these highly *conserved* traits from the current adaptive needs of the organism was essential to
Darwin's use of comparative anatomy to support his theory. Darwin predicted that current adaptations would exist along with traces from former
periods of evolution that create nonadaptive resemblances among living species. These nonadaptive characters are especially problematic for creationists,
for they should expect God to suit each organism for its role in life.

Thus two patterns are discernible in nature, one overlaid on the other.
The first is the match between organisms and the ecological conditions
under which they live. Natural selection accounts for this pattern very well.

The second pattern is formed by the highly conserved traits, which cause organisms with common ancestry to resemble one another. Thus mammals, despite their great differences in lifestyle, all have distinctively shaped ear bones. The ecological variety of the mammals suggests (though it does not prove) that mammal ear bone shape serves no distinct function, as function typically depends on distinctive features of an organism's environment. Yet if these traits have no adaptive value, why don't they disappear? There should be mutations that affect these traits, and nothing to select against them. Many other traits do disappear in this way. Flightless birds lose their flight muscles and their wings become smaller. Cave-dwelling species gradually lose their eyes. But traits like the relative positions of the bones in tetrapods and the fused head segments of crustaceans don't disappear in this way.

One obvious way to explain this is to appeal to developmental biology. Perhaps these structural characters play an essential role in the way organisms grow. Mutations that affected them would disrupt the complex process by which tissues and organs find their proper places in the body. Many anti-adaptationist biologists have stressed the importance of such *developmental constraints* in evolutionary explanation. They argue that some traits do not need natural selection to keep them in existence. Their presence in an organism is explained by its place in the system of nature, not by the specific adaptive pressures generated by the specific environment it faces.

The geneticist C. H. Waddington tried to explain the existence of developmental constraints through his concept of *developmental canalization.* Waddington argued that the developmental system is such that any minor perturbation in a developmental input, such as a gene product, will merely cause a different route to be taken to the same developmental outcome. He compared development to a ball rolling through a landscape. He imagined this landscape as a sheet anchored to many points underneath, representing developmental factors such as genes. Changing one of these factors will not usually change the overall shape of the landscape, and the ball will still roll to the same general place. In some cases this canalization might itself be an adaptation, buffering normal development against some disturbances.

The biologists who have placed the most emphasis on developmental constraints are the so-called process structuralists. In the case of the tetrapod limb, for example, process structuralists appeal to a well-known model of tetrapod limb development. This model dictates that all limb structures will begin with a single bone and that there will be no tripartite branchings. The generic forms of the tetrapod limb are hard to escape because they are dictated by very general aspects of the way in which these organisms achieve

Figure 10.1 C. H. Waddington's representation of developmental canalization. (a) The path of the rolling ball, which represents the developmental trajectory of the organism, is determined by a landscape representing the effects of all the developmental inputs to the organism. (b) The shape of this landscape is determined by genes and other developmental inputs, here represented by pegs pulling the landscape into shape with strings, and by their interactions, represented by connections between strings. Canalization is the idea that many changes in developmental inputs will leave the overall shape of the landscape, and hence the trajectory of the ball, unchanged. Other small changes in inputs may produce radical change by switching the ball from one valley to another. (From Waddington 1957, 36.)

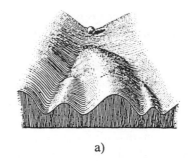

a)

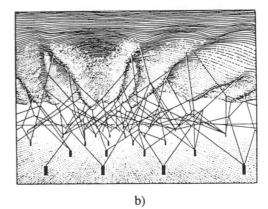

b)

organized growth. In recent years, process structuralism has drawn on chaos and complexity theory to make its case. In the language of those new disciplines, highly conserved traits are *strong attractors* for development.

One way of interpreting process structuralism is to see its defenders as arguing that the space of possible phenotypes—design space—is much smaller than adaptationists suppose. If, for example, there are no six-legged vertebrates in design space—if such organisms are *not possible*—then we do not need to calculate the relative costs and benefits of extra legs to explain their absence. Natural selection *at most* explains why some of the possible organisms are actual and others are not. The complexity theorist Stuart Kauffman, whose ideas we consider in section 15.3, is another who thinks that adaptationists overestimate the extent of design space (Kauffman 1993). Seen in this way, the process structuralists and Kauffman are challenging explanatory adaptationism. There is a striking fact about life—surprising limitations on the range of the possible—about which natural selection is silent. They may be challenging empirical adaptationism as well: many traits

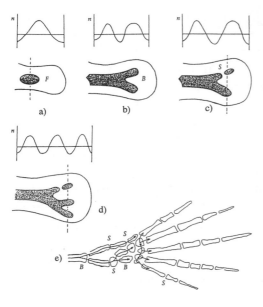

Figure 10.2 In the model of tetrapod limb development proposed by Oster, Murray, and Maini (1985), three processes combine to create the various forms of the tetrapod limb and to limit the forms that can be created. (a) Focal condensation *(F)*: aggregation of cells forms a tightly packed mass that can grow by recruiting more cells. (b) Bifurcation *(B)* of the growing condensate. (c) Segmentation *(S)* into two parts along the length of a limb. (d) The pattern of condensation across the developmental field of a limb growing via these three processes. (e) The role of the three processes *F, B,* and *S* in producing one complete form. The graphs above each drawing show the cell density across the transects indicated by the dotted lines in the drawing. (From Goodwin 1994, 152.)

of many organisms may be explained by developmental constraints and limitations on the possible rather than by natural selection.

Process structuralists hope to return to a pre-Darwinian biology in which explanation by classification was the most important sort of explanation. But we can recognize the importance of both selection and developmental constraints. They are two aspects of the same process. William Wimsatt has shown how natural selection could build organisms with highly conserved characters and strong developmental constraints on their future evolution. He calls this process *generative entrenchment* (Wimsatt and Schank 1988). Wimsatt notes that the key to natural selection is the possibility of incremental design. Very unlikely forms can be produced a piece at a time. Vision starts to evolve as light-sensitive cells appear, then eyeballs, then lenses, then focusing, and so forth. Each stage is selected in its own right because it is better than the last. The improbability of the final design is very large, but the improbability of each stage is quite manageable (2.2). Incremental design has important implications for developmental biology. Each slight modification is generated against the background of the existing developmental system. It makes use of many aspects of what already exists in order to grow correctly. The removal of ancient elements of the developmental system would be likely to remove things that later modifications have made use of and so to disrupt the growth of those modifications. Elements of the developmental system therefore tend to become increasingly generatively entrenched as more is built on top of them. The existing developmental system of the organism comes to shape the space of possibilities available to the organism in its future evolution.

We have no doubt that generative entrenchment is an important idea, and is part of the explanation of the existence of highly conserved traits over long periods of time. However, it should not be regarded as omnipotent. Rudolf Raff (1996) shows that developmental constraints cannot be the whole explanation of the preservation of the body plan. He gives a series of examples of the preservation of the adult form in lineages in which the developmental trajectory to that form has undergone massive modification. Among sea urchins and amphibians, in particular, *direct development* has evolved in many species—that is, those species have evolved developmental trajectories to the adult form of the organism that bypass the usual intermediate stages.

10.6 Optimality and Falsifiability

It's important not to let the rhetoric of the adaptationist debate obscure the fact that some specific adaptationist hypotheses are not controversial. There are cases in which we can read the function of a trait from its complexity and the specific role it plays in an organism's life. For example, vultures have traits that are rightly regarded as adaptations for soaring. First, they have a suite of wing and feather features that are well designed, in an engineering sense, for that particular task. Vultures have broad wings, by virtue of which they have a light wing loading, so that relatively weak thermals will support their soaring. Second, soaring is central to these animals' life histories. Finally, this suite of features is not functionally ambiguous. There are no other tasks in which it plays a critical role. Vultures do not, for example, use their broad wings to shade water more effectively so that they can see into it to hunt, as herons do. In cases like this, an *argument to the best explanation* works. This form of argument claims that if one theory explains the data better than any other, then it is reasonable to accept that theory. Applying it here, we infer that these traits exist because of selection for soaring ability on vultures' ancestors. Equally, no one seriously doubts that the mechanisms that bats now use in echolocation exist because of selection for that function.

However, many other characters are much more problematic. The rapid expansion of brain size in our primate ancestors has been explained as the

effect of an upright stance and the consequent freeing of the hands for complex manual work (Tanner 1981). Alternatively, according to the engaging
"aquatic ape" hypothesis, it is the effect of a period when our ancestors were
supposedly surviving and foraging in shallow coastal waters (Morgan 1982).
The extraordinary "radiator theory" of Falk (1990) suggests that brain expansion is the effect of removing a developmental constraint on the thermoregulation of the brain. Perhaps the most popular current view is that brain
expansion is an effect of the social structure of hominid societies. In these
social groups, as in chimpanzees today, the ability to form and manipulate
personal relationships was the key to success. A person who could form a
more complex system of alliances and remembered favors would do well.
(For more on this "Machiavellian intelligence" hypothesis, see Byrne and
Whiten 1988; Whiten and Byrne 1997.) Like any other science, biology
needs a way of testing such competing theories. One way of doing so is to
turn these stories into rigorous mathematical models of the evolutionary
process and see if they correctly predict the traits that have actually evolved.
The other is through integrating adaptationist and phylogenetic hypotheses—through integrating selection and history.

We begin with the idea of testing hypotheses via rigorous quantitative
models. These models come in two basic kinds. The simplest ones are known
as *optimality models*. An optimality model analyzes an evolutionary problem
the way an engineer would analyze a technological problem. Such a model
has four components: a fitness measure, a heritability assumption, a phenotype set, and a set of state equations. The *fitness measure* specifies the currency
in which the success of various designs will be measured. The ideal measure
would be the number of offspring or grandoffspring an organism produces,
but this is rarely practical. If an optimality model was used to examine different leg designs, it might measure the amount of energy needed to cover a
distance at a given velocity or set of velocities. The model would assume that
the most efficient organisms have the most offspring. The second element of
the model, the *heritability assumption,* specifies the extent to which offspring
will inherit a parent's design. An optimality model of leg design might ide-
alize to an asexual population in which every offspring is identical to its single
parent. This convenient simplification would be unlikely to distort the results of this particular model. The third element, the *phenotype set,* states what
alternative designs are possible. When looking at short-term evolution, the
phenotype set can be restricted to minor variants of types actually observed
in the species under study and in related species. Choosing a phenotype set
for long-term evolution is more difficult. Developmental constraints of the

type mentioned in the last section may rule out many designs. The fourth element, the *state equations,* are the guts of the model. They constitute a theory of the relationship between the organism's phenotype and its environment. In a model of leg design, the equations will come from biomechanics and muscle physiology. The equations determine what result (in terms of the fitness measure) will be produced by each alternative design. When the model is complete, it will show which member of the phenotype set is the optimal design—the one that scores highest on the fitness measure.

Optimality models assume that the fitness of a design depends only on the relationship between the organism and the environment. If hopping is more energy efficient than skipping, it will remain so whether everyone skips or everyone hops. But this assumption is often inappropriate to real-life situations. This problem is addressed by our second variety of quantitative models, called *game theoretic models.* As we noted in section 3.3, selection can be frequency-dependent. The fitness value of a trait can depend on the frequency of that trait in the population. For example, in sexual species, it can be a good idea (in evolutionary terms) to desert your offspring. The deserting parent can devote its resources to having more offspring somewhere else while the other parent looks after the young. But the more organisms that have this habit, the less likely it is to pay off. It becomes increasingly likely that the young will starve as both their parents try to leave the other holding the baby. We have seen this idea before, in considering sex ratios and the hypothetical evolution of water cricket navigation strategies. Evolutionary game theory models the selection of designs whose value depends on how other organisms are designed. Like optimality models, game theoretic models have a measure of fitness, a heritability assumption, and a phenotype set. The fourth element of these models is a *game matrix.* The game matrix describes how the value of each design depends on the designs other organisms use. There are also some terminological differences between optimality models and game theoretic models. In game theory, the score that an organism achieves on the fitness measure is known as a *payoff,* and the different possible phenotypes are usually called *strategies.*

One of the most famous game theoretic models is the "hawk–dove" model. The evolutionary problem it models is how to behave in contests over resources such as food, mates, or nest sites. In the simplest version, the phenotype set contains just two possible strategies: "hawk" and "dove." Hawks fight until one animal is injured. The uninjured animal gets the resource. For simplicity, we assume that every hawk has a 50/50 chance of winning a fight. Doves retreat when a fight threatens and leave the resource to the hawk. If two doves meet, each has a 50/50 chance of getting the resource after a certain amount of posturing and bluffing (perhaps it depends on who runs first!). We assume that the resource is valuable; winning it is worth, say, 50 fitness units. Time costs something, so there is some cost in losing the game of bluff in dove/dove contests; say, 10 units. But that is much less costly than injury; we will suppose that the loser of hawk/hawk fights loses 100 fitness units. (We borrow these numbers from Skyrms 1996.) Then, assuming a hawk has a 50/50 chance in a fight with another hawk and a dove has a 50/50 chance of bluffing another dove, the payoffs will look like this:

	Hawk	Dove
Hawk	$50/2 + -100/2 = -25$	50
Dove	0	$-10 + 50/2 = 15$

Selection in the hawk-dove model is frequency-dependent. When selection is frequency-dependent, it does not make sense to talk of an optimal strategy. Under some conditions one strategy has the highest payoff, but under other conditions another does. Instead of describing the optimal strategy, game theoretic models shows which strategies are *evolutionarily stable*. A strategy is evolutionarily stable (with respect to some set of alternative strategies) if it cannot be *invaded*. A strategy can be invaded if a small number of mutants—would-be invaders using a different strategy—would do better than those organisms using the majority strategy. An evolutionarily stable strategy (ESS) excludes other strategies if it comes to be *fixed*—used by all members of a population.

It is clear that dove is not an evolutionarily stable strategy. The first mutants to follow the hawk strategy in a population of doves would do very well indeed. Hawk is sometimes an ESS. If the value of the resource organisms fight over is greater than the cost of being injured, then even when everyone else is a hawk, it is a bad idea to be a dove. A dove meeting a hawk will get nothing, but if the value of the resource is more than the cost of injury and a hawk wins half its fights, then a hawk will, on average, get a positive payoff. If the value of the resource is less than the cost of injury, however, then hawk is not an ESS. This situation is thought to be common in nature, since for most wild animals any serious injury is fatal. When neither hawk nor dove is an ESS, we expect the evolution of a balanced combination of hawks and doves. The population will be at an evolutionary equilibrium when the average payoff of a hawk is the same as the average payoff of a dove. In this situation, the extra costs hawks bear by fighting other hawks are exactly compensated by the payoffs they get by frightening away doves. The resources doves lose to hawks are exactly compensated by the doves' reduced chances of injury. The proportion of hawks and doves at equilibrium will depend on the value of the resource and the costs of fighting. Notice that a mix of strategies evolves even though everyone would be better off if the whole population consisted of doves. Given the costs and benefits in the table above, at equilibrium, about one-third of the interactions are hawk/hawk fights, so the average payoff per interaction is just over 6. The successful hawk invasion reduces the average fitness of the population. Even so, once an equilibrium ratio of hawks and doves is achieved, selection will keep it in place. If too many hawks are born in one generation, they will find themselves in more fights, and their fitness will be lowered. If too many doves are born, their fitness will be lowered as hawks take more resources without a struggle.

The two strategies can be maintained in a population at equilibrium proportions in several ways. The population can be made up of hawkish individuals and dovish individuals, or every individual can be a hawk on some occasions and a dove on others. Given the costs of fighting and the benefits of resources in the table above, the equilibrium strategy is to play dove five out of twelve times and play hawk seven out of twelve. We could even get a mix of switch hitters, pure hawks, and pure doves.

Optimality models and game theoretic models are tested by comparing their predictions with the way organisms actually are. If an optimality model of leg design is correct, then the legs of real organisms should match the leg design that has the highest fitness score in the model. If an application of the hawk-dove model is correct, then the observed proportions of hawks and doves should be an ESS, given the estimated value of the resource and the estimated cost of injury. If the model is constrained enough to generate precise, quantitative conclusions, a close match with real data is indeed impressive. If, for example, a model of the evolution of clutch size in kookaburras—taking into account the physiological cost of eggs, the risk of foraging for the chicks, the costs of territory defense, and the trade-off between investing in current versus future reproduction—matched actual kookaburra behavior, the model would be very persuasive.

Yet in a model of kookaburra behavior, physiological costs, foraging risk, and the like can all be estimated independently. We can independently test the ecological and physiological assumptions that feed into the fitness measure of the model. It is much less obvious that the same is true of a quantitative model of human brain size evolution. It is very hard to see how we could construct any kind of principled quantitative model in a case like this, for we have no independent access to the ecological information. The Machiavellian intelligence hypothesis, for example, assumes that human groups gradually became larger (and interacted more complexly). But we have no independent information about human group size. So it is less obvious how persuaded we should be by a match between a model of brain evolution and real data about human brain size.

A match is one possible outcome, a mismatch another. What happens if the model fails to predict reality? Gould and Lewontin's complaint about adaptationism is that the failure of an adaptive model is never taken as a failure of adaptationism. The adaptationist assumes that the problem lies in one of the four elements of the model. Perhaps the predicted optimal result is not really in the phenotype set. Perhaps the heritability assumption is too simple. Perhaps the effect on fitness of some action has been overestimated. Perhaps the trait under study is used for two purposes, and represents an optimal compromise between the best design for one purpose and the best design for the other. The possibility that the phenotype is less than perfectly adapted is not considered.

Some defenders of adaptive models admit that they do not consider this possibility. John Maynard Smith, the inventor of evolutionary game theory, insists that when an adaptive model is tested, the assumption that natural selection will choose the optimal phenotype is never under test. Here we see the conflation of the distinct versions of adaptationism we discussed in section 10.4. If adaptationism is treated as a global hypothesis about the biological world—most characteristics of most organisms are mostly the result of natural selection—then the failure to consider nonadaptive hypotheses is worrying. But Maynard Smith's adaptationism is methodological adaptationism (Maynard Smith 1984, 1987) The optimality model is a heuristic device, designed to reveal otherwise unsuspected constraints on adaptation. This heuristic strategy is premised on the idea that we can best find out about restrictions on heritability, or constraints on the array of possible leg shapes, by comparing the actual leg to the best of all possible legs. Suppose, for example, that there are genetic constraints that prevent a potentially adaptive mutation affecting leg shape from becoming fixed. Perhaps the mutation is linked to, and hence inherited with, a gene that is fit only when it is rare. We will discover that the assumptions about heritability in our model were too simple and will modify the model accordingly. Mismatch, not match, is revealing, because mismatch reveals constraints that we would otherwise not suspect—constraints that are not manifested in phenotypes. So the point of testing is to refine the model, adjusting our phenotype set, our fitness measure, and the like until it does correctly predict the observed phenotype. The constraints on adaptation are not ignored; they are incorporated through the fitness measure and the phenotype set.

So it may well be true that though adaptationists test particular theories about adaptation using quantitative models, the basic adaptationist idea is never under test. Does this mean that Gould and Lewontin are correct and adaptationism is unfalsifiable and unscientific? Elliott Sober (1993) thinks that it does not. He argues that their critique depends on much too simple a picture of the way scientific theories are tested. He then goes on to develop an indirect test of adaptationism. We think he is right about the oversimplified picture of hypothesis testing, but we have reservations about his indirect test.

Sober begins by pointing out that adaptationism is not a simple scientific claim, like the claim that kiwis are descended from ancestors that could fly. Adaptationism is a *research program*. The idea of a research program was introduced by Imré Lakatos as a refinement of Karl Popper's falsificationist philosophy of science (Lakatos 1970). Popper's idea is that science makes progress not by proving theories to be true, but by rejecting theories that make false predictions and replacing them with better theories. But simple-minded falsificationism would have been fatal in the history of science. The theory of continental drift suffered from many apparent falsifications—most obviously, its continent-moving mechanism. Darwin's theory clashed with contemporary physicists' calculations of the age of the earth. In the end Darwin was proved right, but to reach this point Darwinians had to tolerate the "anomaly" for eighty years. Faced with examples like these, Lakatos argued that science is organized into research programs. The core ideas of these programs are not tested directly. Instead, scientists spend their time working out how these core ideas can be made to fit the data by elaborating all sorts of extra, detailed theory. This theory comes between the hard core of the research program and the data, just as the four elements of an adaptive model come between the core adaptationist thesis and data about actual organisms. It is only the extra, detailed theory that is tested and perhaps refuted. The core ideas of the research program provide a framework that suggests the detailed hypotheses and makes it possible to test them. If adaptationism is a research program, then one of its core ideas is that natural selection will usually produce optimal phenotypes. This core idea leads to the construction of particular models and also to tests of those models.

Thus, if adaptationism is a research program, it can be tested only indirectly. Lakatos argued that research programs stand or fall on their ability to produce successful detailed results in the long run. A successful research program leads to the discovery of many exciting and unexpected facts. An unsuccessful program spends its time explaining away the continued failure of its detailed research. Orzack and Sober (1994) have discussed how adaptationism could be tested in this global and indirect way. They begin by defining the core of the adaptationist program more precisely. They distinguish three claims about adaptation. The first is that it is *ubiquitous,* meaning that most traits are subject to natural selection. The second is that adaptation is *important.* Adaptation is important if a "censored" model that deliberately left out the effects of natural selection would make seriously mistaken predictions about what sorts of organisms have evolved. Finally, there is the claim that organisms are *optimal.* An organism is optimal if a model censored of all evolutionary mechanisms *except* natural selection could still accurately predict what sorts of organisms have evolved. Orzack and Sober argue that almost all biologists would accept that natural selection is ubiquitous and important. The distinctive feature of adaptationism is its claim that organisms are optimal; that is, that the results of evolution can be predicted reasonably well by models that consider only natural selection.

Orzack and Sober go on to suggest that the real test of adaptationism is whether adaptationist models are successful in predicting how most organisms have evolved. If models censored of all but natural selection correctly predict most of the data, or can be made to predict it with only a few, independently plausible adjustments to their assumptions, then adaptationism is a progressive research program. If such models must be laboriously tinkered with in every case in order to obtain correct predictions, then adaptationism is a degenerate research program and should be abandoned.

10.7   Adaptation and the Comparative Method

At the beginning of section 10.6 we mentioned that "arguments to the best explanation" sometimes make particular adaptationist hypotheses very plausible indeed. Friendly treatments of adaptationism often have great confidence in such arguments. They usually identify two kinds: adaptive thinking and reverse engineering (Dennett 1995). *Adaptive thinking* is the practice of looking at the structure and behavior of an organism in the light of the ecological problems it faces. Adaptive thinking predicts the sorts of features the organism should possess and uses those predictions to guide an investigation of the features it actually possesses. *Reverse engineering* is a way of working out how things actually evolved. One tries to work out what adaptive forces must have produced the existing form by reflecting on the adaptive utility of that form in either the current environment or a postulated ancestral environment. Reverse engineering infers the adaptive problem from the solution that was adopted; adaptive thinking infers the solution from the adaptive problem. Both forms of adaptationist theory can make use of the modeling techniques described in the last section. Adaptive thinking starts with a model and predicts how organisms will be in reality. Reverse engineering starts with how organisms are and constructs a model to explain this. Both forms of adaptive theorizing assume a strong relationship between adaptive forces and the resulting organism, an idea that adaptationists accept and which Orzack and Sober refer to as the claim of optimality. If this claim is correct, a model of evolution censored of forces other than natural selection should predict with reasonable accuracy the trajectory and destination of organisms in the space of possible designs.

The models described in the last section look at the relationship between an organism and its environment or, in game theory, between an organism, its environment, and other competing organisms. The models predict which design should be most successful in competition with others. It is easy to make the mistake of supposing that these models do not involve any particular assumptions about evolutionary history. They seem to involve only general principles about which traits are most efficient. These principles describe the (causal role) functions that certain designs will perform. The hawk design, for example, will beat the dove design in any single conflict. But in fact, these functional considerations cannot make any predictions about evolution unless we specify the particular historical conditions that make up the selective environment. Thus Orzack and Sober point out that optimality is *local,* as even censored models must take some account of the background biology of the lineage. For example, the robust beak of the New Zealand takahe, a large flightless bird, is said to be ideally engineered for feeding on alpine tussocks, but if the bird did not evolve living in the Southern Alps (either as a species or as a locally adapted variant), then this engineering excellence would be irrelevant to its evolution (Gray and Craig 1991). So we need a claim about takahe history conjoined with the engineering claim to generate an explanation of takahe beak structure. History also creeps in when we choose the phenotype set. The range of designs presented for selection will depend on the current state of the organisms facing selection. Without knowing what sorts of ancestors an organism had, it is impossible to say which alternatives competed to produce the form we see today. History has yet another role because evolution is a stochastic process. Only in a very large population can we assume that the fittest traits will be successful. In smaller populations, chance plays a larger role. Conventional evolutionary theory says that many important innovations occur when organisms are isolated in small populations. Chance, referred to as evolutionary *drift,* can be very important in these populations. Taking all these factors into account, the role of particular historical facts in evolution is very large. An adaptive model must make many assumptions of historical fact, although these are often not explicitly mentioned when the model is presented.

Adaptationists have tried to avoid the problem of historical assumptions

Figure 10.3   The adaptationist abduction. This "argument to the best explanation" is supposed to avoid the need to independently test the historical assumptions built into adaptive scenarios. The fit between the model and the observed data provides an argument in support of the historical assumptions that the model requires.

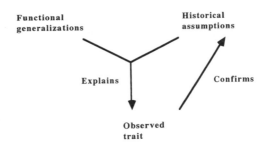

by thinking of an adaptive explanation as a simultaneous abductive argument for the truth of the historical assumptions it requires (figure 10.3). *Abduction,* or "argument to the best explanation," is an important form of scientific reasoning. As we noted at the beginning of section 10.6, it is the idea that if one theory explains the data better than any other, then it is reasonable to accept that theory. Adaptationists argue that if they make certain historical assumptions, then they can neatly explain the actual trait. Therefore, by argument to the best explanation, we have grounds for accepting these historical assumptions.

But for many of the adaptationist hypotheses central to contemporary evolutionary theory, arguments to the best explanation are too blunt an instrument. Optimality modeling, evolutionary game theory, and the like are powerful engines for generating possible explanations. So in considering the evolution of sex, of sexual dimorphism, of strange sex ratios, of reversed sex roles in some bird species, and the like, there are a number of potentially adequate explanations. Argument to the best explanation is not valid when the "best" explanation is just one of several that are equally good. The problem of choosing between several equally adequate adaptive hypotheses is particularly sharp in those many cases—for example, fire resistance in Australian flora—in which uncertainties about past environments meant that the best we can expect is a qualitative fit between theory and data, or—as in brain size expansion in the hominid lineage—in which quantitative prediction depends on ecological features that are not independently known.

There is no methodological magic bullet that solves all the problems of testing adaptationist hypotheses. Requiring a very precise quantitative fit between adaptationist hypotheses and the traits actually observed does something to reduce the proliferation of hypotheses (Orzack and Sober 1994). But, as we have already noted, we think that this requirement is appropriate for a subset—perhaps only a small subset—of adaptationist hypotheses. We think the *comparative method* is more generally promising. This term refers to a range of techniques that infer how one organism evolved by comparing what evolution produced in that case with what it produced in other cases. The comparative method is one of biology's main windows on the past. We think it has three important applications to the adaptationism debate. First, it enables us to directly test the historical assumptions tacit in adaptationist hypotheses. Second, it enables us to test the proposed link between environmental feature and adapted trait. Third, we can use it to make sense of the adaptationist claim about the explanatory priority of selection.

First, let's consider tests of the historical premises that are built into adaptationist explanations. The simplest comparative tests check the actual sequence of evolutionary changes to see if it is the one presumed by the adaptive hypothesis. Jonathan Coddington (Coddington 1988, 10–11) provides a simple example of this sort of test. Living species of rhinoceroses have either one or two horns. This means that both designs were available to the evolving rhinoceros, so it is natural to invent an adaptive scenario in which both horn conditions are evolutionarily stable strategies (Lewontin 1985a; though see Zahavi and Zahavi 1997, 86–87, for a quirky adaptationist explanation of the two-horn design). If horn configuration is important in mate choice or other social interactions, we might suppose that once a population contains a large proportion of individuals with one number of horns, it cannot be invaded by a mutant with the other number of horns. Victory goes to whichever strategy gets established first in a particular population. Some *sexual selection* hypotheses fit this picture. If female rhinoceroses developed a preference, however slight, for one design, then males with that design would be at an advantage. In that case, females that lacked the preference for that design would have both less attractive male offspring and female offspring with their mother's unfashionable taste. The small advantage would thus be reinforced by sexual selection until it became a large advantage. Minor but different female preferences might arise by chance in different populations, leading to the evolution of two rhinoceros designs. However, a cladistic analysis (9.3) of the rhinoceratid group shows that the two-horned condition preceded the one-horned condition in the phylogenetic tree. In some population at some time, the two-horn design was successfully invaded by the one-horn design.

Adaptationist hypotheses often concern the relationship between two traits, and often imply that one evolved before the other. This historical presupposition can be independently tested. Mary McKitrick (1993) provides a simple example. It has been suggested that the low birthweight characteristic of the genus *Ursa*—the bears—is the result of an adaptive trade-off. It is the price bears pay for altering their physiology in order to allow hibernation. But a reconstruction of bear phylogeny shows that this cannot be the case.

Low birthweight emerges before hibernation, and exists on branches of the phylogenetic tree on which hibernation never originated. Tests of this sort have wide application. The "aquatic ape" hypothesis claims as a particular strength its ability to explain a wide range of human characters: upright posture, bipedalism, hair loss, our layer of subcutaneous fat, our diving reflex, and many more. All these are said to have evolved together as an adaptive complex when our ancestors made a return to a semi-seagoing life. Since the hypothesis suggests that these characters emerged together in a single phase of hominid evolution, we can test it by determining when they appeared on the phylogenetic tree for hominids and their relatives. If the traits appeared at different times, they should be inherited by different chunks of the hominid family tree. If the characters emerge at various different points in the tree—if they did not, in fact, evolve together—then however neatly the hypothesis explains them, it cannot be correct.

A second important role for the comparative method lies in directly testing the idea that adapted traits are responses to particular features of an organism's environment. Adaptationist hypotheses can be supported by finding a correlation between certain traits and habitat factors. Such correlations suggest that the habitat factor has something to do with the evolution of the trait. Suppose we are interested in a group of seabird species, some of which nest in burrows, have plain white eggs, and do not remove the eggshells after hatching. Other species nest on ledges, have patterned eggs (camouflage, we suspect), and remove the eggshells after hatching. We reconstruct the phylogeny of the group and discover that (1) the ancestor species nested in a burrow, (2) it had plain white eggs, and (3) it did not remove eggshells after hatching. In case after case, when a descendant species has changed its nesting habit from burrow to ledge, its eggshell pattern and behavior have changed too. Here the inference of an adaptation to the new nesting condition would be enormously powerful. This example is both simple and ideal: real evolutionary data are unlikely to be as clean and as cooperative as our imagined seabird family. But sometimes we can get close. Again and again, rails—a chunky, rather generalist, and widely dispersed group of birds—have become flightless or nearly flightless on islands to which they have dispersed (Trewick 1997). The firm covariation between island life and flightlessness suggests that on islands something about the costs and benefits of flight changes, and that this alteration in the selective regime explains flightlessness.

Adaptationists have always laid great stress on *convergent evolution:* the phenomenon of the independent evolution of the same trait (or set of related traits) in different species. Perhaps the most frequently used example of convergent evolution is streamlining in large marine hunters. The bottle-nosed dolphin, the ichthyosaur, the blue marlin, and the great white shark all have strikingly similar shapes without inheriting them from their (distant) common ancestor. Convergent evolution has played two roles in adaptationist thinking. Sometimes it is taken to illustrate the overwhelming power of natural selection: it has taken widely separated lineages and remade them in the same mold. This is not a persuasive thought: convergence tells us nothing about the relative power of selection and history unless we can somehow count all the possible convergences that have *not* happened—all the times history "won." More reasonably, convergence has played an evidential role in supporting specific adaptationist hypotheses. What else but natural selection to minimize the energetic cost of high-speed travel through water could explain the similarities among these marine predators? Why else would this trait have evolved repeatedly under these particular environmental demands?

Convergence can indeed serve as evidence for an adaptationist hypothesis. But the systematic study of convergence requires an extensive use of the comparative method. For without a proper phylogenetic tree, it is not even possible to tell whether something *is* a convergence. Dennett is struck by the fact that "so many creatures—from fish to human beings—are equipped with special-purpose hardware that is wonderfully sensitive to visual patterns exhibiting symmetry around a vertical axis. . . . The provision is so common that it must have a very general utility" (Dennett 1987, 303). He is impressed by the adaptive hypothesis that this piece of neural hardware is a device for detecting other organisms looking straight at the subject, for then they are, from the subject's perspective, vertically symmetrical. But a phylogenetic tree may reveal that this neural hardware evolved just once, in the ancient common ancestor of all the species that display the trait. If so, then the existence of this cognitive trait in many species is no convergence at all. It has not evolved *repeatedly* in response to some repeated feature of the environment. It could, of course, still be an adaptation. But, equally, having been passed on by descent, it may serve many different adaptive functions in different species, and exist in others merely by "phylogenetic inertia." If so, then seeking an adaptive explanation of why so many organisms are sensitive to vertical symmetry may be as misguided as seeking an adaptive explanation of why humans, birds, and seals all have such similar bones in their forelimbs (the provision is so common that it must have a very general utility!).

The use of the comparative method to test adaptationist claims is widely accepted. We shall conclude this chapter with a more speculative idea. In section 10.3 we discussed the problem of testing empirical adaptationism, but we also noted that it was not an easy idea to interpret. Empirical adaptationists think that selection is the most important force driving evolutionary history. But how could that be, if the evolution of every trait and every organism depends on many other factors as well? The evolution of streamlining in sharks and ichthyosaurs depends not just on selection, but also on history—on the possibilities created by previous evolution in those lineages. Squids are also marine predators, but are not notably shark-shaped. In our view, the comparative method offers a way of interpreting empirical adaptationism.

In a series of recent papers and a book, Robin Dunbar has argued for a connection between group size and cognitive complexity, using brain size scaled against body weight as a rough index of cognitive complexity (Dunbar 1996, in press; Barton and Dunbar 1997). As group size increases, the demands on memory and other cognitive skills increase, because an agent has to learn and remember more individuals, their characteristics, and their social relations. The agent has to learn not only to recognize individuals, but also to keep track of their friends, relations, and enemies. Because the number of relationships increases faster than the number of individuals—each individual has more than one significant relationship—these extra cognitive demands are quite intense.

So group size selects for intelligence: bigger groups, smarter individuals. But it's clear that Dunbar does not expect this relationship to hold in every group of animals. He obviously does not expect ants that live in huge nests to be smarter than ones that live in small nests. It is not clear whether he expects this relationship to hold among birds. Kookaburras are kingfishers, but unlike most of their relatives, they live in social groups consisting of extended families. Does Dunbar's hypothesis predict that kookaburras are smarter than solitary kingfishers? Selective pressure will produce a particular adaptive shift in a population only if that shift is among the evolutionary possibilities created by the previous history of the lineage.

In section 10.4 we remarked that it is hard to evaluate the idea that selection is more important than history, for every adaptive change in a lineage depends on both the history of that lineage and selection acting on it. But if we think of selection in a comparative context, perhaps we can make sense of claims about its relative importance. For selective hypotheses like Dunbar's can be narrow and shallow, intended to apply to only a small fragment of the tree of life. Or they can be wide and broad, applying not just to fancy primates but to bats and kookaburras as well. For the role of history in the explanation of adaptive change enables us to use phylogeny to specify the scope of adaptationist hypotheses. One way of interpreting Dunbar's hypothesis is to see it as nested high in the primate tree. According to this view, depicted in figure 10.4a, the evolutionary preconditions for an adaptive

a)

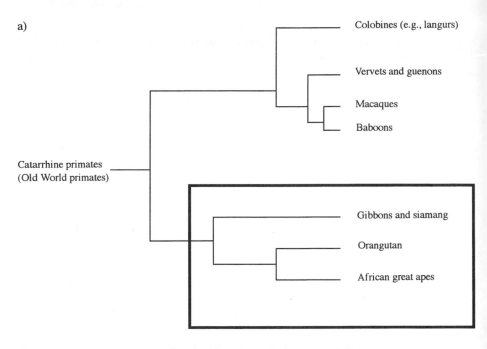

Figure 10.4 Two interpretations of Dunbar's hypothesis. (a) The narrow, shallow-scope interpretation. (b) A deeper, broader interpretation. The heavy box indicates the chunk of the phylogenetic tree in which the preconditions for an adaptive cognitive response to an increase in group size existed.

cognitive response to an increase in group size—getting smarter—have evolved only recently in the primate lineage, in the lineage of the Hominoidea—the lineage of the African and Asian great apes and of our ancestors. Within that small chunk of the primate lineage enclosed by the heavy box, and only there, we predict a correlation between group size and brain size scaled against body weight, for it is only in this clade that the evolutionary preconditions of a cognitive response to group size have arisen.

An alternative, more "history-overriding" version of this adaptationist hypothesis would push the origin of this evolutionary possibility deeper into the tree, and would predict a group size/brain size correlation over more species. So figure 10.4b depicts a less shallow hypothesis. In this version of the hypothesis, the cognitive preconditions for a takeoff in intelligence in response to group size evolved early in the primate lineage, just after the deepest and oldest split in the lineage. In this reading of the hypothesis, we would expect group size and weighted brain size to covary in all primate species except those few survivors of the ancient lemur/loris/bushbaby

b)

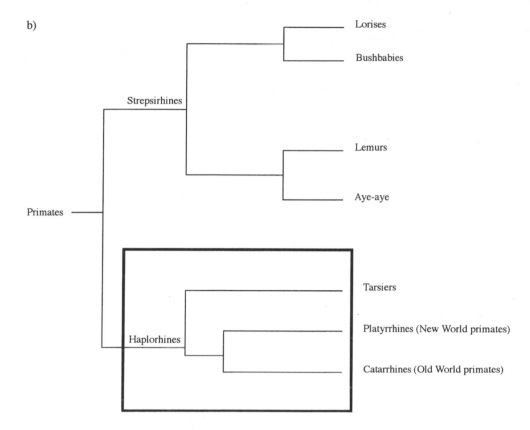

branch. Dunbar himself cites the fact that social bats have larger brains (scaled to body size) than their less social relatives as evidence for his idea, so perhaps he would push the origin of the takeoff point still deeper into the tree, perhaps early in the mammal lineage. Note that this way of interpreting adaptationist hypotheses is insurance against cheating. If social bats with big brains (for bats) count in favor of the hypothesis, then any other social animal with a normal-sized brain in the clade that includes bats and primates counts against the hypothesis.

Seen in this way, empirical adaptationism does not downplay the causal importance of history. Without the developmental and phenotypic possibilities the evolutionary history of a lineage creates, selection for cognitive sophistication would be ineffectual. Rather, adaptationists emphasize (or should emphasize) that the explanatory salience of selection over history depends on the fact that historical factors remain relatively constant, whereas the role of selection changes. We focus on selection in, say, explaining a cognitive change in primate evolution because selection is the *varying factor* and is primarily responsible (along with drift) for explaining *variance* of this feature in the primate tree. Were it not for selection (and to a lesser degree, drift), all the primates would be the same. Adaptationists, then, are those who develop and defend deep-scope hypotheses, hypotheses about large chunks of the tree of life. Nesting Dunbar's adaptationist hypothesis deeper in the tree would make it apply to social living bats such as vampires; still deeper—much deeper—and it would apply to social birds like the kookaburra. A defender of deep-scope hypotheses expects many of the historical and developmental constraints on evolutionary change to remain relatively constant over large chunks of the tree of life. Skeptics of adaptationism so understood are those who expect these constraints not just to be important (on this, all are agreed), but to be variable. According to this way of reading adaptationist ideas, Dunbar's hypothesis would be in trouble if the evolutionary possibilities—the range of evolutionarily possible phenotypes—differed significantly from orangutan to chimp to gibbon to siamang. For then, even if group size were important, so too would be the different possibilities of response made available by the evolutionary histories of each of these lineages since their divergence from one another. So there would be no general pattern to capture in the response to selection for behavioral adjustment to living in larger groups.

We see this conclusion as enjoyably ironic. For Gould, in particular, is not just one of the arch-critics of adaptationism. He is also one of the defenders of the idea of stable constraints—of the idea of the conservation of what is evolutionarily possible for a lineage, and what is not, over time. As we see it, he is the defender of the critical empirical presupposition of empirical adaptationism.

Further Reading

10.1 As is often the case, Keller and Lloyd 1992 is a good entrée to the literature, with entries on adaptation and teleology. Rose and Lauder 1996 is an impressive recent collection on many of the topics covered in this chapter. Hull and Ruse 1998 has good sections on both adaptation and function; so too does Sober 1994. Many of the important recent papers on function are collected in Allen, Bekoff, and Lauder, 1998, and Buller, in press. A collection of new papers on adaptationism edited by Orzack and Sober (in press) is about to emerge.

There is a voluminous literature on adaptation, adaptive traits, fitness, and related concepts. For accounts of the development of the contemporary concept of adaptation, see Burian 1983, 1992 and Amundson 1996. Belew and Mitchell 1996 has a good selection of early classics on adaptation. The contemporary concept is discussed by Brandon (1990, 1996) and by West-Eberhard (1992). Gould and Vrba (1982) introduce the adaptation/exaptation distinction, which is criticized by Griffiths (1992), Reeve and Sherman (1993), and Dennett (1995). However, only Reeve and Sherman question the more basic distinction between being an adaptation and being currently adaptive.

The concept of fitness has also been the focus of much interest, as it has evolved from an intuitive notion of "fit" between organism and environment into an array of more precise but more technical concepts. The three essays on fitness in Keller and Lloyd 1992 are probably the best introduction to this difficult topic. The majority view of fitness is to treat it as a reproductive propensity that depends on the other features of an organism. A standard formulation and defense of this view is given by Mills and Beatty (1994). It is criticized by Byerly and Michod (1991). For a good introduction to the different uses of the notion of fitness in evolutionary theory, see Dawkins 1982, chap. 10.

10.2   The etiological account of function is usually credited to L. Wright (1994). We prefer the more biologically informed and better developed version of the basic idea found in Millikan 1989b, Neander 1991, and Godfrey-Smith 1994b. The propensity view can be found in Bigelow and Pargetter 1987. The causal role view of functions is often credited to Cummins (1994). Godfrey-Smith (1993) and Amundson and Lauder (1994) present very clear and intelligent defenses of the need for distinct function concepts in different areas of biology. The consensus view that functions in evolutionary biology are explained by the etiological theory has recently been called into question by Walsh (1996), Walsh and Ariew (1996), and Schlosser (in press).

10.3, 10.4   Gould and Lewontin's original attack on adaptationism (1978) is reprinted in Sober 1994. Dupré 1987 is a very important collection on this issue. Godfrey-Smith develops his distinction between different kinds of adaptationism most fully in a forthcoming paper (Godfrey-Smith, in press-c). Amundson 1998 is an insightful exploration of the relationship between adaptationism and developmental constraint. Adaptationism is vigorously defended by Dennett (1983, 1995) and Cronin (1991). In addition to the three lines of criticism we discuss in the text, Lewontin also argues that adaptationism misstates the relationship between organisms and their environments. We discuss this issue in chapter 11, and give references there.

10.5   Goodwin (1994) offers a very simple introduction to process structuralist research. Kauffman (1993) has written a very important but *extremely* difficult book on the role of complexity in evolution. The best introduction to Kauffman's work is Depew and Weber 1995; their introduction to structuralism is also very helpful. Kauffman provides his own introduction in Kauffman 1995a,b. Smith (1992), Dennett (1995), and Griffiths (1996a) all argue for the compatibility of these ideas with conventional Darwinism. Wimsatt is another important but difficult author; his ideas are most accessibly presented in Wimsatt and Schank 1988. Gould's views on the relationship between development and form are explored most fully in his *Ontogeny and Phylogeny* (1977), a most impressive combination of history and theory. His views have continued to develop since that work, and we discuss them extensively in chapter 12. The empirical literature on vestiges is surveyed by Fong, Kane, and Culver (1995). There has been some debate on whether there is a real conflict between developmental and selectionist explanations of a trait; see Sherman 1988, 1989; Jamieson 1989; Mitchell 1992; Sterelny 1996a.

10.6 Maynard Smith 1982 is a fairly accessible introduction to evolutionary game theory, but an even better introduction is Sigmund 1993. Maynard Smith's take on the philosophical issues can be found in Maynard Smith 1984, 1987. For a recent review of issues on optimality, see Seger and Stubblefield 1996. The idea of a global test of adaptationism is defended by Sober (1993) and by Orzack and Sober (1994). Brandon and Rausher (1996) present a critical response, arguing that Orzack and Sober's suggestion is biased toward adaptationism. They reply in Orzack and Sober 1996. Gray (1987) and Pierce and Ollason (1987) present detailed critiques of optimality theory.

10.7 The significance of the comparative method for the study of adaptation is discussed by Taylor (1987), Horan (1989), Griffiths (1994, 1996b), and Sterelny (1997b). There are two very good book-length surveys of the modern comparative method and its application: Brooks and McLennan 1991 and Harvey and Pagel 1991. Eggleton and Vane-Wright 1994 is an important recent collection on the use of phylogenetic methods to study adaptation; the first four papers are general discussions of the issues discussed in this section. Lauder, Armand, and Rose (1993) discuss the limitations of these methods. Finally, for a wonderful parody of all these debates, see Ellstrand 1983. In a similar vein, see Shykoff and Widmer 1998 for the application of the comparative method to the vexed question of the temporal order of eggs and chickens.

11

Adaptation, Ecology, and the Environment

11.1 The Received View in Ecology

Ecology and evolutionary theory usually deal with organisms on strikingly different temporal and spatial scales. While ecologists might survey the seasonal changes in abundance of the Polynesian rat on an island over a few years, evolutionary theorists might treat of the evolution of the rat over entire continents and tens of millions of years. Moreover, the conceptual tools of ecology and of evolutionary theory seem very different. Ecologists are interested in local populations of organisms, for these are parts of local communities, and the central concern of ecology has typically been the structure of these communities and the abundances of the organisms within them. These local communities, in turn, are parts of ecosystems, and ecosystems may be nested in larger units. In contrast, evolutionary biology nests organisms in breeding populations, species, and higher taxa. These groupings are defined by descent and by their reproductive boundaries, not by their ecological properties. Marsupials, for example, are a great mammalian clade spread out over vast distances in space (New Guinea, Australia, South America, North America, formerly Antarctica) and time. They are ecologically extremely heterogeneous. It is true that none fly and none have become exclusively aquatic, but they live anywhere from rainforest to desert, and they eat everything that grows or moves. The same ecological heterogeneity is evident at smaller scales: many species have members in a range of different communities. In Australia, introduced foxes live quite successfully in urban and suburban environments, probably by scavenging. But they also are effective predators—too effective—in the quite different communities of semi-desert wilderness. So the units that evolve are not usually ecologically cohesive: they are not parts of communities or ecosystems.

Yet despite these differences of scale and classification scheme, ecology and evolutionary biology are intimately connected. Ecological interactions among organisms and between organisms and environment explain the evolutionary fitnesses of organisms. In discussing evolutionary theory, Sober (1984b) usefully distinguishes *source laws* from *consequence laws*. Consequence laws explain the *effects* of fitness differences. For example, if we know that a dark variant of a moth is 1% more likely to survive to reproduce than a white variant, consequence laws tell us how many generations it will take for the fitter variant to predominate in the population. Although they are, in a sense, only mathematical derivations from fitness values, consequence laws are far from trivial; they must take into account heritability, the size of the population, and much else. Source laws are an essential supplement to consequence laws. Source laws explain the *origins* of fitness differences; they explain *why* the dark variant is fitter than the white variant by a 1% margin. Ecology clearly must play the central role in the formulation of evolutionary biology's source laws. So ecology feeds into evolution. Equally, the features of organisms that determine their ecological fate are themselves the product of evolutionary history.

The relationship between ecology and evolution once seemed clear. Darwin had talked of a "struggle for existence," meaning both the struggle between organisms and the struggle of an organism to cope with its environment. His German disciple Ernst Haeckel coined the word *ecology* to mean "the science of the struggle for existence." The network of relationships among organisms and between organisms and physical environment provides the context of evolutionary change and hence generates the selective forces that drive that change. Ecology describes the environment within which evolutionary change takes place, and provides the conceptual tools for describing the complex, interactive relationships that drive evolution. These concepts include general coarse-grained descriptions of an organism's role in a community. For example, some animals are carnivores, whereas others are insectivores or herbivores. Among the herbivores, some are browsers specializing on foliage and others are adapted to eating grass. Ecology has also developed concepts to describe the overall structure and dynamics of environments. For instance, ecologists have developed a body of theory about food chains and food webs that helps to explain energy and nutrient cycling through particular ecosystems.

Ecology at its most traditional made use of Elton's concept of a niche to understand the structure of communities and the sources of fitness. In Elton's conception, a *niche* is a particular way of making a living in an organic community (Elton 1927). Niches are like the career opportunities in a human community. A striking and important feature of this classic conception of the niche is that it is independent of its particular occupant. Communities widely scattered in space and time are nonetheless sometimes very similar, the idea runs, because they are functionally similar. Each community contains different organisms, but in each community the organisms play similar *causal roles* (10.2). The same niches are available in, say, grassland communities in Africa, Asia, North America, and South America. Hence in each of these places we will find an array of grazers, predators, scavengers, seed eaters, dungivores, and so on. For instance, we can characterize a "small warm-blooded grassland carnivore niche," which might be occupied by a bird, a marsupial, a placental mammal, a dinosaur (perhaps!), or by nothing at all. Different occupants of the same niche make similar contributions to the functioning of the community. All of the different dungivores make similar contributions to the overall organization of the community by breaking down dung and recycling the nutrients in it. In many different forest communities there are roles for fruit-eating species. These species vary widely: bats, birds, mammals, and lizards all eat fruit and hence disperse seeds through the landscape. While the example of fruit eaters shows that there are many exceptions, organisms playing the same causal roles in different communities are often physically rather alike, even though they may not be close relatives. Old World and New World vultures, for example, are not particularly close relatives, but they are physically and behaviorally very similar.

This conception of ecology is readily linked to a conception of evolution that sees change in a population over time as an adaptive response to the demands the environment makes on organisms. In section 10.4, we characterized empirical adaptationism as the idea that selection is by far the most important force driving evolution. Empirical adaptationists are often "externalists" as well, although there is no necessary connection between the two ideas (Sterelny 1997b). *Externalism* is the view that selective pressures are determined by the environment of the population, and adaptive change is a response to the problems the environment poses. Features of the environment explain features of the organism. Organisms are shaped by selection to fit their environments. The classic conception of a niche encapsulates this picture. Classic niches exist independently of their occupants; they constrain their occupants, and their occupants can vary in the degree to which they fit their niche. Selection, of course, will prefer those that fit their niche well over those that fit it less well. So, over time, if the environment is not disturbed by external shocks, a niche's occupants will become ever better fitted to it. They will come to fit their niche as a key fits a lock. The environment drives adaptive evolution, and the niche identifies the aspects of the environment doing the driving.

Hence classic ecology fitted one version of the adaptationist program by providing a functional analysis of biological environments that explained their adaptive dynamics. The niche structure identified the key adaptive pressures in an environment. Thus an ambitious ecological program—an account of community structure that emphasized the functional similarities between communities independent of particular taxa—and an ambitious version of adaptationist evolutionary biology were (literally) made for each other. In combination, these two programs were intended to explain striking patterns on both ecological and evolutionary time scales.

The organization of a biological community into niches determines the *rules of community assembly;* that is, it determines which kinds of species can find places in the community and which cannot. These rules explain *ecological convergence*—for example, the ecological similarity of different tropical rainforests despite the fact that most of the species in, say, Borneo are not closely related to those in the Congo or South America. The community assembly rules impose constraints on species coexistence and thus determine whether invasions by new species will succeed or fail. In particular, the principle of *competitive exclusion* states that two species that compete for the same niche in the same community cannot both survive indefinitely, since one is bound to be at least a little better suited to it than the other. So competitive exclusion will exclude some potential invaders, whereas others will find vacant or poorly defended niches, which they may then occupy. The rules of assembly also explain the division of resources among species. Two species with overlapping niches may each become restricted to those areas to which they are best adapted. Australasian harriers are restricted to swampy regions of Australia, for Australia has a rich array of birds of prey. In Australia, other birds of prey restrict the harrier to the regions to which it is best suited. In New Zealand, where the only other bird of prey is the New Zealand falcon, the harrier's range is much broader.

Over evolutionary time, these very same processes explain *parallel evolution,* as related species respond independently but in similar ways to a change in their environment, and *evolutionary convergence.* One example of convergence is the multiple evolution of "saber-toothed tigers" in response to the evolution of very large herbivores. The "signature" saber-tooth is a true cat, *Smilodon.* But versions of the same basic feeding apparatus also evolved in the extinct mammalian carnivore lineage of nimravids more than once, and even in a South American marsupial lineage (Janis 1994). Such convergences occur, the idea goes, when unrelated species become adapted to the same niche in different communities. As well as explaining similarities among unrelated species in different places, the rules of community assembly explain resemblances between different times. They explain adaptive radiation and the reestablishment of broadly similar communities, though with different components, after regional or global mass extinctions.

There are many examples of closely related and very similar bird species living in the same community. But though similar, these species are not identical. In some cases, there is variation in size. In other cases, the different species have become specialized for foraging in different layers of the forest, or for using slightly different resources. Over evolutionary time these related species have subdivided what was originally a single niche between them, and have differentiated enough behaviorally and physically to coexist. This process is called *character displacement,* and it is the evolutionary signature of the action of competitive exclusion. These species groups thus come to form *guilds* of related and similar tribes (for a classic series of studies in this genre see Lack 1971).

So the idea that communities are organized into functional niches is the analog in ecology of a particular version of adaptationism in evolutionary biology. In the next three sections we discuss recent developments in ecology that have undermined this picture of the fit between ecological and evolutionary theory. In the next section, we rather skeptically consider the idea that the structure and dynamics of communities can be explained by ecological theories that abstract away from the particular species that inhabit them. In section 11.3, we take up the problem of equilibrium in ecology. Only if a community is at or near equilibrium will the niches available in a particular habitat determine the presence and abundances of species. If the community has suffered significant disturbance, niches that are available at equilibrium may be absent, and species that would be squeezed out at equilibrium may be present. So the idea that the network of niches in a community explains the selective forces on a population depends on an equilibrium picture of ecology, and as we shall show, that picture is decidedly controversial. In section 11.4 we turn to the concept of the niche itself. Lewontin, in particular, argues that the classic concept of the niche fundamentally misunderstands the relationship between organism and environment. It does so precisely because it is committed to the idea that the explanation of evolutionary change runs from "outside" to "inside"; from environment to organism (Godfrey-Smith 1996). Lewontin claims that organisms construct environments every bit as much as environments make organisms. We think that Lewontin's critique has considerable force, but we argue that it overlooks the kernel of truth in the classic conception. So in section 11.5 we sketch out a possible compromise candidate: our own reconstruction of the idea of the niche.

11.2 History and Theory in Ecology

Many important biological generalizations depend on the shared history of the specific species group on which they focus. These generalizations are about groups of species linked by common descent, and they are true because of this common descent. Like most principles in evolutionary biology, many of these historically based generalizations have exceptions. Some do not apply to all members of the historical group. Most generalizations about the diets of birds of prey ("the raptors") will not be true of the vegetarian African palmnut vulture. Other generalizations do not apply *only* to a group of species linked by common descent. The Australian black shouldered kite is a bird of prey, not an owl. But it has many of the owl clade's distinctive adaptations for nocturnal hunting.

There is no deep mystery about these exceptions. The divergence of the palmnut vulture from her relatives explains why she is so different in diet from most raptors; the convergence of the black shouldered kite onto night hunting explains why she is owl-like, though no owl. As Dennett (1991) has put it in another context, patterns can be real, but "noisy." History is sending us a signal through time in the form of jointly inherited features. Convergence and divergence are noise, exceptions that degrade that historical signal and make it harder for us to receive. Given enough time, some signals will be lost entirely, completely obscured by evolutionary differentiation outward and convergence inward. We do not know the nature of the first multicellular animal because divergence away from its pattern in its descendants has obliterated that historical signal. That, however, is a long way from happening with the birds of prey and the owls, clades whose members we can easily recognize despite their changes.

So when we are interested in reconstructing history from the traces it has left in the present, the effects of ecology are noise, static in the signal that partially drowns it. Perhaps it's equally true that there are ecological patterns in which historical inheritance is "just noise," obscuring patterns that are the consequences of similar ecological processes found in historically unconnected communities. Big fierce animals are rare, whether they are tigers in an Asian rainforest, marsupial "lions" in pre-aboriginal Australia, or great white sharks in the Great Australian Bight (Colinvaux 1980).

One particularly important tradition in ecology has been built around the hope of turning ecology into a precise, mathematical, and general science that excludes the "noise" produced by history. This program is associated with Robert MacArthur, and in his work the ahistorical program is explicit:

We are looking for general patterns, which we can hope to explain. There are many of them if we confine our attention to birds or butterflies, but no one has ever claimed to find a diversity pattern in which birds plus butterflies made more sense than either one alone. Hence, we use our naturalist's judgment to pick groups large enough for history to have played a minimal role but small enough so that the patterns remain clear. (MacArthur 1972, 176)

A qualitative example of the general patterns MacArthur had in mind is plant succession. Cleared or otherwise disturbed lands return to forest in broadly similar ways. First annual weeds appear: plants with many small seeds that invest in fast growth and reproduction. These are quickly replaced by perennial shrubs, which invest more in deep roots and stronger, more resistant structures. These, in turn, are slowly replaced by trees, which are slower-growing, invest even more heavily in growth and in chemical defenses against herbivores and other enemies, and which often produce fewer but larger seeds. Plant succession looks similar in many different regions, even though the particular weed, shrub, and tree species differ greatly from community to community. There are other ecological patterns that are supposed to be identifiable across different ecosystems and which are supposed to apply independently of the particular species composition of those ecosystems. Predator/prey ratios, for example, are supposed to be relatively constant just so long as the predators are all warm-blooded. Hence one way in which the "warm-blooded dinosaur hypothesis" has been investigated is by the use of fossil evidence to determine the carnivore/herbivore ratios in dinosaur communities. One of the flagship theories of ahistorical ecology is MacArthur's own theory of *island biogeography.* MacArthur developed a theory of equilibrium species diversity that essentially depended only on the size of the island (for size determines intrinsic extinction rates) and its distance from immigration sources.

There are some major threats to MacArthur's dream of an ecology that transcends history. One potential problem is *contingency:* the sensitivity of community structure to small variations in the factors that impinge on it. Sharon Kingsland's fine history of population ecology (1985) largely revolves around the tension between those who hope for a general theory of ecology and those who think that the particularity of individual ecosystems puts a rich and informative general theory out of reach. Kingsland is inclined to picture this controversy as a dispute between mathematical modelers and those who think ecosystems are just too complex or too sensitively dependent on initial conditions to be treated this way. Consider, for example, island biogeography. In 1883 a series of savage volcanic explosions purged the island of Krakatoa of all life and physically reshaped the island itself. This island has become a paradigm of a reassembled ecosystem. When life was reestablishing itself on Krakatoa, the island's particular location, the species present in the local bioregion, their capacity to survive in a grossly disturbed habitat, and their dispersal capacities jointly determined the list of potential colonizers. But chance determined the order of arrival of those potential colonizers and no doubt played a role in determining the survival of those that turned up.

The theory of island biogeography is committed to the idea that the basic ecosystem structure that develops on an island—its richness or diversity—is independent of accidents of arrival and timing. Those accidents may determine which member of (say) the goanna guild establishes itself, but they do not determine the fundamental structure and richness of the ecosystem. A contemporary critique of island biogeography rejects this idea in favor of the idea that particular initial conditions profoundly influence ecosystem development. In this view, the particular goanna, rat, or fruit bat that becomes established first may make a major difference to the downstream community. The nature of the ecosystem that will develop is not predictable to any decent approximation from general facts about local geography and regional biology. In the next chapter we will encounter a similar contingency thesis applied to life as a whole. Just as, it is alleged, we cannot predict Krakatoa's biological character from knowledge of its size, location, climate, and the surrounding biota, perhaps the history of life on earth is a one-off occurrence not dictated by any general facts about the planet.

So MacArthur's program—the development of qualitative ecological theories that apply across many different types of ecosystems—might be derailed by the sensitivity of ecological processes to detail. His program depends on the viability of robust process explanations of the assembly of island ecosystems from scratch. It requires that every island that is physically similar and which draws on a similar pool of potential colonists end up with similar communities. We have no guarantee that that is true. It will not be true if, say, the order of arrival of potential colonists is important, as it may well be.

However, another challenge to the program—and we think it a successful one—is the inescapable importance of history. Consider the interactions between dingos, foxes, and Tasmanian devils (a miniature marsupial hyena) in Australia. Foxes have never become established in Tasmania, a fact of great consequence for the small native mammals. That is not because there is no fox niche, but apparently because they are excluded by Tasmanian devils

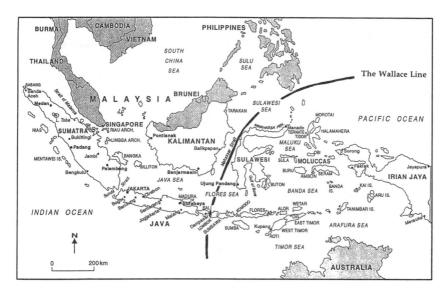

Figure 11.1   The Wallace line. (Adapted from Grant 1996)

preying on their kits. Devils, on the other hand, have probably been excluded from mainland Australia by invading dingos. So the biological character of Tasmania substantially depends on two accidents. First, the dingo
invasion took place when sea levels were rising, so by the time they reached
southeastern Australia, the Bass Strait had re-formed, and the sea blocked
their route to Tasmania. Second, it depends on a particular quirk of fox/devil
interaction. So the biological character of a community depends not just on
what niches are occupied, but also on the species that occupy them. Perhaps
we can predict from general considerations the ecological roles a community
will make available. There are no large carnivores on small islands. Even so,
there will be much that matters about the community that depends on the
occupants of those roles. The character of communities depends in part on
particular taxa. Not every devil-sized carnivore would exclude foxes.

This point is absolutely critical. If the ecological character of a community
mainly depends on the causal roles of the organisms in it—on, for example,
the fact that some organism or other is recycling dung—then the prospects
are good for general theories of community types. But if the particular
species composition is critical to the character of a community, then history must play a central role in explaining the nature of biological communities. For no one denies that history determines which particular species are
found in a community. Many organisms are not found in communities in

Figure 11.2 Vicariance. At A, a new species has come into existence. Since species typically form from small, isolated populations, its range at speciation is small. At B, it proceeds to spread through the geographically continuous habitat that is available to it. At C, geological change has fragmented the species' range. By D, as the result of further change, the species' distribution has become disjunct. It has gone extinct in some parts of its original range (including its point of origin).

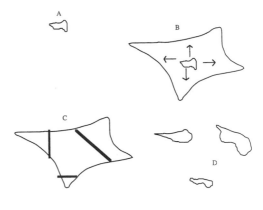

which they could live. A wonderfully striking example is the biological difference between Lombok and Bali, two similar-sized islands separated by 30 kilometers or so. Alfred Russel Wallace, the co-discoverer of natural selection, noted that Bali, to the west of Lombok, has or had tigers, monkeys, hornbills, and bears. Lombok lacks these species, but has cockatoos, honeyeaters, birds of paradise, and possums. Bali is biologically part of Asia; Lombok, of Australasia. The very striking difference in the biological communities of these two islands—one on each side of the "Wallace line" separating the Australasian biota from the Asian biota—is not explained by the different niches available in the two regions. Their difference is a difference in history.

There has been an important controversy in biogeography about the role of history in explaining the distributions of organisms. As we shall shortly see, *vicariant* explanations downplay the organism's own power of getting around, and emphasize instead the role of geological change in explaining distributions. *Dispersalist* explanations, as the name suggests, emphasize the organism's own ability to get itself or its gametes to new places. This contrast in views is important, but it does not contradict our main point here about the importance of historical ecology. Both views are historical explanations of organism distributions; they just invoke different historical mechanisms. Also, as we shall see, for a very important group of cases, vicariant explanations are uncontroversial. Figures 11.2 and 11.3 schematically illustrate these two mechanisms of distribution.

Vicariant explanations appeal to the geographic history of the region in question. Shared geological and evolutionary history sometimes explains rather surprising *disjunct distributions:* species or groups that are scattered in unconnected chunks. For example, shared history explains the distribution

Figure 11.3 Dispersal. At A, a new species has come into existence. Since species typically form from small, isolated populations, its range at speciation is small. At B, it has dispersed from its point of origin to many locations in the geographic vicinity of its origin, but without ever having established a large, continuous distribution across this space. These may be real islands, which cannot be reached without some lucky accident, or they may be "habitat islands": regions suitable for the species surrounded by areas that are typically, but not always, unsuitable. So successful dispersal events may be widely separated in time. Some dispersal may be indirect, with previously colonized islands seeding others. At C, the conditions of life have changed. Some of the habitat islands have disappeared, and the species has suffered many local extinctions, including at its point of origin.

of the southern beech family, a family of large trees that is scattered in odd bits of New Zealand, New Guinea, Australia, and Chile. Biogeographers believe that an originally continuous distribution in Gondwana was fragmented and shifted by geological forces acting on the group's original range. So vicariant explanations of species distributions depend not on the intrinsic dispersal abilities of the species in question, but on their histories and the histories of their homes. Southern beeches and leiopelmid frogs (an ancient frog lineage) are two of the many groups of organisms that have similar distribution patterns around the Pacific Rim. This is not because frog spawn and beech seed tend to be dispersed together, but because these groups and many others originated in the same region and have shared the subsequent history of that region's fracturing and motion. In figure 11.4 we show some of the Pacific Rim distributions that are biogeographic traces of the ancient continent of Gondwana.

These vicariant explanations contrast with explanations appealing to the dispersal of a population radiating out from its point of origin. In many cases there is no difficulty in choosing our mechanisms. No one doubts that vicariant processes explain the distribution of southern beeches throughout the surviving suitable fragments of Gondwana. No one doubts that dispersal explains the distribution of the Norway rat. There has been much debate—at times heated—in the biological literature on the importance of vicariance and dispersal in the explanation of discontinuous distributions. But no one

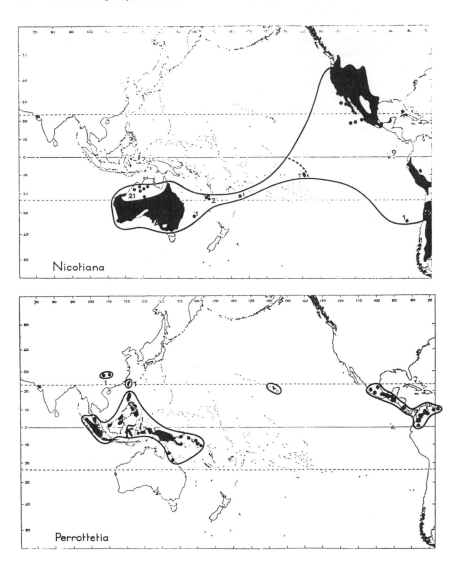

Figure 11.4 The Pacific Rim distributions of plants of the genera *Nicotiana* and *Perrottetia,* two of the many groups whose biogeographic distribution shows traces of the ancient continent of Gondwana. (From Nelson and Platnick 1981, 537-38.)

doubts the importance of vicariance in the explanation of continuously distributed species in adjoining communities. As the Bass Strait divided Tasmania from mainland Australia, their shared geological and biological history ensured that there would continue to be great similarities between these two ecosystem complexes. The shared history of these adjoining, though now separated, ecosystems plays a significant role in explaining their continuing similarities. Many of the ecological similarities between Tasmania and mainland Australia are the ecological equivalent of homologies. They are inherited from the larger system whose fragments they are. There are also, of course, considerable differences due to differences in climate, habitat, and chance.

These elementary considerations of history's importance have important implications for the way we think about some of ecology's paradigm examples. Studies of, say, the regeneration of life on Krakatoa have the potential to be seriously misleading if they are taken to illustrate typical mechanisms of community assembly. Studies of community reestablishment on literally barren ground are potentially misleading. Assembly from scratch will be very different from reassembly after habitat fragmentation or disturbance, for it is assembly after the historical signal has been canceled—a signal that is typically of great importance. The utter destruction of life on Krakatoa meant that the new Krakatoan ecology inherited nothing directly from the pre-eruption ecology (as far as is known, nothing whatsoever survived, though we would bet that some bacteria made it through).

Though a striking case, Krakatoa is surely a very unusual one. We have no doubt that the mechanisms of interest to ecologists who try to identify the causal roles sustaining an ecosystem play a role in the explanation of community organization. The size of an island, its physical structure, and its habitat diversity will surely influence the plants and animals that become established there. But our guess is that these considerations will play their greatest role in explaining the differences between historically related communities, not their similarities. So the fact that Tasmania is cooler, wetter, and (much) smaller than mainland Australia is central to the explanation of the differences between Tasmania and mainland Australia. These ecological processes will help to explain the presence of some vagrant species from outside the parent community, and losses that one but not the other daughter community has suffered. There are fewer niches on Tasmania for nectar-eating birds because the cooler, wetter landscapes have prevented eucalypts (whose flowers are designed for bird and insect pollination) from taking over completely from more ancient wind-pollinated trees. So there are far fewer Tasmanian honeyeaters. But vicariant processes dominate. For the same reasons in reverse, Lombok and Bali are different because they do not share a history. Despite their geographic proximity, they are chips off quite different tectonic blocks.

In sum, ecologically important properties of ecosystems have historical explanations. A habitat might be resistant to invasion, be at a species equilibrium, or show a robust oscillation between predator and prey numbers not because of its intrinsic characteristics (area, climate, number of trophic levels), but because of its particular species composition. Its history might explain its ecological character. So we agree with those who think that ecology must have a historical dimension. In section 10.7 we suggested some ways in which historical and adaptive thinking in evolutionary theory might be integrated. In section 11.5 we shall try out a similar, admittedly speculative, idea along the same lines in ecology.

11.3 The Balance of Nature

There is no doubt that the popular idea of "the balance of nature" has a powerful and seductive appeal. Left to herself, nature regulates herself. The environment and the particular communities within it are at or near equilibrium. Storm, fire, and drought will from time to time push a community away from equilibrium, but intrinsic mechanisms will restore the community. Even after a major fire, for example, the process of succession will eventually restore a climax forest community. Human activities are often wrong because they (almost alone) subvert this natural balance.

The scientific discipline of ecology has often hosted a more sophisticated and less mystical version of the same idea. Much scientific ecology has been equilibrium ecology. The principle of competitive exclusion is an equilibrium principle. It states that at equilibrium, no two species can occupy the same niche in the same community. Arguably, so too is the classic theory of the niche, for the rules of community assembly—what species can and cannot coexist—are determined for the community in equilibrium. In their review of ecological ideas, May and Seger (1986) distinguish two central explanatory concerns of ecology. One is community structure: the variety and interrelations of the species in a community. The other is the mechanisms regulating the population size of particular species. Though these are distinct problems, the idea of a structured community with determinate rules of community assembly fits well with equilibrium conceptions of population regulation. If the different populations in the community interact strongly with one another through competition, predation, parasitism, and the like, those interactions both help to determine the niche of each population and help to regulate its size. Community organization will determine both the number of slots—the niches—available in the community and the size of those slots. Community structure specifies, though no doubt within fairly rough limits, the carrying capacity of each niche in that community. If communities have rich and powerful internal checks and balances, then unless they are devastated by outside forces, they will remain close to equilibrium conditions.

Equilibrium models in ecology are clearly problematic. There is no single idea of equilibrium, nor any single idea of its disturbance. Rather, there is a family of perhaps loosely related equilibrium ideas, none remotely uncontroversial in their application to real ecosystems (Pimm 1991). Even more to the point, the idea that communities are typically regulated by the internal checks and balances of the biological interactions within them may simply be wrong. It may well be the case that communities out of equilibrium are the norm, not the exception. Reice (1994) has argued that "biological communities are always recovering from the last disturbance." In his view, whatever equilibrium is, biological communities are rarely close to it. In most communities, the "return time" from disturbance is shorter than the life span of the community's dominant elements. So communities are typically disturbed again before they return to equilibrium. Fire returns again to a forest patch before recovery from the last fire is complete.

If this alternative view of repeated disturbance is right, perhaps we should think of communities as assembled through the interplay of two different sets of rules. One set derives from the type of disturbance typical for the community. In a fire-prone landscape, some of the plants present are there by virtue of their ability to establish themselves rapidly in freshly burned-over ground. The other set derives from the character of the community as it moves towards equilibrium. Other plants are present because of their ability to compete successfully for light and nutrients in a crowded landscape. These plants might exclude the first type entirely were it not for the frequent disturbances. So in frequently disturbed communities, we would expect two different suites of adaptations to evolve, one suited to rapid invasion of disturbed habitat and the other allowing successful competition as the system recovers (and perhaps resistance to the typical disturbing agent).

According to Reice, the fact that communities are not at equilibrium leads to increased community diversity. First, different members of the community can vary in fitness with respect to the distinct types of selective forces they will face. A species that is competitively inferior in near-to-equilibrium conditions might persist through being competitively superior in response to disturbance. This, presumably, explains the persistence of some of the plants that are the typical first colonizers of cleared ground in ecological succession. Second, since communities do not reach equilibrium, competitively inferior organisms are not forced to local extinction before the next disturbance. Reice illustrates this idea through the *intermediate disturbance hypothesis*. According to this hypothesis, species richness in a community is maximized when the community experiences disturbance at intermediate levels of severity. That is, species richness is maximized when the disturbance is not so severe as to remove the competitively superior occupant species, but is severe enough to reduce their numbers sufficiently (or create enough short-term environmental heterogeneity) to allow the persistence of inferior competitors.

If these nonequilibrium perspectives are right, then the niche organization of a community is only one factor explaining membership and abundance in the community. As the effect of a disturbance fades, biological interactions within the community—the niches within that community—will become increasingly important in explaining the numbers and distributions of the taxa within the community. But if disturbance is important, such interactions do not play the only significant role.

11.4 Niches and Organisms

In section 11.1 we linked the classic theory of the niche to a version of adaptationism that takes adaptive change to be a response to problems posed by the environment. The arrow of explanation runs from "outside to inside" as organisms are shaped to fit their particular niche. Lewontin, in particular, has argued that this view of evolution is misconceived. He claims that the arrow of explanation runs both ways: organisms construct environments as much as environments construct organisms (Lewontin 1979, 1982b, 1985a,b, 1991). Organisms select their environment—most spectacularly in the case of migratory birds that fly north to breed in the Arctic and take advantage of that region's seasonal boom in insect production. Organisms also determine the features of the environment that are relevant to them. For example, plants do so by evolving metabolic mechanisms for dealing with aridity, salinity, low soil nutrients, and so on. And organisms physically alter their environment. Herbivores, for example, can transform their environment through their feeding. Beavers engineer ponds in which they live and to which they retreat from danger. In ecological succession, the successional vegetation creates the environment every bit as much as the environment sorts organisms into those that can establish themselves in it and those that cannot. The environment is reshaped by its successive occupants as the plant community changes from the initial wave of short-lived invaders to climax forest. If organisms both make and are made by their environments, then we need a transformation of the idea of the niche.

In an insightful discussion of Lewontin's ideas, Godfrey-Smith has distinguished between ecological properties like territory and properties like temperature (Godfrey-Smith 1996). Organisms may accommodate themselves to temperature in various ways; for example, they can grow thicker coats. Organisms may even act on the world to change the temperature of their immediate environment. Humans, bees, and termites all do so. Organisms determine the relevance of a wide range of physical environmental parameters of their world. Even so, temperature is an objective feature of the world that we can identify independently of the organisms that experience it. The same is not true of territories. A magpie family's territory does not exist when you remove the magpies from the landscape.

Robert Brandon has addressed similar issues by defining three different concepts of the environment. The *external environment* is simply all the physical factors, biotic and abiotic, surrounding the organism. The *ecological environment* is the subset of those factors that affect the organism's reproductive output. Finally, the *selective environment* is the subset of the ecological environment that *differentially* affects reproductive output across a range of individuals. The selective environment is the critical notion for evolutionary theory, because it is what two organisms must share if they are to be subject to a single selection process (Brandon 1990; Brandon and Antonovics 1996). The claim that the environment is constructed by the organism means very different things for these three different senses of *environment*. It is an empirical truth that some populations construct some aspects of their external environment over time. It is, however, a conceptual truth that all organisms "construct" their ecological and selective environments. A factor counts as part of the ecological environment of a population if the organisms' reproductive output is sensitive to that factor. It is part of the selective environment if the population is differentially sensitive to it, so that it can be more favorable for some individuals than for others. Whereas the external environment can be described without knowing which organism it surrounds, the ecological and selective environments cannot. Somewhat counterintuitively, this means that the external environment is the hardest of the three to describe: there is simply too much of it. What we humans think of as a simple description of what is "out there" typically focuses on the ecological environment of humans. There are innumerable factors of importance to plants and microorganisms that we do not notice. All these are part of Brandon's external environment.

In Lewontin's eyes, niches are more like territories than temperatures. Niches describe the ecological and selective environments of organisms, not their external environments. Lewontin interprets the history of the niche concept as a move from the idea that niches exist and can be identified independently of the organisms that fill them to the idea that niches are defined by their occupants. Lewontin adapts Hutchinson's conception of the niche for this purpose. Instead of following Elton in defining niches in terms of functional roles in a community, Hutchinson defined niches as volumes in

Figure 11.5 A simple example of the distinction between fundamental and realized niche. Consider just two dimensions of the niche space representing factors affecting a species of plankton: water salinity and predator density. The light stippling represents the species' fundamental niche with respect to those factors; the solid black represents its realized niche. When a particular competitor becomes a factor, the species retreats to areas of its former habitat that are highly saline, but have fewer predators.

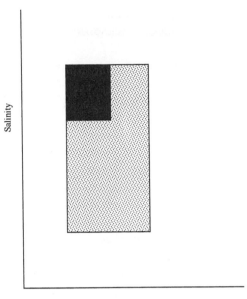

Predator density

an "abstract space." The dimensions of these volumes are the environmental quantities relevant to the population of interest. So for, say, a plankton species, the niche dimensions might include water temperature, salinity, sunlight, oxygen content, and the densities of parasites and predators. In Hutchinson's terminology, a species' *fundamental niche* is the region within this space in which that species could maintain itself indefinitely. The fundamental niche measures the species' tolerance for physical conditions, food supplies, predation, reproductive requirements, and parasites, but it ignores the effects of competition. The species' *realized niche* is the region within this space that it actually occupies, for competition may exclude it from some part of its fundamental niche. To a first approximation, New Zealand has only two sorts of endemic birds: endangered ones and extinct ones. One of the former is the takahe, a giant flightless rail. Its fundamental niche encompasses much of New Zealand, but since the arrival of placental mammals it has become restricted to an otherwise less than optimal corner of this niche space in some tussock grasslands of the Southern Alps (Gray and Craig 1991).

There were powerful reasons for switching from Elton's conception of the niche as a role in the community to Hutchinson's conception of the niche as a volume in a multidimensional space. It's a notion with much wider applicability. First, Hutchinson's notion of the niche makes it much easier to compare the ecological roles of similar but not identical organisms. So long as the same dimensions are relevant to the two types of organisms, both their differences and their similarities are clearly captured. Second, it fits better with those conceptions of ecology that emphasize the importance of the physical environment. Elton's picture emphasized the relations between the organisms in the community, whereas Hutchinson's formulation is neutral on this question. The niche dimensions can be physical parameters of the environment or the densities of predators, prey, and other biological factors. Third, we can define a niche even for those organisms whose daily or seasonal rhythms take them through many ecological communities, playing different roles in each. Fourth, Elton's conception is committed to the idea that there are natural boundaries to "a community." But small invertebrates, large invertebrates, reptiles, small mammals, large mammals, and birds all have greatly differing ranges, and often interact with very different samples of vegetation. Given the very different scales on which organisms live, and the great differences in their mobility and in the means through which their young are dispersed, we might very well doubt that there is any such thing as "the community" of which a particular ant nest, eucalypt, or bird is a part. Those doubts are irrelevant to Hutchinson's definition, for we can choose different dimensions for each different kind of organism.

In Lewontin's view, we derive our niche dimensions from our knowledge of the organisms of interest. We might define the niches of a warbler group using the dimensions of food size, foraging height, and foraging location: whether the warbler forages close to the main trunk of trees or on the outer leaves. For a goanna, our dimensions will be quite different. Hence Lewontin does not think that there is any interesting sense in which niches exist independently of their occupants; like magpie territories, they are not found, but made. So he and others regard Hutchinson's concept as abandoning anything like a "lock and key" conception of adaptation. In their view, we cannot explain adaptive radiation in evolutionary time, or invasion in ecological time, through the idea of unfilled niches. The physical universe pre-exists its particular organic inhabitants, but its physical structure—the external environment—does not explain the diversity of organic form. There are an infinite number of ways of partitioning the universe into potential niches, corresponding to all the possible Hutchinsonian dimensions we could choose. "Unless there is a preferred or correct way in which to partition the world, the idea of an ecological niche without an organism to fill it loses all meaning" (Lewontin 1985b, 68). If there are infinite numbers of unfilled niches, then we cannot explain adaptive radiation as filling vacant niches, for the vast majority remain unfilled. So all the genuine explanatory work—our account of the filling of the few that are filled—remains to be done. It makes

no difference whether we think of the notion of "an empty niche" as unde-
fined, or whether we think of empty niches as well defined but inordinately
numerous—like all the nameless beetle species, another instance of the
Deity's excessive creative zeal.

11.5 Reconstructing Niches

If we had to choose between the classic concept of the niche and Lewontin's
concept, we would certainly choose Lewontin's. His main points on the
reciprocal relations between organisms and environments are well taken.
Moreover, his view of niches looks very plausible when we think about
some particular ecological problems. We shall illustrate its usefulness through
Eldredge's approach to a famous puzzle: the variation in species number
between tropical and temperate habitats. Tropical habitats are proverbially
diverse in comparison with temperate ones. Why is this so? Does the species-
making engine run at a higher rate in tropical habitats, or do incipient, fresh-
made species survive to be counted more often?

Eldredge (1995) uses Lewontin's conception of the niche to explore the
hypothesis that species diversity is explained by ecological *invariance*—that is,
by environmental simplicity or predictability. The Tropics show little seasonal change, and this stability allows organisms to specialize, for they are not
compelled by physical change—climatic heterogeneity—to be able to live
in different regimes of temperature and moisture. So they are free to specialize, and hence to subdivide their fundamental niches more finely:

> The greater equability of the tropics allows species to be more special-
> ized, focusing more narrowly on habitat parameters. In ecological
> parlance, species in the tropics tend to perceive their habitats in a more
> fine-grained manner. In a very real sense, Stevens is providing a variant
> version of the claim that there are more "niches" in the tropics. (Eldredge
> 1995, 163–64)

Thus, the Tropics do have more niches available, but only because of species' "perception" of niche space. The organic response to invariance *generates* those niches—Lewontin's niches. Rather than thinking that tropical
communities provide many niches, and hence allow ancestral species to differentiate into finely subdivided roles within ecosystems, we can think of
organic response as creating niche structure. In this picture, very stable
environments—invariant ones—allow for specialization. Slightly differing
stable environments, and slightly differing inheritances, will select for differing specializations. So some of the diversity of life in tropical rainforests is

generated by the ecological stability that allows specialization. Then, once a phylogenetically complex community has developed, that itself creates a complex environment for other organisms. We do not, of course, know whether this explanation of tropical diversity is correct, but our examination of the hypothesis illustrates the fruitful nature of Lewontin's idea.

Nevertheless, Lewontin's view is surely too skeptical. There seems to be something right about the classic concepts of niche and ecosystem. Experience teaches us that broadly similar ecological communities are reassembled after extinction events from new components. We find broadly similar ecological communities in widely separated places, again made from distinct components. The different adaptive radiations of birds really do yield striking similarities. New World vultures are very like Old World ones, even though they are not closely related. Many Australian bird species are strikingly like Northern Hemisphere forms, a fact reflected in their names. Australian flycatchers, warblers, wrens, and magpies are not especially closely related to their Northern Hemisphere counterparts, but these pairs have many physical and behavioral similarities. There does seem to be something to the idea that there are functional similarities—similar roles—in different communities.

We think the solution may be to construct an account of niches intermediate in generality between Lewontin's niches, whose dimensions are generated by our knowledge of the life history of the particular species, and the classic idea of a niche as a role in a community of a particular type. We can extract niche dimensions not from the species in question, but from the larger clade to which it belongs. Hence Lewontin's dilemma is avoided: niche dimensions do not have to be chosen once and for all for each community type, but neither do they have to be chosen anew for each species. We shall use an example from Vrba to illustrate this idea. In her macroevolutionary work, Vrba contrasts the generalist impala clade with its specialized sister group, the wildebeests and their allies. There are fewer species in the impala clade, but they are all generalist and long-lasting. The wildebeest clade is specialist and species-rich. The pattern of inheritance within these sister groups includes their ecological characteristics. Thus, Vrba claims, "within broad limits, components of habitat specificity can be heritable and characteristic for entire clades through millions of years" (Vrba 1995, 17).

If Vrba is right, then we can identify a common set of niche dimensions for the species-rich group, and compare the actual occupants of the niche space so characterized with their potential occupants. Is there a missing gnu? Perhaps a comparison of this kind might throw some light on the distinct evolutionary strategies of the two sister groups: specialization and subdivision on the one hand and broad ecological tolerance on the other. MacArthur's famous work on warblers (1958), which showed that different species of war-blers forage in trees of the same kind but exploit different regions of them, seems to fit this general picture. For the same set of dimensions describes the different niches within this clade.

So just as we think it wrong to oppose history and selection in evolution, we think it wrong to oppose history and community organization in ecol-ogy. If this historical conception of the niche is defensible, then the notion of a vacant niche—with respect to a particular clade—is not vacuous. In section 10.7 we suggested that the empirical adaptationist's bet is that his-torical and developmental constraints are relatively constant within a clade, so that they can often be treated as fixed background conditions. We are suggesting a similar bet about the *ecology* of a clade. Australasian robins vary, one from another, in how they live in and use their environment. But per-haps these differences are variations within the same set of niche dimensions. The aspects of the environment that are relevant (the amount of ground cover, the density of the tree canopy, the richness of invertebrate life) are consistent across the contemporary robins, and change relatively slowly over evolutionary time. So in viewing a particular woodland community, we may be able to identify vacant niches, poorly defended niches, and niches that have been divided between two or more species. However, we can identify these niches only relative to clades. The general biogeography of the region determines the clades that are relevant, and hence the niche structure that is relevant, in a given community. Parrot niches are relevant to explaining the community organization of Australasian woodlands, but not North Ameri-can woodlands.

One important problem is determining the depth of the nesting within clades of ecological invariants. Are the dimensions that determine a taxon's niche typically conservative, inherited from deep in the tree? Or are they typically shallow? If they are shallow, evolutionary change often invents new ecological dimensions rather than just changing the values of old ones. These are empirical questions, and the answers will surely vary from group to group. Cats, for example, are a strikingly homogeneous clade, and we would not mind betting that the dimensions relevant to one cat species will be rele-vant to most. We would be less confident in making the same bet about Darwin's famous finches of the Galápagos, some of which seem to have taken on quite unfinchlike ways of life. Moreover, there will be many individual exceptions. Clades, especially old and species-rich clades, will not be per-fectly ecologically conservative with respect to the aspects of the environ-ment that are relevant to them. The panda and the palm nut-eating African palmnut vulture are very different from the rest of their clades. Our suggestion can work only if these are exceptional rather than typical cases.

We began this chapter by recognizing the pervasive interconnections of ecology and evolution. The taxa out of which communities are built are the products of evolution. Those communities form the local environment within which evolutionary change takes place. However, our view of this local environment has been transformed. We suspect that communities are not best represented by the Eltonian model, as networks of interlocking niches that are fairly constant in physically similar habitats. Rather, communities achieve their composition through a mix of chance, inheritance, and ecological opportunity. In this mix, as vicariance biogeographers have noted, inheritance will often be of central importance. Moreover, it helps to specify the *relevant* ecological opportunities. The niches that matter are niches for particular clades. On the Torres Strait islands (between North Queensland and New Guinea) there are goanna niches, fruit bat niches, but no niches—in any way that matters—for genets, civets, and their relatives. The relevant niche space helps to explain the pattern of local migration and survival on ecological time scales. On evolutionary time scales, it helps to determine the adaptations we might find in local populations. In this view, there is no useful notion of (say) the small carnivore niche on the Torres Strait islands. The differences between goannas and hawks are too important. Nor is there a fruit-eating niche. But there are fruit bat niches, fruit-eating pigeon niches, and a single, rather lonely, cassowary species in the large ratite fruit-eating niche.

How should we think of adaptive radiation from this perspective? Consider, for example, the spectacular radiation of cichlid fishes in the East African lake system. The nature of those lakes surely plays an important role in this radiation. The lakes are vast and varied, and their coastlines have many local discontinuities. But the nature of the lineage matters too. The males are territorial, and they disperse rather poorly. Moreover, there is a critical redundancy in the jaw design of the No-Name Base Model cichlid, which has a second set of jaws functionally decoupled from the oral jaws. So though the lakes do present the lineage with an array of many different niches, they do so only through the way in which that lineage reacts in that lake environment. The independent and pre-existing structure of the lakes matters—not just any lake would fuel such a radiation. But features of the lineage make features of the lake system relevant to ecological and evolutionary patterns in the cichlids (Goldschmidt 1996).

In section 11.1, we noted Sober's distinction between source laws and consequence laws (Sober 1984b). Source laws are ecology's job. In this picture, they will not be fully general. They will not be written once and for all for each causal role in a community. But if niche dimensions are inherited with high fidelity from deep within a clade, then they need not be wholly local either. With luck, we can hope for ecological and evolutionary theory to produce source laws of intermediate generality. In ecological and evolutionary modeling, the trade-offs between realism, detail, and robustness are notorious. We may not be able to expect robust general principles about medium-sized carnivores, still less robust detailed ones. But equally, we may not need a new set of source laws for every goanna population in Australasia. There may be source laws about goanna niches that are both robust and which include a modest amount of detail.

11.6   Unfinished Business

The conceptual problems posed by ecology have not been as intensely debated as those posed by evolutionary theory. In this chapter we have focused on the idea of the niche, using it as our stalking horse in exploring the connections between evolution and ecology. This issue is certainly not the only one in ecology worthy of attention. As a hint of the extent of the unfinished business in philosophy of ecology, we will wind up this chapter by noting two interesting and very open questions.

First, the nature of ecology's units is rather problematic. In this chapter, we have blithely written about biological communities as if it were clear that these are real units in nature and that we can recognize them. On reflection, this may not be so. Communities on small islands are easily recognized. Some other boundaries are relatively sharp as well; for instance, the treeline-subalpine shift. But other habitat shifts are very smooth, such as changes from closed forest to woodland and from woodland to grassland. Moreover, the membership of such communities is not well defined. For example, do birds coming through a woodland and roosting for a while on a tree count as members of that ecosystem, or as vagrants passing through it? Furthermore, the unit—the community—seems to vary with the nature of the ecologist's investigation. For some purposes, the entire headwaters of a drainage system might be the ecological unit. For others, it might be a single tree. Perhaps this problem does not undercut the objective existence of the community. We do not have to define a community's boundaries the same way for every causal factor that affects them. The boundaries of a human suburb can differ for different purposes. The boundary for the purposes of local government may not be the same as the one for the purposes of the school system or of bus transport. Similarly, we might define the boundaries of an Australian woodland community one way if we are interested in the effect of eucalyptus dieback, and another way if we are interested in the effect of rabbit calcivirus. The boundary of the community may be objective (though obviously not precise) once we have specified the relevant causal factor. Even so, it's not obvious that extraterrestrial ecologists would discriminate the same communities, and community types, that we do.

Second, there is a problem of scale in testing ecological theory. Indeed, this is the conceptual problem that has most worried ecologists themselves. Some of their worries seem to derive from an excessive reverence for Karl Popper (10.6), but there are clearly real issues as well. First, if the fine-grained detail of a community matters, we obviously have the problem of generalizing from one case to another. Second, as Pimm in particular has argued, the scale of practical experiment fails to fit with the scale of explanatory interest. Ecologists are very often both theoretically and practically interested in the ecology of ecosystems and still larger units. What would be the effect of eliminating rabbits from Australia? Would predators, both endemic and introduced, that now largely eat rabbits successfully switch to rarer and more vulnerable native species, with tragic consequences? Yet the experiments ecologists can run and control are on shorter time frames and much smaller spatial scales. So ecologists are always struggling to test their theories.

Further Reading

11.1   Eldredge has written frequently on ecological units, and in particular on the relationship between ecological units and evolutionary ones (see Eldredge 1985a,b, 1995).

Both Worster 1994 and Kingsland 1985 are good introductions to the historical development of ecology. Also valuable, but more specifically focused on the concept of the ecosystem, is Golley 1993. Real and Brown 1991 is a fine collection of classic papers in the discipline. There are many texts that give a broad overview; two are Krebs and Davies 1981 and Rosenzweig 1995. Two less technical, more popular introductions are Colinvaux 1980 and Ehrlich 1988. For a short and snappy overview, see May and Seger 1986. Cooper, in press, is a general philosophical examination of ecology.

11.2   For a general defense of the importance of historical ecology, see Brooks and McLennan 1991. For introductions to biogeography that emphasize the importance of history, see Wiley 1988, Cracraft 1983, and Grehan 1991. For an attempt to apply these principles to the specific instance of Australian marsupials, see Duellman and Pianka 1990. Our account of the Wallace line understates some of the complexities; for an excellent treatment of this issue, see Oosterzee 1997.

11.3 The history of the balance of nature concept is briefly outlined in Egerton 1973. For more on contingency in the reestablishment of a community after major disturbance, see Del Moral and Bliss 1993. For a discussion of ecological theory in a conservation context, where it has been applied to the design and management of national parks, see Budiansky 1995 and Quammen 1996. Quammen is impressive, interesting, and readable, but sometimes annoyingly smart-arsed. Budiansky is very skeptical (more so than Quammen) about such uses of the theory. Both rely on Simberloff, whom Quammen in particular cites extensively. The general problem of equilibrium and disturbance is discussed insightfully by Pimm (1991). The intermediate disturbance hypothesis as an explanation of diversity is discussed in an Antipodean context by Wilson (1990, 1994) and Padisak (1994).

11.4, 11.5 As we mentioned, there is less work on the conceptual and theoretical structure of ecology than there is on evolutionary theory. But the idea of the niche is an exception, on which there is a rich literature. It is introduced well in two essays in Keller and Lloyd 1992. Elton (1927) introduced and popularized the idea of the niche. For an overview of the problem of deciding whether there are common patterns of community organization in different communities, see May 1984, an introduction to a collection on just this issue. Hutchinson (1965, 1975, 1978, 1991) developed his views in a series of publications. He himself seems to have thought that he was just making his predecessor's intuitive idea more precise, but his interpreters (Schoener 1989; Colwell 1992; Griesemer 1992b) think otherwise. For Hutchinson does not define niches independently of their occupants: a niche is a volume occupied by some population. So Colwell claims that, in Hutchinson's view, "the niche is an attribute of the population (or species) in relation to its environment" (241). Similarly, Schoener argues that Hutchinson's formulation is revolutionary, for "it defined a niche strictly with respect to its occupant, a species population, and not at all with respect to a place or 'recess' in the community" (90).

For Lewontin's critique of a "lock and key" concept of adaptation and the view of niches that comes with it, see Lewontin 1982b, 1985a,b, 1991, Griesemer 1992b, and Odling-Smee, Laland, and Feldman 1996. Lewontin's constructivist view of the organism/environment relation is discussed critically but sympathetically in chapter 5 of Godfrey-Smith 1996. Brandon

(1990) develops an account of environments, and discusses practical strategies for studying the coevolution of organism and environment in a more recent paper with biologist Janis Antonovics (Brandon and Antonovics 1996). In an interesting recent paper, Leibold (1995) develops an alternative interpretation of the Hutchinson/Elton contrast. He argues that Hutchinson's niche definition focuses on the environment's effect *on* a population—on the population's habitat requirements, which limit the population. In contrast, Elton was concerned with what the population did—on the effect *of* the population. Furthermore, he argues that, despite a formalism inherited via Hutchinson through MacArthur, the most recent developments in niche theory are best seen as a continuation of Elton's ideas, for these developments are preoccupied with the effect of a population on its depletable resources and with the consequences of that depletion.

Eldredge's ideas on tropical diversity serve as an illustration of the way Lewontin's views on the niche might play a role in biological explanation. These ideas are by no means unproblematic, however; see Platnick 1992 for a skeptical view of the whole problem. He thinks the greater species richness of the Tropics is largely a myth. Plant diversity in many nontropical Southern Hemisphere habitats is very impressive indeed. For more on the evolution of specialist versus generalist organisms, see Vrba 1984a,b, 1993, Sultan 1987, and Godfrey-Smith 1996. For the connection between specialization and the definition of the niche, see Price 1984. On the cichlid adaptive radiation, see Meyer 1993, Rossiter 1995, and Goldschmidt 1996.

11.6   There has been within ecology itself a long-standing dispute over the relationship between communities and ecosystems on the one hand, and the individual organisms and populations that make them up on the other. In what sense do communities reduce to their individual components, and in what sense do they have properties that are in some sense distinct and additional to those of their components, so that they are "more than the sum of their parts?" For debate within ecological circles, see Simberloff 1980, Levins and Lewontin 1985, O'Neill et al. 1986, and Underwood 1986. The historical dimension is well treated in Kingsland 1985 and in Real and Brown 1991. For a recent philosopher's discussion of reduction, with an emphasis on ecology, see Dupré 1993. For a first pass at some of the methodological problems of testing ecological theory, see Conner and Simberloff 1986. Cooper (1990) offers a philosophical perspective on testing ecological models. A recent and extensive treatment of these problems, also with a conservation orientation, is Shrader-Frechette and McCoy 1993.

12

Life on Earth: The Big Picture

12.1 The Arrow of Time and the Ladder of Progress

The history of life has directionality. If a time machine deposited you on earth and you were able to examine the life around you, within rough limits, you would know *when* you were. The fact that history has a direction depends only on the ubiquity and irreversibility of evolutionary change. Species do not survive forever, and once extinct, they stay extinct. So-called *Lazarus taxa*—taxa like the coelacanth that appear in the fossil record, then disappear for a very long period, only to reappear—are famous in paleobiology. But no one suspects that the coelacanth actually went extinct, only to re-evolve in recent times. Hence fossils have been considered reliable ways of dating rock strata since the mid-nineteenth century (Rudwick 1985). But life's history has an arrow of time in a stronger sense: there is a time line not just in the particular taxa that have evolved, but in the *kinds* of taxa (the *grades*) that have evolved. Early in life's history, only prokaryotic organisms existed. A billion or so years later, eukaryotic organisms existed too. About 600 million years ago, complex animals appeared on the scene, and few hundred million years or so after that, plants became more complex as they colonized the land and diversified.

So life's history has a direction. There are, however, much more ambitious, controversial, and conceptually murky claims about life's overall history. Those claims are the topic of this chapter. We have already prefigured a few of them. In section 2.1 we noted Gould's distinction between disparity and diversity, and in section 2.3, his idea that life's disparity has decreased over the last 500 million years or so. We take up that idea in sections 12.2 and 12.3. It is developed in conjunction with two other ideas, one about the contingency of life's history and a matched claim about the role of mass extinction in reshaping life. We focus on these claims in sections 12.4 and 12.5.

As we shall see, all of these debates pose both conceptual and empirical problems.

In this section we take up the idea of progress, a long-standing theme in thinking about evolution. Ruse (1996) has shown that over much of its history, most evolutionary theory has assumed that the history of life is the history of progressive improvement. Gould (1989, chap. 1) has shown that this idea remains central to popular thought about evolution. In response to this work, we consider the following questions:

1. Is there a trend in the history of life toward increased complexity? Is life as a whole more complex now than it was, say, at the base of the Cambrian?

2. Is the history of life progressive?

3. If the answer to (1) or (2) is yes, what explains that trend? How does evolutionary history generate progress and/or complexity?

Both (1) and (2) raise difficult conceptual questions. It is not obvious that complexity is a feature of an organism itself, rather than a feature of our conception of that organism. Dawkins accepts the claim that complexity is a feature of our description of the world rather than of the world itself, but he suggests that even so, we can develop a reasonably objective account of relative complexity. We can do so by comparing the length of the descriptions of two organisms. A cat is more complex than a fly if it takes longer to describe a cat than a fly. To apply this test, of course we have to ensure that we use the same language in each case and that the level of detail (the "grain" of the description) is the same (Dawkins 1992, 1997).

[margin annotation: complexity a "feature of our description of the world"?]

This idea seems defensible when we consider structures that are fundamentally similar. So, to take an important example, we can compare the complexity of vertebrate backbones. McShea is a central figure in both defining complexity and testing the idea that it has increased over time, and backbone complexity is one of his key examples. He develops both a definition of backbone complexity and a test to see whether "the average backbone" has become more complex over time. McShea defines complexity by the number of parts in a system together with their degree of differentiation. A maximally simple backbone is one in which each unit—each vertebra— is the same, so that the backbone as a whole is a series of similar segments. Backbones become more complex as the units become increasingly unlike one another. An animal with differentiated vertebrae is more complex, at least in this respect, than one with a uniform structure (McShea 1991, 1994, 1996a,b).

Thus we could reasonably claim that a description *at the same level of detail* of the giraffe backbone would be longer than a description of the trout backbone. Since these are descriptions of the same structure modified in different ways in the different animals, we would have an objective account of "an equivalent level of detail." But if we were comparing radically different organisms, it would be much less clear that we had an objective specification of "the same level of detail." In comparing, say, the complexity of a fish to the complexity of a fish parasite, what are the equivalent levels of detail? Though parasites are (to our eyes) morphologically simple, they often have both extraordinarily complex life cycles that take them through many physical transformations, and many physiological and biochemical specializations for their parasitic existence.

So the idea of complexity is far from unproblematic. Nonetheless, even someone as skeptical as Gould about progressive pictures of evolution is prepared to concede that there is a minimal sense in which life has increased in complexity over time. The change from a world in which the most complex organisms were bacteria to a world in which large organisms abound, each made from many millions of cells each of which is more complex than a bacterium, is a change to a more complex world. So perhaps complexity really has increased over time. What of progress?

If there are problems with the idea of complexity, we have those problems in spades with the idea of progress. In the tradition of thinking about evolution as an engine of progress, "progressive change" has often just meant "change in the direction of humans." The shift from mammal-like reptiles to mammals is progressive because it is a step on the road toward us, or something like us. There is no doubt that the idea of progress has been important in evolutionary thinking. But if that is all progress means, then it is an anthropocentric notion of no real biological interest. "Progress" so defined is a human-focused notion. As many have pointed out, there is no more reason to be interested in progress in that sense than in progress toward worminess or toward the characteristics of any other organism. Evolution has produced humans, but it has also produced tens of millions of other species. In every current view of evolution, there is little reason to suppose that our evolution was any more fated, any more to be expected, than that of any other species.

But what else could "progress" mean? Does it have any more objective definition? One possibility is that progress might be identified with a *progressive increase in complexity,* thus turning our first two questions into one. Indeed, Gould is interested in complexity precisely because he regards it as an objective and empirically tractable surrogate for progress. Dawkins, in contrast, regards this definition as just disguised anthropocentrism. In his view, we think increased complexity is progress only because we happen to be complex organisms ourselves (Dawkins 1997). We will return to complexity, but we shall first consider Dawkins's alternative, the idea that evolution is progressive because over time life is becoming *better adapted*. As time goes on and natural selection grinds away, living creatures become better designed; they become better adapted to their niches.

This hypothesis faces both conceptual and empirical challenges. Its central idea is that organisms differ in their degree of adaptedness to their worlds. Organisms at some time period are typically better adapted than earlier ones and not as well adapted as later ones. In a very limited sense, this idea is uncontroversial. If two organisms are members of the same population responding to selection, we can certainly compare their fitness. As we shall see, we can extend these comparisons to a population evolving over time. But if we define progress as increasing levels of adaptedness over millions of years, then we are required to compare organisms of disparate morphologies, physiologies, and environments. This would require us to be able to identify a general property of adaptedness—of the degree of "fit" between an organism and its environment. The idea that there is such a property has great intuitive force; we have met something like it before in considering explanatory adaptationism (10.4). However, despite its plausibility, it has turned out to be very difficult to cash out this idea. To the best of our knowledge, no one has done so.

Natural selection generates adaptation to the local conditions of life. As we have just seen, while those local conditions remain the same, and while they are defined the same way by the evolving population, we have a good intuitive grip on an increase in adaptedness. As bittern camouflage patterns improve, bitterns become better adapted. But the force of this example depends on the fact that the niche of the bittern (in the sense defined in section 11.5) remains the same. Both the features of the environment and the relevance of these features to the bittern population are constant over the evolutionary transformation of poorly camouflaged to well-camouflaged bittern. So we may well be able to use the notion of adaptedness to give content to the idea of evolutionary progress over the short run—progress in a single lineage. But the conditions of life are not stable over the long term, both because the physical parameters of environments change and because organisms stop perceiving their environments in the same way.

Dawkins disagrees with this pessimistic assessment of our capacity to identify progress. He argues that evolutionary *arms races* between competing lineages define an arrow of progress, a trend of improvement over the long term, though not the very long term. Arms races between lineages are cut short by mass extinction events, but while they are in progress, each lineage is objectively improving (Dawkins 1997). We are not convinced. First, we suspect that prolonged evolutionary arms races may reconstruct the environment in which evolutionary change takes place. If primate cognitive evolution were driven by a race to manipulate others and to avoid manipulation oneself, the upshot might well transform the selective environment. Second, arms races may involve a rock/paper/scissors evolutionary shuffle. This is no idle possibility. If Hamilton is right, the evolution of sex involves a rock/paper/scissors game. As a genome becomes common, parasites become well adapted to it because it is common, and less well adapted to rare ones because they are rare. The function of sex is to change the target at which parasites must aim. Within limits, the intrinsic features of the target matter less than avoiding presenting a stationary target at which parasites can aim. So a genome that is now well adapted because it is rare might have been common and ill-adapted several evolutionary moves ago.

Thus the idea that evolution is progressive is particularly problematic both conceptually and empirically. The idea that evolution results in a directional change in complexity is much harder to deny, despite lingering doubts about the objectivity of our complexity measures. So let's return to it. Gould notes that we cannot escape the conclusion that evolution has its arrow of time (Gould 1996a). For the living world was once represented wholly by prokaryotic cells: bacteria and other organisms of similar complexity. In our world we can hardly escape noticing the existence of gigantic and complex animals, for we are such animals ourselves. In the face of this apparently inescapable fact, it is natural to suppose that there is a very broad-scale trend in the history of life: complexity has increased on average.

Gould argues that while in a certain sense this is true, it's a very misleading formulation of the facts. It invites us to ask the wrong questions and overlook the right ones. For the trend toward increasing complexity is nothing but the spread of variation. Life starts off as simple as life can be. Mostly, it stays that way. Most living things have always been as simple as the first living things, for nearly every organism is a bacterium. Occasionally lineages split and a species appears that is more complex than its parent. No global evolutionary mechanisms make this impossible, but none make it more likely. Complexity increases by passive diffusion from a point of origin close to minimum complexity. The real change is an increase in the total variance. If life originates close to the point of minimum complexity, then wholly undirected, stochastic mechanisms will increase the variance, and that variance must include a bias in the direction of increased complexity. Mechanisms that are blind to complexity suffice to produce an upward drift in average complexity. The fact that there is no bias in the mechanisms of adaptation, speciation, or extinction that favors increased complexity, together with the persistence of bacterial domination of the living world, is fatal to any robust version of the idea that evolution over time has generated increased complexity (Gould 1996a).

In 1995, Maynard Smith and Szathmary published *The Major Transitions in Evolution*. On the face of it, there seems to be a real contrast between Gould's conception of life's history and their conception of life's history, which involves a series of major transitions and hence an inherent directionality. About half of the transitions on which Maynard Smith and Szathmary focus took place on the road to the invention of the bacterium. These include the shift from independently replicating structures to the aggregation of codependent replicators into chromosomes and the shift from RNA to DNA as the main replicator . But they also identify the invention of eukaryotes, cellular differentiation, and the invention of plants, animals, and fungi, colonial organisms, and even human language as major transitions. So can we reconcile the idea of life's history as a series of major transitions with Gould's alternative of complexity drifting upward undirected?

We think the difference between Gould on the one hand and Maynard Smith and Szathmary on the other is a difference in how they picture variation. We have illustrated Gould's picture in figure 12.1. He sees variation as a curve. The "left wall" of the graph represents the minimal complexity of life, but complexity has no intrinsic maximum to the right. So there is a lower limit to complexity, but no upper limit. When we consider a series of these graphs over evolutionary time, we see that not much happens. The curve stays pretty much the same, with the peak close to the left wall. The only difference is the gradual spread of the tail to the right, as maximal complexity creeps up over time.

Maynard Smith and Szathmary do not formulate their ideas in Gould's language, but if they did, the difference would be that they do not regard the walls of the graph as fixed over time. The major transitions in evolution are *movements of the walls*. Until the foundations of eukaryotic life were gradually assembled, there was a right wall, the intrinsic limit on the size and structural complexity of prokaryotes. In a certain timeless sense—the sense in which Gould operates—the right wall was open, but for much of its history, bacterial evolution was confined within two walls. Similarly, after the evolution of eukaryotes, there was another shift of the right wall, but only a shift. The invention of the organism required a complex series of evolutionary innovations. Until these came into existence, there was a right boundary to complexity set by the limits on a single eukaryotic cell. Maynard Smith and Szathmary argue that colonial and social life, too, have evolutionary

Figure 12.1 Average complexity increasing by passive diffusion in a series of snapshots from the history of life. The vertical dimension graphs the number of species in existence at a time at a given degree of complexity. The horizontal dimension graphs complexity, from minimum complexity at the left to increasing complexity at the right. Physics and chemistry impose constraints that define the least complex possible form of life; the heavy bar represents that "left wall." The basic pattern remains the same in all four snapshots: most species are (relatively) uncomplex, for the typical species is a bacterium, and they are simple (for a living thing). But over time, the right tail spreads: the complexity of the most complex species increases by passive diffusion away from the point of origin near the left wall. Relative to bacteria,

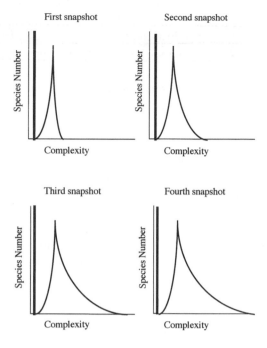

there are never many of these organisms, but the level of complexity of the most complex creeps up. Interestingly, the left tail of the distribution probably creeps out a little over time as well. The least complex organisms are those that use the complexity of others to survive, and these cannot evolve first! Viruses do without machinery for replicating their own genetic material, replying on their host to provide this. If current hypotheses about prions (the infective agents in scrapie, bovine spongiform encephalitis (BSE), and Creutzfeldt-Jakob disease) prove to be correct, prions rely on their hosts for "their" genetic material as well as for their metabolism.

preconditions. Until these are met, a wall remains to the right. The left wall has moved too: bacterial life makes viral life possible. So where Gould sees a drama played out within unchanging boundaries set by physics and chemistry, Maynard Smith and Szathmary see evolution as transforming these boundaries irreversibly. Once the eukaryotic cell comes into existence, once sexual reproduction comes into existence, once cellular differentiation comes into existence, the theater in which the evolutionary drama takes place is changed irreversibly. The boundaries change over time, and mostly in a direction that increases the maximum possible complexity.

In sum, as we see it, there has been an important shift in evolution over time: an evolution of evolvability. The mechanisms of evolution have changed, and changed in ways that have opened up new evolutionary possibilities (and perhaps shut down others). So while it's true that bacterial evolution has continued, and that, as Gould notes, this age and every other is the age of the bacteria, that is not the whole truth.

12.2 Gould's Challenge

Complex animals first appear in the fossil record about 600 million years ago. We might naturally expect the complexity and diversity of animal life to increase gradually over time, as the first multicellular lineages differentiate, becoming more complex and more adaptively sophisticated under the influence of selection and competition. Beginning in his book *Wonderful Life: The Burgess Shale and the Nature of History* (1989), Gould argued that this expectation is dramatically mistaken. Multicellular life diversified with extraordinary rapidity soon after its invention. The *Cambrian explosion* of evolutionary experimentation began about 530 million years ago and lasted 5–10 million years (Gould 1996b, 96–98). This explosion, Gould argues, generated greater disparity in animal life than we find today. Today's faunas in particular, and post-Cambrian faunas in general, are no more than remnants of these disparate forms. So the received view of the pattern of life is mistaken, and this mistake about patterns in history carries with it implications about the processes that generated those patterns. We here pick up themes from sections 10.3 and 10.5, where we introduced the idea that natural selection fails to explain the *persistence of type,* the preservation of the characteristic organizations of life's great clades. In this chapter we broaden the issue to include the establishment of those distinctive organizations and their extinctions.

In Gould's view, we are in the midst of a radical reinterpretation of life's history. This reinterpretation begins with a bed of fossils discovered by Charles Walcott in Canada in the second decade of the twentieth century, the Burgess Shale fauna. These fossils preserved the soft parts of some remarkable animals. These animals were metazoans, members of the lineage of all surviving multicellular animals. But they were from the Cambrian period, about 530 million years ago, so they provide us with a picture of early animal life. Walcott described a number of these organisms, for the most part regarding them as earlier and simpler members of lineages that are still with us. There is now a reasonable consensus that many of these judgments were mistaken. Beginning in 1972, Whittington, Briggs, and Conway Morris redescribed most of these organisms. In some cases they could not fit the animals into existing major groups. They were metazoans, but radically

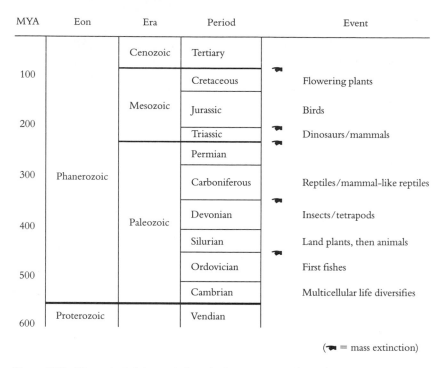

MYA	Eon	Era	Period	Event
		Cenozoic	Tertiary	
100			Cretaceous	Flowering plants
		Mesozoic	Jurassic	Birds
200			Triassic	Dinosaurs/mammals
			Permian	
300	Phanerozoic		Carboniferous	Reptiles/mammal-like reptiles
			Devonian	Insects/tetrapods
400		Paleozoic	Silurian	Land plants, then animals
			Ordovician	First fishes
500			Cambrian	Multicellular life diversifies
600	Proterozoic		Vendian	

(🐾 = mass extinction)

Figure 12.2   The geological time scale from the first appearance of complex animals.

unlike any surviving members of that lineage. Others were indeed, as Walcott thought, arthropods. Arthropods are segmented invertebrates with exoskeletons and jointed appendages. Flies, beetles, crustaceans, spiders, and that emblematic fossil, the trilobite, are all members of the arthropod group, one that dominates multicellular animal life even today. But the arthropods Walcott had discovered were unlike any of the surviving major branches of that lineage, and unlike the trilobites as well.

Gould has turned these empirical discoveries into an apparent conceptual revolution in our understanding of the nature of evolution and the shape of evolutionary history. First, he argues that the Burgess Shale fauna overthrows the orthodox conception of the shape of the tree of life. When Walcott first described the Burgess fauna, it was common to believe that life has become more complex, better adapted, and more disparate. The idea of evolutionary progress has since largely dropped out of professional evolutionary biology, but the idea that life has become more disparate has not suffered a similar fate. The conventional iconography of the history of life pictures it as a tree, narrow at its earlier stages close to its roots and becoming bushier as it

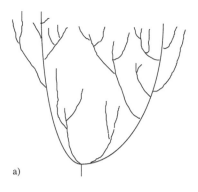

a)

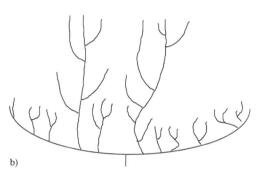

b)

Figure 12.3   Two models of the change in disparity over time. (a) Gould's interpretation of the received view, with disparity creeping up gradually over time. (b) Gould's own view: disparity was greatest early on. Much of life's initial disparity has disappeared into extinction, and little new has replaced it. (Adapted from Gould 1989, 46.)

approaches the present. According to Gould, the Burgess discoveries challenge the idea that the history of life is a history of increasing disparity. Instead, life is an irregularly shaped, straggly bush, thick with diversity around the period when multicellular life was invented, but with only a few tendrils representing those groups pushing up into the future (Gould 1989, 46).

In section 2.1 we introduced Gould's distinction between *diversity*—mere species numbers—and *disparity*—difference in basic body organization or basic body plans. Since Cambrian life has invaded the land and the range of climatic zones life occupies has increased, no doubt there are more species around today. *Diversity* has increased. But although there are, for example, fabulously many beetle species, they are all recognizably beetles, precisely because they all share a basic beetle body plan: beetles are insects whose forewings have hardened into protective sheaths *(elytra)* that serve to protect their more delicate hindwings. This body plan has been conserved since its invention, at least 265 million years ago. Gould thinks that, compared with the Burgess fauna, contemporary species are variations on fewer themes—they are less *disparate*. We are faced with the surprising conclusion that disparity has decreased since the Cambrian. Even if it has not increased, that would be surprising enough. As we shall see, though, there are serious doubts both about the idea of disparity and its measurement.

With Gould's view of the shape of the tree of life has come a further development of his views of the evolutionary process. As we saw in section 10.3, he is one of those evolutionary biologists who think that the importance of natural selection within evolution has been exaggerated. He has used his views on life's history as a vehicle for reformulating those thoughts. If animal life was at its most disparate in the Cambrian explosion, this is at least unexpected to those with an adaptationist conception of evolution. Moreover, Gould suggests that selection has not played the central role in either generating disparity or reducing it. For the history of life is *contingent*. If we replayed the tape of life's history and made minor changes at one time, they would be likely to ramify into great differences at a much later time. For example, the Burgess fauna includes early chordates, the ancestors of all the vertebrates and hence our ancestors as well. Perhaps only a very minor change in those shallow Cambrian seas would have made those early chordates go extinct, and some other group that has disappeared would instead have survived and flourished down the eons. If those first chordates had gone under, perhaps the whole chordate body plan and the array of evolutionary possibilities dependent on it would have disappeared forever. In this view, evolutionary outcomes—who makes it and who doesn't—are sensitively dependent on initial conditions. Though life's history is intelligible in retrospect, at least on large scales it is unpredictable in prospect.

If small changes could make big differences, presumably so could large external shocks to the system. Mass extinction looms large in this view of life. The path of evolutionary history is profoundly influenced by catastrophe. Dominant groups often do not wither away, but are cleared out by these great calamities, and that opportunity, and only that opportunity, allows the lucky survivors to diversify. The variety of life *over time* depends on catastrophic shocks making room for new experiments in life. These ideas on contingency and mass extinction are linked to natural selection. The more we think of the history of life as contingent, and the more important we think mass extinction is to its general course, the less we think of it as being determined by selection. The decimation of disparity since the Cambrian has depended not on the adaptive excellence of the various body plans, but on the details of history, on the contingencies of life. The adaptive radiation of the surviving lineages has depended on the mostly accidental disappearance of other groups. At the largest scale of life, we see survival of the luckiest, not survival of the best adapted.

12.3   What Is Disparity?

Gould argues that life at the high point of the Cambrian explosion was more disparate than it has ever been again. But what is disparity, and in what ways was Cambrian life disparate? In section 9.3 we explained how systematic biologists group organisms into more and more inclusive categories, from species at the least inclusive through kingdoms at the most inclusive. Though this scheme is now somewhat controversial, until recently, multicellular animals, the Metazoa, were considered one of five kingdoms (the other four are plants, fungi, prokaryotes—bacteria and their allies—and protists—mostly single-celled eukaryotes). In turn, the Metazoa have traditionally been divided into about thirty-five phyla (for a good quick review see Raff 1996, chap. 2). The *chordates*—all the animals with their nerve tissue in a dorsal column—constitute a single phylum. Mollusks make up another phylum; so do arthropods, and so do jellyfishes. This big picture is now being revised in ways that recognize the vast importance of the bacteria and the depth of their divergence from other prokaryotes, but even so, phyla remain major groupings. So in classic taxonomy, a phylum is a major division within the living world. In *Wonderful Life,* Gould (1989) argues that we find in the Burgess fauna seven or eight phyla—fundamental body plans—that are no longer with us. So Cambrian life had greater disparity than contemporary life because more body plans were exemplified then than now. But most of Gould's discussion centers around disparity within the arthropods, the most species-rich, diverse, and successful group of multicellular animals.

Gould's unclassifiable "weird wonders" do look profoundly strange. But many of them have been reclassified and placed in existing phyla. *Hallucigenia,* one of the weirdest of them all, turned out to have been reconstructed upside down. Right side up, it was recognizably an onychophoran, a segmented worm-like animal, whose relatives survive in the Southern Hemisphere. Moreover, it is hard to know what to say about these creatures, so we will follow the usual practice and focus on disparity within the arthropods. How might we tell whether arthropod disparity has gone up or down since the Cambrian? To answer this question, we need to skim the standard systematics of the arthropods. Arthropods have been sorted into major subdivisions (known as *classes*) on the basis of the type and distribution of appendages on the body. The hypothetical common ancestor, the Mother of All Arthropods, was a segmented creature with a pair of appendages on each segment, one on each side. These appendages themselves ramified into two branches. One of these branches was thought to be primarily for walking, the other a gill-like organ for breathing. As arthropods evolved, some of these original double-branched appendages became specialized, leaving only one branch. Arthropod design, then, is a mix-and-match business. You can fuse segments together to form larger body sections, and you can stick various sorts and numbers of appendages on the sections. So, for example, *chelicerates* (spiders, scorpions, and others) have a two-part arrangement. They have six single-branched pairs of appendages on the front body section, typically with the first jawlike pair specialized for manipulation, the second for sensory purposes, and the remaining four pairs for locomotion, giving spiders their excessively leggy look. The back section has various numbers of single-branched breathing tubes.

It's easy to imagine lots of variations on arthropod anatomy that don't exist. Break out your Construct-a-Pod kit and make a crab with three pairs of antennae. But, sadly, there are no such crabs. The *crustaceans* (true crabs, barnacles, lobsters, and their relatives) have two, not three, pairs of antennae. Today's arthropod patterns were exemplified in the Burgess, but it also contained arthropods with segment and appendage patterns different from any found today, and different again from the trilobites. *Yohoia tenuis,* for example, in Whittington's reconstruction (Whittington 1974), turns out to have a large pair of grasping manipulative appendages at the front of its head, followed by three pairs of walking legs, as well as configurations of the rest of its segments that depart from contemporary models. There are many other Burgess arthropods that do not fit into today's patterns. So if the segmentation patterns that we use to distinguish between the four great groups of arthropods—trilobites, crustaceans (crabs, lobsters, and the like), uniramians (insects and their allies), and chelicerates (spiders, ticks, and the like)—measure disparity among arthropods, then disparity has decreased. But do they measure disparity?

Gould, like many others, has used spatial models to express his ideas. He conceives of the organic world as located within "morphospace," the space that represents the physical forms of all actual and possible organisms. Similar organisms, like humans and chimps, are clumped close together in that space, because a similar set of physical dimensions describes us both, and we even have similar values on most of those dimensions. The disparity of life at some time is the amount of morphospace through which life is scattered at that time. Once large, bipedal, forward-looking, omnivorous primates had evolved, the addition of hominids would add little to disparity, however species-rich that lineage might be. But what measures *distance* in morphospace? What makes it true, say, that the hairlessness of humans, or the different shape of our jaw, represents a trivial increase in disparity? Does eye number matter? We would be deeply impressed by a six-eyed primate, and even

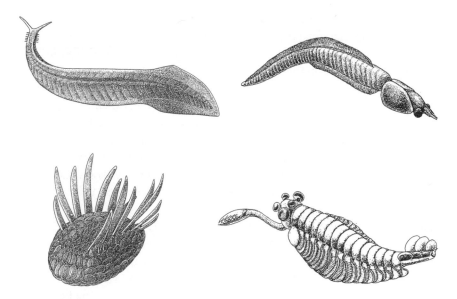

Figure 12.4 Some of the Burgess Shale fauna. Clockwise from top left: *Pikaia, Nectocaris, Opabinia,* and *Wiwaxia.* Only *Pikaia,* an early chordate, is a member of a taxonomic group known to exist at a later date. (From Gould 1989, 126, 146, 192, 322.)

the near-functional third "eye" of some reptiles strikes us as a bit weird and spooky. But eye number is no big deal among the arthropods: different spider species differ in eye number. The region of morphospace sampled by life will depend on what we count as traits, and how we weigh their importance.

An Australasian and a South American flycatcher really do seem very like one another, and very unlike an emu. It's hard to believe that these judgments of similarity and difference are wholly subjective. Yet there may be no objective dimension of disparity, of the degree of morphological differentiation of a group of taxa. For it turns out to be very hard to vindicate these appealing intuitions. The last thirty years of systematics has seen a vigorous controversy between cladists and evolutionary taxonomists (9.3). Evolutionary taxonomy fits in well with Gould's ideas, for its classification of organisms is sensitive both to the branching order of lineages and their degree of differentiation. In this view, barnacles are very aberrant crustaceans, and that can legitimately be taken into account in their biological classification. In contrast, many cladists doubt that we can objectively measure the differentiation of barnacles and clams. These judgments represent human projections onto nature, not objective features of nature (Ridley 1986). For example, in judging the relative similarity of spiders to ants or trilobites, does the ratio of eye number to leg number count as a similarity? Cladists think that in principle there is no way of answering this question. So they think that the problem of "quantifying morphospace" is intractable, not for technical reasons, but because there are no facts to discover.

These cladistic arguments are hard to answer, so there may be no such property as disparity. However, even if disparity is a real and biologically important property, the standard traits used to classify arthropods are unlikely to measure it. Systematic biologists are historians of the tree of life, so they are interested in traits that are informative about the history of organisms rather than traits that are important for measuring disparity. Traits that are *too conservative* are genealogically uninformative. If the Mother of All Arthropods has a trait—for example, a segmented body—that has been inherited unchanged in all the descendant forms, that trait tells us nothing about relationships within the arthropods. As we pointed out in section 9.3, traits that are primitive for a group are uninformative about relationships within that group. Traits that are *too evolutionarily labile* are also uninformative. If arthropod eye number changes at the drop of a hat, then eye counts will be genealogically uninformative. Arthropods with the same eye number will often not be closely related, for that similarity will often be a product of parallel and convergent evolution. Arthropods with different eye counts will often be sister species. So the traits we want are traits that are fairly conservative relative to the group of interest, but not too conservative. Segment number might, for example, be such a trait. It would be if, say, two segments that are fused on a certain insect are also fused in the taxa that descend from that insect.

Let us apply these considerations to disparity among the Burgess arthropods. First (and most important), genealogy is one property and disparity—if it is a property at all—is another. We need shared derived traits to reconstruct evolutionary relationships, but there is no reason to suppose that these have any special significance for disparity. Second, segment and appendage patterns may not tell us much about the genealogical relationships among the Burgess arthropods, because characteristics that are labile at one period can become more conservative. Finger number is now, and has been for many millions of years, fairly conservative within the vertebrates. One of the important and fairly widespread homologies among tetrapods is being pentadigital. This character stabilized quite early. But, as Gould himself points out, early tetrapods were not largely pentadigital (Gould 1993a). So it is quite likely that very early in tetrapod evolution, finger number would not have been a good trait to use to map evolutionary relationships. It later became a good index as the alternative versions of tetrapod limbs began to disappear and variation from the standard pattern became informative. Presumably, though, the value of finger number as an index of disparity has not changed.

In the same way, features of living arthropods that now index long histories of separate evolution may not have the same significance in early arthropods. If we found a living arthropod that differed in structure from all the other living groups, we would think that we had discovered a whole new limb of the tree of arthropod life. We would be reasonably sure that that branch had a deep history, for the features that now identify the great arthropod clades do not change without world enough and time. Many of them, very likely, have become developmentally entrenched over the last 540 million years (10.5). So variation from the standard mix-and-match arthropod anatomy found in a living arthropod would be good evidence of long genealogical separation from the other clades. But though the discovery of an arthropod alive today with the limb configuration of *Opabinia* might indicate a very long period of distinct evolutionary history, we do not know that *Opabinia* was more than a speciation event or two (or even a few hundred) away from, say, a Burgess crustacean. In an exceptionally bold version of this idea, Ohno (1996) argues that the animals of the Cambrian explosion all had more or less the same genome, varying in their developmental expression of common genetic resources.

Gould's critics think that arthropod segmentation and appendage patterns are like finger number in the tetrapods. Since the Cambrian shakeout of arthropod design, these patterns indicate membership in the four great arthropod clades of chelicerates, crustaceans, trilobites, and uniramians. But it does not follow that early in arthropod evolution, these traits would be good guides even to the branching pattern among the early arthropods, let alone their disparity. Thus, as Ridley (1990, 1993a) argues, it would be a "retrospective fallacy" to use features that now index major arthropod lineages to infer greater Cambrian disparity. There are more segmentation patterns in the Burgess fauna, and a more eclectic distribution of legs and feelers on those segments. But the fact that segmentation pattern *now* serves as a means of distinguishing chelicerates from trilobites does not show that it is a trait important in itself for measuring disparity, nor that it was then genealogically informative.

These ideas have sparked a lively debate. Some of the weirdest Burgess organisms have been redescribed. But some are still decidedly odd—odd enough, Gould suggests, to make the idea of greater initial disparity stick. So he still hankers after a "diversification and decimation" model that retains the idea that the Burgess fauna really did occupy more morphospace than the descendant fauna. But he accepts that nothing in the early Cambrian corresponds to the great clades—trilobites, chelicerates, uniramians, crustaceans—of the subsequent arthropods: "all characters are similarly labile in the Cambrian. . . . Nothing defined stable clades back then, because nothing had stabilized in this great era of experimentation and lability" (Gould 1991, 416). In an important sense, there were no "higher taxa" in this experimental period. We can see the changes since the Cambrian as the invention of "the basic arthropod body plan." Developmental programs became entrenched, and hence formerly labile morphological patterns have been frozen. The broad sweep of evolutionary history is the result of developmental stabilization, which constrains the ways in which morphospace can thereafter be explored. Reduced disparity is a consequence of developmental freezing plus extinctions. It is not the result of the extinction of early-evolving higher taxa, early great branches in the tree of life. In the initial period of diversification, there were many twigs, but no great branches.

If disparity really was maximal early in the Cambrian, the received view would face a potential challenge: What role did selection play in generating that early disparity, and what role did it play in reducing it? Gould thinks that selection played a relatively minor role both in generating and in reducing disparity. So he thinks that the explanation of a critical feature of the tree of life is essentially independent of selection. However, the idea that evolution since the Cambrian has resulted in a loss rather than a gain of disparity remains problematic. It is not at all obvious that disparity is an objective feature of life at a time. Even if it is an objective property of a biota, we do not know how to measure it. In particular, we cannot simply use either classic taxonomy, with its higher taxa, or cladistic systematics for the job. At best, the shape hypothesis of figure 12.3 remains unproven.

12.4 Contingency and Its Consequences

A central theme of Gould's recent work has been the contingency of life's history. He emphasizes the importance of particular events in shaping the history of life, and of the unpredictability of their consequences. Though some features of life are predictable from general laws of physics, chemistry, biomechanics, and the like, many are not. Thus he writes:

> The question we face is one of scale or level of focus. Life exhibits a structure obedient to physical principles. . . . I suspect that the origin of life on earth was virtually inevitable. . . . Much about the basic form of multicellular organisms must be constrained by the rules of construction

and good design. The laws of surfaces and volumes . . . require that large organisms evolve different shapes from smaller relatives in order to maintain the same relative surface area. Similarly, bilateral symmetry can be expected in mobile organisms built by cellular division. . . . But these phenomena, rich and extensive though they are, lie too far from the details that interest us about life's history. Invariant laws of nature impact the general forms and functions of organisms; they set the channels in which organic design must evolve. But the channels are so broad . . . (Gould 1989, 289)

Even if we add to our arsenal of general principles those internal to biology, the importance of particular events is not lessened.

Contingency, however, is more than the importance of unique and unpredictable events—meteor impacts, volcanic upheavals—on the growth of the tree of life. Gould frequently uses the metaphor of "replaying the tape" to force home the importance not just of vast upheavals and cataclysms, but also of the apparently insignificant. History can turn on particularity at fine grains: "any replay, altered by an apparently insignificant jot or tittle at the outset, would have yielded an . . . outcome of entirely different form" (Gould 1989, 289).

For instance, perhaps tetrapod evolution required a precise sequence of mutations in ancestral vertebrates, each of which was independently improbable. It depended on a chain of lucky breaks. If that were so, then tetrapod evolution would have been fragile; a cosmic ray in the wrong place at the wrong time could derail it. So we take the contingency hypothesis proper to be the idea that important features of the history of life are not *counterfactually resilient*. Within evolutionary biology there is no robust process explanation of their evolution. Their existence depends on the precise details of the history that produced them—on the actual sequence of events in their evolution. If a minor deviation from the actual circumstances under which tetrapods evolved would have derailed that evolutionary process, then the evolution of the tetrapods is not counterfactually resilient. In contrast, Gould thinks that the origin of life *is* counterfactually resilient. In any broadly similar environment, life would have evolved.

These ideas are intended to defend the importance of paleobiology as a discipline by insisting on the critical importance of particular events that only that discipline can reveal. But too strong an insistence on contingency dooms paleobiology too. Historical narrative and historical explanation are selective. In giving an account of some particular episode in evolutionary history, we suppress some details as unimportant and highlight others. If literally everything matters—if every element in the chain of causation is of in-eliminable significance—then the paleobiologist's job is hopeless. If human evolution depended on every meal eaten by our actual ancestors—if we would not have evolved if these meals had not taken place *exactly* as they did—then we would never be in a position to explain human evolution. Those lost details would be every bit as critical as the general features of paleoecology and biogeography that, with luck, might be preserved. Obvi-ously, no one defends a contingency hypothesis that strong. So there is no single "contingency hypothesis," but rather a family of related hypotheses. We think there are at least three "contingency hypotheses," and that they have different implications for evolutionary theory.

The Contingency of Specific Taxa

In one common view of species formation, new species are born from the isolation and then divergence of small chunks of their parent populations. Small, peripheral, isolated populations are the founding stock of most spe-cies. But most small, peripheral, isolated populations do not turn into new species. Most of them go extinct. For such populations to survive, they need a lot of luck. If this view is right, then the existence of specific taxa is contin-gent. If we trace the history of the malarial mosquito, *Anopheles maculipennis,* there will be a time when its population was small and a minor bit of bad luck—a storm, a dry spell, a short-term dearth of hosts—would have fin-ished it off. The birth of specific taxa is not counterfactually resilient. Of course, from the fact that the evolution of *Anopheles maculipennis* is contin-gent, it does not follow that the existence of some very similar species is contingent. Perhaps its ancestor was widely dispersed, and populations at its periphery frequently became isolated. Sooner or later one of the fragments was likely to survive to become a new species. This brings us to a second contingency hypothesis.

The Contingency of Adaptive Complexes

A second hypothesis is the idea that the evolution of adaptive innovations is contingent. Bats, probably, are a monophyletic group: there is a single an-cestral species from which all the bats derive, an ancestor of all the bats and none but the bats. But the bats are also united by having a distinctive adaptive complex: powered flight based on a distinctive wing structure. One branch of the bats has another distinctive adaptive complex: echolocation. The evolution of true bats is contingent, for the evolution of the ancestral bat species is contingent. But it by no means follows that the evolution of the bats' distinctive adaptive complex is contingent. Perhaps some other twig of the mammalian branch of the tree of life would have evolved batlike characteristics had that ecological space not already been occupied by the bats. Bats probably evolved from tree shrews (or possibly lagomorphs—the group that eventually gave rise to the rabbits and hares: Bailey, Slighthorn, and Goodman 1992). But suppose that this evolutionary development had never taken place. Suppose that instead, in a batless world, a flying echolocating insectivore evolved from a night-hunting rodent. That would not be the evolution, in that world, of a bat. But it would be the evolution in that world of the same adaptive complex that the bats invented in our world. So the contingency of specific taxa does not imply the contingency of adaptive complexes. Conway Morris (1998) is very skeptical about Gould's ideas on contingency because he thinks that convergence is pervasive. So the contingency claim he objects to is this one, the contingency of the evolution of adaptive complexes.

Contingent Explorations of Morphospace

If there are powerful historical and developmental constraints on evolution, then once those constraints evolve, the persistence of (say) the crustacean body plan without radical revisions is *not* contingent. In this "constraints view," the volume of morphospace that ever will be explored was largely determined by its initial exploration in the Cambrian explosion. However, it might be that the particular products of that period of experimentation were contingent. The Burgess experimentation with arthropod design was exuberant, but perhaps even in the Cambrian there were a lot more possible arthropod patterns than actual ones. In this picture, the experiments tried were a tiny fragment of the ones that might have been tried. Perhaps each early arthropod pattern was an unlikely accident, easily derailed by any trifling change in the causal sequence that created it. So the tree of life has explored only a small fraction of the ways of being an arthropod, a mollusk, a chordate, and so on, and those explorations depend on the exact causal history of the initial experimentation period. Contingency infects the initial laying down of the trails through which morphospace has subsequently been explored.

The contingency hypothesis about specific taxa is probably true, but it is an implication of the received view, not a revision of it. On the other hand, a contingency hypothesis about the evolution of adaptive complexes does threaten a version of evolutionary thought. It undercuts the adaptationist idea that traits are primarily or robustly explained by their selective environment. To illustrate, let's consider Jared Diamond's hypothesis that chimpanzees' large testicles are an adaptive response to sperm competition (Diamond 1992, 72–75). Of course, other factors play a causal role—the historical and developmental inheritance of the chimp lineage are necessary too—but within limits, the adaptationist expects that though variations in history and development might affect the details of the evolutionary trajectory through which large testicles have evolved, they would not change the qualitative outcome. In sections 10.6 and 10.7 we discussed a number of ways of unpacking this adaptationist idea. Orzack and Sober aim to make it precise through their notion of a censored model, one that excludes all forces but selection in predicting the evolutionary trajectory of a population yet still gets chimp testicle size right (Orzack and Sober 1994). Presumably, though, if this evolutionary response is contingent, then small changes in the environment, in the population structure, in the genetic variation that happened to be available, or in the developmental mechanisms that previous history had made available could have derailed this evolutionary response. If tiny changes in these factors would have resulted in very different testicles, then a "censored model" of chimp testicle evolution, one ignoring all but natural selection, could lead to wildly inaccurate predictions about testicles. The evolutionary trajectory would not be robust.

A similar message emerges from the comparative interpretation of adaptationism we explored in section 10.7. According to that interpretation, the adaptationist conjecture is that the developmental and historical preconditions for many evolutionary changes remain stable over considerable periods of a lineage's history. But, of course, the fine-grained details of developmental mechanisms, genetic variation, and population structure do not remain stable. So the contingency hypothesis about adaptive complexes is inconsistent with that adaptationist conjecture. Furthermore, if adaptive change were sensitively dependent on all these factors and more, we would not expect the comparative method to provide good evidence of adaptive evolution. Imagine, as we suggested in section 10.7, that flightlessness on islands is adaptive for rails. Even so, if that adaptive evolutionary change is contingent—that is, if the evolution of flightlessness is sensitively dependent not just on selection, but on much else as well—we would expect at best a very poor correlation between island life and flightlessness. Too much would have to be just right for flightlessness to evolve for there to be a clear covariation between flightlessness and island life.

So the idea that adaptive evolution is contingent really does seem inconsistent with any version of empirical adaptationism. But it is far from obvious that this hypothesis is true. The truth is quite likely to be heterogenous: some adaptive complexes might depend on a very lucky sequence of accidents, without that being true of others. The soaring scavenging lifestyle has evolved at least twice in bird groups: the Old World vultures and the New World vultures evolved their very similar adaptations independently. The distinctive beak, reinforced head, and peculiar tongue structure of wood-peckers has evolved only once, even though in Australasia there are other birds that nest in tree hollows and make their living gleaning insects from tree trunks.

Unfortunately, it is hard to test the idea that the evolution of a particular adaptive complex is contingent. Parallel evolution put some limits on contin-gency. The fact that many species of eucalypts were able to evolve distinctive adaptations to drought and fire shows that this adaptive shift is fairly robust, for it did not depend on a very specific set of factors that applied to only one population. It is tempting to draw a similar conclusion from convergent evo-lution. Vision has evolved so often that it is tempting to suppose that the preconditions for its emergence cannot be very specific. But we need to be very cautious about this line of thought. First, no lineages are fully indepen-dent of one another. The double invention of soaring and scavenging in birds might therefore depend on the prior invention of developmental or genetic resources deep in the bird tree and inherited by both Old and New World vultures. The conservation of genetic systems over hundreds of millions of years makes this a real possibility, not an idle speculation. There is increasing evidence to suggest that the invention of eyes in many different lineages does depend on conserved and common genetic resources (Gould 1994; Quiring et al. 1994). Second, convergence differs from the paradigm cases of parallel evolution in that only the victors are visible. In our eucalypt example, we have some prospects of at least estimating the number of eucalypt species that failed to evolve adaptations to aridity by counting the species that exist only as relicts in moist areas. Perhaps many bird lineages would have been advan-taged if they had been able to respond to selection for soaring and scaveng-ing. But only the New and Old World vultures did. If so, they might exem-plify contingency without our knowing it.

So a contingency hypothesis about adaptive complexes really does chal-lenge an important conception of evolution, but not in a way that lends itself to empirical testing. What would we expect the tree of life to be like if a contingency thesis about the evolution of adaptive complexes were true? Is the world we have the one we would expect? It is obvious that this question is extraordinarily difficult to answer. Similar epistemic problems haunt the idea that the exploration of morphospace is contingent. In sum, we think that Gould's ideas about evolutionary contingency are close relatives of his earlier criticisms of adaptationist evolutionary biology. One contingency hypothesis is relatively uncontroversial, but it is not a challenge to the received view. Other contingency hypotheses seem to be relatives of well-known worries about the importance of adaptation in evolution. We think those worries about adaptation are important, but we do not see how revisiting them in the language of contingency makes them theoretically or empirically more tractable.

12.5   Mass Extinction and the History of Life

The explosive diversification of animal life early in the Cambrian was one of the most dramatic events in life's history. Equally dramatic were several great disruptions of that history. For those unfamiliar with it, geological history is divided into a baffling array of incomprehensibly named eras, periods, and epochs (see figure 12.2 above). These distinctions are not arbitrary. The units of geological time are identified by their characteristic biotas—the distinctive assemblies of life forms revealed by their fossil records. One era ends and another begins when that biota is transformed, with many of its most characteristic elements disappearing and other forms appearing. Extinction in itself is not unusual; species come and go in the normal course of nature. Presumably, though, the extinction that occurs as one era ends and another begins is not routine. It is characterized by a comparatively rapid and more or less coordinated shift from one biota to another, marked by the disappearance of many species. The major transitions in life's history are defined by *mass extinctions,* not routine or *background extinction.*

The nature and importance of mass extinctions is a subject of deep debate within paleobiology. In one view, mass extinctions have profound historical significance, for the disparity of life over time depends on the extirpation of dominant groups in mass extinctions. If the ancient rugosian corals had not perished at the end of the Permian, modern corals would never have evolved. Erwin (1993a) argues that the reduction of the snail and echinoderm lineages at the end of the Permian permanently changed the character of those groups, for their post-Permian radiation depended on the few surviving lineages, and hence all modern lineages inherited the distinctive characteristics of those survivors. The most famous of all these ideas is the hypothesis that only the extinction of the dinosaurs at the end of the Cretaceous period (often known as the *K/T boundary*) allowed the mammals to diversify.

This view of the importance of mass extinctions confronts us with two questions. First, is it right? Perhaps this line of thought exaggerates the significance of these events. Second, supposing that it is right, does it undercut the received view of evolution? Gould, for one, has argued that the received view is committed to the idea that the history of life is just an accumulation of evolutionary episodes at the scale we can observe. Nothing new characterizes life at large scales in long runs. What we see there is just an extrapolation from, and many iterations of, life at our scale. First we consider the claims about the effects of mass extinctions on life's history, then we turn to their theoretical significance.

The importance of mass extinction in life's history has been challenged in two ways. The first looks at the contrast between mass and background extinction. The more a mass extinction can be shown to be *sudden,* a result of sharp and unpredictable changes in conditions, the more robust is this contrast. So it matters whether the meteorite impact at the K/T boundary was just one of many factors involved in that transition. It matters whether the groups that went extinct—pterosaurs, dinosaurs, ammonites, and the like—were already showing a reduction in range and diversity before the impact. If they were already declining, and if the impact was just one of many factors involved in their demise, then the contrast between mass and background extinction loses its sharpness. The factors acting during the period of transition may have been qualitatively similar to those acting before the transition, and unusual only in their intensity, or in that several were operating at once. In that case, a mass extinction episode might just accelerate or accentuate evolutionary business as usual, rather than radically altering the course evolution would have taken in its absence. On the other hand, if the K/T extinction was the result of the meteorite impact and nothing but the impact, and if the groups that disappeared were flourishing until the time of impact, then the distinction between mass and background extinction is robust.

A second line of thought, while accepting the reality of mass extinctions, suggests that they only accelerate a process that is already under way. Sepkoski (1984) explores the possibility that the great shifts in life over the last 600 million years might be a reflection not of great environmental upheavals, but of factors internal to lineages. Imagine two families reaching an island at the same time. One, living by the maxim "live hard, breed fast, die young," might initially spread faster than the family characterized by a more conservative lifestyle, but that second family might well come to dominate the island further down the track. Sepkoski has shown that if two lineages differ in their intrinsic rates of speciation and extinction, a fast speciation/fast extinction biota can give way to a slower speciation/even slower extinction biota without that change having any unitary external cause. The turnover of lineages is thus a consequence of intrinsic features of the lineages. He has developed models that suggest this might be the case for the replacement of the Cambrian by the Paleozoic and then by the Modern fauna. In each case, he suspects that the faunas that dominated the next phase of the earth's history were already becoming more significant before mass extinction intervened. For example, the Modern fauna was expanding before the end of the Permian, so the turnover may have been accelerated by the extinction, but not caused by it.

We are skeptical, for we are unconvinced that there is a distinct chunk of the tree of life that can be thought of as the Paleozoic fauna, or the Mesozoic fauna. If we could see the diversity of Permian life before the Permian mass extinction, is there anything that would unite the living forms that we retrospectively label "the Modern fauna"? Sepkoski's examples of the Cambrian fauna include trilobites and inarticulates (a kind of brachiopod; these, in turn, are members of a shelled invertebrate phylum that externally look like bivalves). His Paleozoic fauna is exemplified by articulates (another kind of brachiopod), cephalopods, crinoids (a stalked, immobile kind of echinoderm), and ostracodes (a kind of crustacean). The Modern fauna continues to include a wide range of marine invertebrates, but takes in various fish groups, mammals, and reptiles as well as gastropods (snails), bivalves, and lobsters. These different groups do not seem unified in any obvious way. So it's hard to see how the members of the Modern fauna could have any biologically significant property in common that explains their future dominance. Rather, they seem to be just a collection of lineages that happened to survive and diversify at the same time. If the Modern fauna or the Paleozoic fauna were a clade, or something close to a clade—a well-formed branch of the tree of life—then differing speciation and extinction rates might characterize those two groups despite their dispersal through many distinct geographic and ecological zones. For a property internal to the lineage might be inherited through the tree: differing turnover rates might result from genetic and other developmental mechanisms projected through the tree from an ancestor to a unified class of descendants. But since the Paleozoic fauna and Modern fauna are polyphyletic, there is no way a distinctive set of intrinsic speciation and/or extinction rates could be inherited by those groups alone. So if their extinction/speciation rates differed, it would have to be because of some common ecological feature of the Modern or of the Paleozoic fauna. But those faunas are no more ecologically unified than they are genealogically unified.

The mass extinction debates are of great intrinsic interest, but we doubt that they are critical to the central theme of this book, the adequacy of the received view of evolutionary theory. Gould would disagree. He thinks these catastrophic external shocks to the system are important for three reasons. First, in mass extinction episodes, the rules of the game change. Species go extinct or survive for reasons unconnected to their histories of selection. We might suppose that mass extinction episodes are periods in which selection acts with special intensity, but in the normal direction—that the race is still to the swift and the hunt to the strong, though only the very strongest and swiftest survive. To the contrary, Gould, Raup, Ward, and others argue that mass extinctions differ from normal periods not just in the number of extinctions, but also in their causes. Second, episodes of mass extinction have profound effects on the world's biota. If the dinosaurs did not fade away, but instead were slaughtered in some unpredictable ruin, then only that catastrophe allowed mammals to flourish and diversify. The variety of life *over time* depends on catastrophic shocks making room for new experiments in life. Third, and most important, mass extinctions mean that the history of life cannot be understood by extrapolating from evolutionary change over ecological time frames in local populations. The importance of these large rare events dooms this *extrapolationist* strategy. Gould (1995) argues that extrapolation is central to the received view, which is undercut by a recognition of the significance of mass extinction.

We agree that a demonstration of the significance of mass extinction would doom a very ambitious version of extrapolationism. For that demonstration would reveal the significance of rare events: events not visible at ecological temporal and spatial scales. The importance of mass extinction thus vindicates paleobiology's essential role within the discipline of evolutionary biology. If there is no understanding life's history without understanding the role of these huge but rare events, then large-scale change is not just an accumulation of small-scale change, and cannot be studied in the same way.

Nonetheless, a fall-back idea is still alive. In thinking about evolution, we can distinguish between the mechanisms of evolution and the environment in which those mechanisms operate. The mechanisms include mutation and recombination, mechanisms that generate diversity. They also include selection and drift, mechanisms through which diversity is shaped. And they include migration and the other mechanisms that impose structure on populations of organisms, often dividing populations into subpopulations. We saw the importance of this structure in chapter 8. These mechanisms also include, as we noted in chapter 5, the many interactions between genetic and nongenetic mechanisms in heredity. One natural way to understand extrapolationism is to see it as the claim that the mechanisms we can observe in natural (and experimental) populations on ecological time scales give us a complete inventory of the mechanisms of evolution. In different environments—including the very exceptional ones characteristic of mass extinctions—those mechanisms interact with one another and the environment to generate *outcomes* very different from those we are in a position to observe. But they do so by the operation of normal mechanisms in an abnormal world. So conceived, extrapolationism is consistent with the idea that mass extinctions fundamentally reshape the tree of life.

In section 9.4 we discussed the idea of species-level selection. There can be species selection if properties of the species itself (rather than its component organisms) are relevant to extinction and survival. If there are such selection processes, then extrapolationism is indeed false. But, as we noted in section 9.4, the existence of such processes remains controversial. Mass extinction by itself, however important, is no threat to the received view of evolution. Mass extinction undercuts it only if it provides crucial evidence for species selection or some other high-level mechanism.

12.6 Conclusions

In our view, the debates opened up by Gould in *Wonderful Life* remain far from settled. It remains to be seen whether the concept of disparity can be tamed well enough for Gould's hypothesis on the shape of the tree of life to be formulated precisely. For only then can it be tested. If it were tested and confirmed, explaining that shape would then be a major challenge to every view of evolution. The ideas involved in Gould's contingency hypotheses are less well developed, and perhaps even more empirically intractable. The most progress, we suspect, is being made on mass extinction. There seems little doubt that something extraordinary closed out the Permian, and equally little doubt that a severe asteroid impact played at least some role in the K/T extinction. So the evidence for mass extinction's importance strikes us as persuasive, and the considerations in favor of some kind of high-level selection are at least suggestive. So extrapolationism, even as we understand it, may be in trouble.

However, we would like to close our discussion of these issues with a general caution. Gould's case for disparity and contingency depends on his history of multicellular life—primarily of animal life. Single-celled organisms and plants are left out. Yet, as we saw in section 12.1, Gould himself has rightly cautioned us against confusing life-like-us with life. If our concern is with typical evolutionary patterns and the processes that generate them, we should be very cautious in resting our case on metazoan evolution. Vascular plants evolved on land after the Cambrian, so plant diversity did not peak in the Cambrian and then decline. Moreover, the typical organism is a bacterium. The different basic metabolic systems found among the bacteria suggest that their disparity is very great indeed, and we suppose otherwise only because of the bias our own size induces. If the received view fits bacterial evolution well, one might well argue that it is a good picture of evolution, whether or not it fits the sideshow of metazoan evolution.

Further Reading

12.1 For a very readable and lively take on the role of progress in evolutionary thought, see Ruse 1996. Nitecki 1988 is an important collection on these issues, and they are covered well in section 9 of Hull and Ruse 1998. The issue of complexity is taken up in the context of progress by Gould; see especially Gould 1996a. Gould and Dawkins have recently exchanged blows on these issues; see Gould 1997 and Dawkins 1997. McShea (1991, 1994, 1996b) discusses complexity, but without using it as a surrogate for progress.

12.2 In discussing the Cambrian explosion of multicellular life and its significance, we have focused on the theoretical and conceptual issues, but the empirical ones are no less contested. The status of this "explosion," and the relation of the Cambrian fauna to its immediate predecessor, the Ediacaran fauna, remain matters of acute controversy. For no superb Burgess Shale-quality fossils of the Ediacaran fauna exist, and their biological affinities remain a matter of debate. In some views, the Ediacaran fauna is ancestral to some surviving multicellular lineages. In others, it represents a different and extinct experiment in complex life. Gould (1996b,d) reiterates the case for the sudden and unique character of the Cambrian explosion. Erwin (1993b) and Knoll (1994) reinforce this conclusion by arguing that the period saw an expansion in the diversity of microorganisms as well. But Bell (1997) and Cooper and Fortey (1998) review molecular evidence that suggests that the phyla came into existence well before the Cambrian explosion; the earliest known fossils would then be much younger than the earliest members of the lineages they supposedly represent. Moreover, Knoll (1996) suggests that the paleontological evidence is not so clear-cut either; in particular, the Ediacaran fauna may well have overlapped in time with its supposed successors.

Gould's main defense, both of the revolutionary impact of the Burgess Shale fauna on our conception of the overall shape of evolutionary history and of its implications for evolutionary theory, is his *Wonderful Life* (1989). For other recent reviews of this critical period in the evolution of animals, see McMenamin and McMenamin 1990, Briggs, Erwin, and Collier 1994, and Signor and Lipps 1992, which has a good introductory chapter to a rather technical collection on the topic. The McMenamins accept Gould's view that disparity after the Cambrian explosion was indeed very great, suggesting that the Cambrian fauna might represent as many as 100 distinct phyla (168), but they offer a more adaptationist explanation of its origin. They take it to be the result of both a great increase in nutrient availability and the evolutionary fallout of the invention of predation. So for them the explosion is caused not by a relaxation in competition (as both Gould and Whittington tentatively suggest), but by the evolution of fear, panic, and loathing as predation bit into a pre-Cambrian idyll. This scenario, the "Garden of Ediacara," is insightfully discussed in Raff 1996, a well-developed alternative synthesis. Whittington (1985) gave his own take on the Burgess somewhat earlier. His views on the evolutionary significance of the Burgess are similar to those of Gould, though without defending a contingency claim about the way disparity shrank. Conway Morris (1998) has recently given his overview.

12.3 For a very succinct table showing the disparity of Burgess arthropods given contemporary standards for measuring it, see Baird 1990. See Ridley 1986 for a forceful statement of the idea that judgments of similarity and dissimilarity are ineliminably subjective. For a more recent review, see de Queiroz and Good 1997. For a review of these debates, see Sober 1994 and the other works listed in the further readings for section 9.3.

The problem of defining disparity is well aired in a series of papers in the journal *Paleobiology* (Gould 1991; McShea 1993a; Ridley 1993a; Gould 1993b). Of these, the most important are Gould 1991 and McShea 1993a. Wagner (1995b) returns to the same issue with an attempted solution in his very technical paper. In a series of papers, McShea tries to tame one element of disparity, namely, morphological complexity, with the aim of rendering empirically tractable the idea that complexity has increased over time (McShea 1991, 1992, 1994, 1996b). For a crisp overview of the changes in species number over time, see Signor 1994.

Gould is certainly sensitive to the problem of counting or measuring disparity. In a footnote to *Wonderful Life,* (Gould 1989, 209), he worries about the problem without doing much more than relying on the educated judgment of evolutionary taxonomy, but he returns to it later (Gould 1991; Foote and Gould 1992). In two papers in *Eight Little Piggies* (Gould 1993a, chaps. 23 and 24; see especially the diagram on page 339), he continues to defend the idea that we can define the disparity of a group of lineages just in terms of their morphological variance. Briggs himself is rather skeptical about claims that the Burgess fauna shows great disparity (see Briggs, Fortey, and Wills 1992a,b). Conway Morris (1993) also gives his views, but his focus is fitting the Burgess fauna into a coherent phylogeny of the early animals, rather than disparity as such. The same is true of his discussion of disparity (1998), in which, for example, he argues that since *Anomalocaris* probably resembles an ancestor of the arthropods, it cannot be very disparate from them. This argument seems to conflate phylogenetic questions with those of disparity. However, in an earlier paper (1989), he defends a view very similar to Gould's, emphasizing both the disparity of the Cambrian fauna and the contingency of survival and extinction. Wagner 1995a is an interesting but very difficult attempt to resolve problems of measuring diversity created by the uncertain, confusing, and inconsistent taxonomies of early animals. Another important but difficult attack on these issues is by Foote (1993).

12.4, 12.5 The idea that evolutionary biology is an essentially historical discipline has received a good deal of attention in the literature of philosophy of biology, often through the problem of whether there are "laws of nature" that are specifically biological. For recent works in this genre, see Mayr 1988, Dupré 1993, Rosenberg 1994, Depew and Weber 1995, and a 1996 symposium of the Philosophy of Science Association, which featured papers by Beatty, Brandon, Sober, and Mitchell and has been published in its entirety in *Philosophy of Science,* in the supplement to volume 64, no. 4 (1997). But the idea of contingency—that evolutionary history sometimes depends on fine-grained details of its particular historical causes—is a stronger claim. This claim has not received very much attention in the literature, perhaps because it is sometimes conflated with the problem of the importance of mass extinction. Not for the only time, Lewontin (1966) anticipated an important debate. Oyama (1995) takes up Gould's ideas with the dual aim of (1) arguing for a continuum between more and less constrained processes rather than a dichotomy and (2) arguing that developmental biology, not just evolutionary biology, needs some appropriately refined descendant version of the concept of contingency. One of us (Sterelny 1996a) takes up the idea of contingency in relation to adaptationism. Dennett (1995) discusses it very skeptically, but treats it as the hypothesis that failure and success are due to chance, not adaptation. We do not adopt this reading because chance extinction, strictly interpreted, does not make sense of the claim that the patterns of failure and success are retrospectively intelligible. For more on the issue of convergence and its interpretation, see Mueller and Wagner 1996 and Burian 1997.

In contrast, a great deal has been written on mass extinction and its importance. For a splendid review of the largest of all extinctions, see Erwin

1993a, especially the concluding chapter, in which he defends the idea that mass extinctions have reshaped biological history. See Glen 1994 for a collection focused on impact theories of mass extinction. Ward (1992, 1994) has written two good popular reviews of mass extinctions and their effects (the second, however, uses ancient mass extinctions as a means of focusing attention on human-caused extinctions). Leakey and Lewin 1996 is in the same genre, and is heavily indebted to Gould and to Raup in its conception of extinction and its significance. Raup (1991) presents an extensive defense of mass extinction's importance and of the idea that "different rules" govern survival through such periods. Jablonski (1986a,b, 1991) defends the same idea. For Gould's recent defenses of the difference between background and mass extinction, and of the central importance of the latter, see chapters 21–24 of Gould 1993a. Chaloner and Hallam 1989 is a fine survey of different opinions on extinction and its significance: see especially the papers by Maynard Smith, Hoffman, Jablonski, and Benton. Sepkoski (1989) and Jablonski (1991) review the empirical issues concerning the intensity and periodicity of mass extinction. Hoffman (1989) has been continually skeptical about mass extinction and the alleged difference between mass and background extinction; see also his contribution to Chaloner and Hallam 1989. For the latest on the K/T extinction and the end of the dinosaurs, see Ward 1995 and MacLeod 1996; for an excellent review of this extinction, see Archibald 1996.

Evolution and Human Nature

13

From Sociobiology to Evolutionary Psychology

13.1 1975 and All That

In 1975, E. O. Wilson published *Sociobiology: The New Synthesis,* a sweeping overview of the evolution of social behavior. It finished with a bold, speculative, and ambitious attempt to apply adaptationist reasoning to human behavior. The response was extraordinarily varied, ranging from high praise to venomous attack. On the face of it, the venom of the negative response was puzzling. We are evolved organisms. Understanding our evolutionary history should help us to understand both what we have in common as humans and our differences. So the attempt to apply evolutionary theory to us cannot be just wrong. Yet human sociobiology has a dark reputation in many circles. We think there are some intrinsic problems impeding the development of sociobiology, but its development has also been impeded by some serious misunderstandings and misapplications of evolutionary theory. So we begin with an initial sketch of this landscape, outlining both the unavoidable difficulties and the avoidable confusions.

Humans as Experimental Animals

Homo sapiens would be a very poor choice as an experimental organism. We are long-lived, so our generations turn over very slowly. We contrast badly with fruit flies in this respect. Humans are expensive to keep in captivity; worse still, there are very considerable restrictions on the experimental regimes to which they may legally be subjected. So, though field data about our species is rich (though perhaps not always reliable), experimental data is restricted in many important ways. This is a real impediment to the ambitions of sociobiology.

Humans as Evolutionary Orphans

Most of our immediate relatives are extinct. That makes it hard to test evolutionary hypotheses about our psychological and social traits by the comparative method, which uses the distributions of traits across a group of related species to infer when and why those traits arose (10.7). This is another way in which the study of human evolution really is more difficult than the study of many other species.

The Problem of Changing Environments

Virtually all humans now live in environments that differ in important ways from the environments in which we evolved. The foods most of us eat are unlike those yielded by hunter-gatherer lifestyles. Selective breeding has greatly changed the food species we use, so our food is very different even from that consumed by early farmers. No doubt our biochemical environment has changed in many ways. The wide availability of artificial light has changed our daily life rhythms. The social groups in which we now live differ in size, and perhaps in composition, from those in which we evolved. Of course, there was no single ancestral human environment. For much of human evolution we have lived in a wide variety of physical and social environments. But the range of our ancestral environments probably overlapped very little with the current range.

This difference between our past and present environments has at least two important consequences for the application of evolutionary theory to human nature. First, it makes the link between selective history and current utility fragile. Suppose that we discover that the fantail, a New Zealand flycatcher, chooses nest materials and nest sites that result in well-camouflaged nests. The birds' current nesting practices contribute to their fitness by making their nests difficult for predators to find. We would be tempted to infer that those nesting habits evolved because they helped protect fledglings from searching predators. But even in this case, such inferences are somewhat risky. A behavior can evolve for one reason and be adaptive now for another (10.1, 10.3). But the inference from current utility to evolutionary cause is especially chancy if the environment has changed in important ways. Zoos, botanic gardens, and late-twentieth-century human societies are risky settings in which to study evolution. Even if we have a good understanding of the effect of behavior on fitness in these new settings, it is very hard to use that understanding to confirm a hypothesis about previous evolutionary history.

Second, environmental change can change developmental outcomes. According to the interactionist consensus (5.3), our development depends on a complex matrix of developmental resources, not just on our genes. So alterations in the social and physical environment can result in new phenotypes. Taking the simplest example, European populations are generally much taller today than 100 years ago, because they eat better. The question of whether their current height is evolutionarily optimal simply does not arise. This lesson applies to behavior as well. Bonobos immersed in a sign language-using environment spontaneously learn to use signs themselves. They develop new behaviors—and there is every likelihood that we have done the same. So not only is it unsafe to assume *adaptive stability* over significant environmental change, it is even unsafe to assume *phenotypic stability* (Alexander 1990; Turke 1990).

These three problems pose unavoidable, intrinsic problems for theories of the evolution of human behavior. Indeed, *evolutionary psychology,* the modern descendant of sociobiology, is to a significant extent a response to these problems stemming from the change in human environments.

What Should We Study?

What features of an organism are its traits? That is, what features of an organism have an evolutionary history to call their own? We have already met the mandrill and its color scheme (10.3): males have electric blue muzzles and a matching blue on their buttocks and genitals. Should we consider these colors part of a single evolving trait—the overall mandrill color scheme—or do these colorings of these particularly salient parts of the male monkey have evolutionary histories to call their own? This is no simple question. One of the worries about adaptationist conceptions of evolutionary history is that they underestimate the extent to which different properties of an organism form a linked evolutionary system (10.3, 10.5).

Nonetheless, we conjecture that educated biological judgment is often quite sufficient. No biologist would think of treating the several orange stripes and black stripes on a tiger as separate traits with separate adaptive and phylogenetic histories. The whole striping pattern is a single trait. With human behavior, however, it becomes very difficult to specify the appropriate grain of analysis. Should we think of human aggression as a single trait? Perhaps, instead, there are many forms of aggression with different histories, just as in chimpanzees there are two quite different dominant and subordinate threat displays. If so, then "aggression" names a bundle of traits with no more than a superficial similarity. In that case, evolutionary hypotheses about aggression will fail by being at too coarse a grain of analysis. Different patterns of aggression will have evolved at different times for different reasons, and may develop in very different ways as well. Hypotheses can also fail by being too fine-grained. If aggression occurs as part of a tit-for-tat strategy of doing as you are done to, then it is part of a single trait that will produce cooperation on some occasions and revenge on others. As we will see in this chapter and the next, this problem is a very difficult one.

Adaptation and Development: Distinct Issues

Sociobiology has often been accused of genetic determinism— of supposing, that is, that human behavioral patterns are insensitive to the life experiences of individual humans (1.4). This has led many critics to allege that socio-biological explanations are restricted to "instinctive" behaviors. But little, if any, human behavior is instinctive. Human behavior depends in complex, subtle, and sensitive ways on the environment in which humans develop. Therefore, the critics conclude, sociobiology has little to say about most human behavior.

Sociobiologists, especially in the early days, were often most incautious in expressing their views. Dawkins, for example, wrote of our "selfish genes": "Now they swarm in huge colonies, safe inside gigantic lumbering robots, sealed off from the outside world, communicating with it by tortuous and indirect routes, manipulating it by remote control. They are in you and in me; they created us, body and mind; and their preservation is the ultimate rationale for our existence" (Dawkins 1976, 21). Dawkins is no genetic determinist, yet this passage makes him sound like one. The determinist reading is so irresistible that his critics misquote him, turning "create" into "control" (Dawkins 1989b, 271).

Though some sociobiologists have drifted into genetic determinism, they need commit no such error. Randy and Nancy Thornhill, for example, have conjectured that rape by human males is a facultative adaptation to sexual exclusion, an adaptive behavioral trait, though one that is exhibited only under certain conditions (Thornhill and Thornhill 1987, 1992). To put it mildly, this conjecture faces very serious empirical problems, as the accompanying commentaries on the 1992 paper make clear. Rape is very dangerous for its perpetrators, and probably was once even more so. The chance of fertilization is low, and in the less tender-minded environments in which this adaptation is thought to have evolved, the chance that the resulting child would actually be raised would have been smaller still. But genetic determinism is not among the vices of this hypothesis. The Thornhills' claim stands or falls on selective history. They are committed to the claims that: (1) our male ancestors differed in their propensity to commit rape in certain circumstances; (2) this difference was heritable; (3) those having the propensity to commit rape in those circumstances had greater expected fitness and have thereby maintained that trait in the general population. The conjecture says nothing whatsoever about the proximate mechanism by which the trait develops. Some adaptations require very specific inputs from the local environment for their development; others do not. Social deprivation of rhesus monkeys during infancy can entirely eliminate normal play. The adult then develops without the ability (if female) to care for an infant or (if male) to successfully copulate. Maternal care and copulation are adaptations, but their development requires a richly structured social environment. Sociobiological conjectures have often been used to argue that certain social changes are impossible: to defend, for example, the idea that sex roles are fixed. But even if the adaptive conjectures are right, they lend little support to these claims about inevitability.

The link between adaptation and developmental stability is weak in the other direction, too. Evidence of developmental stability does not show that a trait is an adaptation. Insensitivity to environmental factors can result from general features of the developmental system (10.5) as well as from adaptive evolution buffering a valuable outcome against environmental disturbance. "Genetic diseases" are developmental outcomes that are insensitive to environmental change; that is, roughly, what makes a disease genetic. But they are not adaptations. Hereditary breast cancer, for example, is no adaptation. The mythical link between adaptation and unchangeability has done a great deal to muddy the waters in the debate over sociobiology.

The Fact/Value Swamp: Danger—Keep Out!

No doubt, in most research, what researchers want to be true plays a role in what they believe to be true. It's uncharitable to say so, but we suspect that hope has been rather too fecund a father to belief in the debates on evolutionary theories of human behavior. Too few of those involved seem able to resist ideology and moralizing. Sad to say, this moralizing tradition continues to this day, especially in popular presentations of sociobiology. In *The Moral Animal,* for example, Robert Wright both presents the newest brand of sociobiology and waves a disapproving finger at many aspects of contemporary life (Wright 1994). Darwin reappears as Granny. The enemies of sociobiology often agree that sociobiological hypotheses have direct implications for social policy. Wilson's sociobiology in particular was seen as threatening liberal social ideas by defending the current social order as natural and inevitable. With so much at stake, the resulting spectacle has not been edifying. Speculative evolutionary explanations—"just so stories"— have abounded. Claims about adaptive history have been conflated with ones about developmental fixity and used to denounce programs for social change. In response, sociobiology's critics have erected absurd standards of proof for any claim about human behavioral adaptation and embraced extraordinarily strong versions of environmental determinism. Too much mud has been slung; it's hard to construct good theory while mud wrestling.

In summary, some of the barriers to the application of evolutionary theory to our species are based on misapplications of evolutionary ideas. But some of these barriers are intrinsic to the topic. The loss of diversity in the hominid clade, for example, is an irretrievable loss of historical information. We should expect an evolutionary theory of the behavior of behaviorally complex organisms, ecologically released from their ancestral environments, and with no similarly complex living relatives, to remain decidedly conjectural. In the rest of this chapter and the next, we attempt to see what can be salvaged.

13.2 The Wilson Program

E. O. Wilson and his various co-workers originally attempted a fairly direct extension of evolutionary models of animal behavior to humans. Their driving concept was the idea that some human behaviors are adaptations—that human behaviors are molded by natural selection for some function. They had in mind behaviors such as incest avoidance, male sexual promiscuity and female coyness, infanticide, rape, and hostility to strangers. Rape and infanticide are important examples, for they illustrate the idea of a *facultative,* as opposed to an *obligate,* behavioral adaptation. Selection can produce conditional as well as unconditional behavioral rules. Just as some fishes have been selected to change sex under certain circumstances, perhaps we have been selected to murder our children when necessary.

The Wilson program is based on the idea that behavioral differences are just like any other phenotypic differences. They can make a difference to fitness. Moreover, populations are just as apt to differ in behavioral profile as in morphology, so there are behavioral differences among individuals on which selection can work. Moreover, there is good reason to expect that behavioral differences are heritable. Evolutionary histories reconstructed from behavioral traits agree well with those reconstructed from morphological and genetic ones (Paterson, Wallis, and Gray 1995; Kennedy, Spencer, and Gray

1996). If behavioral profiles vary within a population, but are inherited from one generation to the next, then selection can choose among them.

We will work through the most plausible case to illustrate both the temptations and the problems of the Wilson program. Consider sex role differentiation. There are some animals with "reversed" sex roles. In fishes, it is not particularly rare for the male to take primary responsibility for care of eggs. Male seahorses brood eggs in a special pouch, and in other fish species the male defends a nest. In some bird species (especially jacanas) the male takes responsibility for brooding the eggs and caring for the chicks. But, in general, females, and especially female mammals, have a higher initial investment than males in any reproductive act. First, eggs are much larger, and hence more biologically expensive to produce, than sperm. Second, the costs of pregnancy and postnatal care are very significant. By accepting a sexual partner, a female mammal commits a serious fraction of her total lifetime reproductive resources. That need not be true of males: they don't bear the costs of pregnancy and lactation. Hence females are the "limiting resource" for reproduction: healthy females are not normally in danger of failing to find a mate.

Despite the fact that sperm is metabolically cheap, sex is not without costs to males. They risk interference by other males, increased danger of predation while distracted, and the exchange of parasites and pathogens. Still, the female bears these costs too, together with those of pregnancy. This asymmetry of mating costs suggests different strategies for males and females. Male are likely to be more promiscuous; females, more coy. So we expect gender differentiation in mating decisions. We might also expect some gender differentiation in parental care decisions. If her young are to live, a female mammal has no choice but to engage in parental care. Not so the male, who is unconstrained by physiology. For him, all options, from outright desertion of the pregnant female, through diversion of some of his resources to other mates and their young, to full participation in parental care are possible.

There seems to be some evidence that this general picture fits the human case. First, there is evidence that men have, by and large, more promiscuous inclinations than women. Second, while men normally play a considerable role in child care, both directly and by providing resources, in a fair range of human societies, diversion of resources to other mates takes place. Polygamy is a fairly common human social arrangement; polyandry is rare.

Two cautionary notes should be sounded when applying this general pattern of explanation to humans. First, human sexuality is very unusual. In women, ovulation is concealed, at least in terms of visual cues, and there is no special breeding season. So it's quite likely that in humans sexual relations have functions additional to fertilization. The more human sexual decisions are divorced from reproductive decisions, the less of a grip the general model (females are more cautious because unwise reproduction is more expensive for them) has on the human case. Second, in arguing that female coyness is an adaptation, Wilson and others may have overlooked the possibility that female sexual behavior is a primate inheritance rather than a specific human adaptation. Such oversights invite the general criticism that adaptationism tends to focus on short-term adaptation to the exclusion of other known biological processes (10.3). No doubt Wilsonian sociobiology should have more readily considered evolutionary but nonselective explanations of human behavioral characteristics.

However, even if Wilsonian sociobiology tended toward adaptationism, that was not its main failing. To see its central problem, we need to make a distinction between *mosaic* and *connected* traits. A mosaic trait is one that can evolve relatively independently of the rest of an organism's phenotype. Human skin color is a mosaic trait, for it can evolve with relatively little change in the rest of the organism. When that trait changes as the result of selection, we can identify the selective forces involved and the adaptive function of the change. So mosaic traits are evolutionary atoms with specific adaptive characteristics. Connected traits, as you will guess, are precisely those that are tied intimately to many features of an organism or its development. These traits cannot change without profound alterations in development and phenotype. A plausible example of a connected trait is our having two lungs. Why do we have *two* lungs? Our lung number may well be a consequence of the general bilateral symmetry of our bodies and of the developmental mechanisms involved in that symmetry. In the language of section 10.5, lung number may be developmentally entrenched. There may never have been any variation in the primate line in lung number. Moreover, a change in lung number would involve a cascade of other changes. It would be part of a bigger package. So it's not at all clear that we should think of lung number as an evolutionary unit, a feature of our phenotype that has a more or less independent explanation. There are, of course, many traits less changeable than skin color but more changeable than lung number. *Mosaic* and *connected* name endpoints of a continuum.

Some behaviors may be mosaic traits. Hygienic behavior in bees—the removal of dead larvae from the hive—is a good candidate. We know that hygienic species have close nonhygienic relatives. Perhaps the whole behavioral repertoire of the bee results from a bundle of independent behavioral programs. Such an organism would have a set of distinct behavioral modules that could be taken out and replaced by variants without disrupting the others. The human behavioral repertoire, however, is not an aggregation of independent units. Our behavior is produced by mental mechanisms that play a role in many different behaviors. Some of the mental mechanisms used in hunting are also used in storytelling. So speculations about the adaptive significance of rape, xenophobia, child abuse, or homosexuality seem to be at the wrong grain of analysis. Such behaviors might be alterable only by altering the underlying mental mechanisms, and since these mechanisms are used for many different purposes, any change in them would have many other consequences. Hence individual behaviors are unlikely to have histories to call their own, or to have independent adaptive significance.

13.3   From Darwinian Behaviorism to Darwinian Psychology

The idea that we should not be looking for adaptive hypotheses about specific behaviors has gradually become part of the accepted wisdom of human sociobiology. In one form or another, the idea that the psychological mechanisms that generate behavior are the proper focus of evolutionary theorizing is now widely accepted. The comparison with the "cognitive revolution" in psychology is compelling. In that earlier revolution, psychologists turned from the idea that each behavior develops because it is rewarded to the idea that behaviors are caused by a small set of cognitive mechanisms. The full gamut of our social actions may depend on the interaction of just a few distinct cognitive devices, and it is these mechanisms that develop as an individual grows up. Likewise, sociobiologists have turned from the idea that each behavior evolved because it was selected to the idea that many different behaviors are caused by a relatively small number of cognitive mechanisms. It is these mechanisms that have evolved.

Biological anthropology played, and continues to play, a curious role in this transition from Darwinian behaviorism to Darwinian psychology. A key figure in this transition was Richard Alexander (1979, 1987). Alexander did not expect to find human behavioral uniformity, and he did not believe that specific behavioral patterns had adaptationist explanations. He accepted that many human behaviors were novel. They were learned on the spot in response to unusual circumstances, and hence were not in themselves adaptations. Even so, they could be understood adaptively. Some cephalopods can camouflage themselves by altering their color and pigmentation pattern to match their background. A particular animal, matching itself against a discarded diver's mask, might generate a pattern unique in that species' history. The pattern itself is not an adaptation, but it is a direct result of adaptive mechanisms that have the job of producing the animal/environment match of which this pattern is a specific instance. Alexander, and the biological anthropologists he influenced, defended an analogous conception of the diversity of human behavior. That behavior is genuinely diverse, but it's the manifestation of a naturally selected learning rule. For what we learn to do is to maximize our inclusive fitness. We will find enormous variation in social arrangements, economies, and political organizations as differing groups of people adapt themselves to differing physical and social environments. But we will not find societies in which people typically lavish their resources on second cousins or strangers, but not on their own children or siblings. The research program in Darwinian anthropology has been the attempt to confirm this hypothesis through example, by showing that surprising social behaviors—often ones that at first sight seem to disconfirm the idea—actually turn out to be confirmations of it after all.

We will consider one of Alexander's own examples before offering our critical discussion of this project. In the avunculate social system, a man directs his resources to his sisters' children rather than his wife's children. This system seems to contradict the idea that humans maximize their expected inclusive fitness. Men are less closely related to their nieces and nephews than to their own children. Alexander argues that the avunculate system arises when societal organization forces husbands and wives to live separately, and when this leads to "a general society-wide lowering of confidence of paternity [which] will lead to a society-wide prominence, or institutionalization, of mother's brother as an appropriate male dispenser of parental benefits" (Alexander 1979, 172). At least your sister's children are some kin of yours, for their maternity will never be in dispute. Your wife's children may be no kin at all. So while humans are extraordinarily good at adapting to new situations, what they learn is how to maximize inclusive fitness in whatever circumstances they find themselves. In the extraordinary situation of lowered confidence in paternity, males respond by caring for their sisters' children.

We have four criticisms of this example and the general program it represents. First, biological anthropology has its own "grain problem." As Kitcher (1985, 299–307) points out, according to Alexander's own analysis, the avunculate system should be unstable. It could be subverted by a female strategy. Consider the richest man in an avunculate society. He supports his sister's children, but this is only a second best option for him. If his sister is only a half-sister, her children may share only one-eighth of his genes. Since he is the richest man in the group, it would pay any other woman in the group to make him the following offer: to guarantee his paternity (say, by living with his mother) in return for his support. It would pay the richest man to accept. The same strategy is then open with the second richest man, and so forth. The avunculate system should collapse as women offer fidelity to all men of above average resources. As more and more men accept, their sisters will lose their resources and be forced to strike deals with whomever else is available.

We hope Kitcher's instability argument strikes you as unrealistic, because therein lies the grain problem for the Alexander program. The argument is unrealistic because the avunculate system is an organized social system, not the result of each individual deciding on a reproductive strategy. It is a society-wide organization for the rearing and support of children. It is stable, we imagine, because social mechanisms prevent women from following the subversive strategy Kitcher describes. It's likely that attempts by women to form socially abnormal households would be punished by the rest of the society. Perhaps males attempting to withdraw the usual support from their sisters would be punished too. These sociological factors, and the more general human motivation of avoiding punishment, are sufficient to explain human behavior in avunculate societies. It is unnecessary to postulate a mental mechanism designed to choose the best reproductive strategy for the circumstances.

Second, we think that it is often hard to measure the effect of behavior on fitness. Economic resources are typically used as a measure of fitness benefits. Even in hunter-gatherer societies, this measure is probably too crude. Fitness probably does not vary as a linear function of economic resources. There are likely to be thresholds below which fitness is zero, ranges in which marginal additions of resources have little effect, and thresholds at which fitness increases dramatically, as that of a man does when he can support a second wife. A more complex function of this kind from resources to fitness may be applicable to hunter-gatherer societies, but in many societies, like our own, economic resources are apparently unrelated to biological fitness. So the inference from the economic returns of a behavior to its genetic returns is dodgy.

Third, even if the effects of a behavior can be measured, to assess whether the actual behavior is optimal we need to understand the space of possible alternative behaviors. That can be extraordinarily difficult. What were the realistic alternatives of, say, a nineteenth-century Chinese woman faced with having her feet bound? Combining this problem with the last suggests that the most we may be able to manage are crude qualitative judgments of the relative values of different behaviors.

Finally, and most importantly, we doubt the significance of even such correlations as it is possible to establish between behavioral traits and inclusive fitness. Finding that a behavior increases inclusive fitness does not tell us much about the proximal mechanisms that produce that behavior. The data on the avunculate system, for example, do not discriminate between these two hypotheses:

- Humans possess adapted mechanisms specific to resource distribution that are sensitive to degrees of kinship.
- Humans make resource distribution decisions on the basis of cognitive and emotion structures that are relatively unspecialized. These unspecialized mechanisms are involved in mate choice, reciprocal interactions, bargaining, and many other social activities.

If resource distribution behavior tracked inclusive fitness differences in an extraordinarily sensitive and accurate way, then we might be driven to posit a specific resource distribution mechanism. Perhaps only such a mechanism could explain a precise covariance of behavior and inclusive fitness. Unfortunately, such precise data are not to be expected. Moreover, what should we say about disconfirming data? Perhaps it is reasonable to dismiss failures to maximize fitness—drug abuse, celibacy, falling birthrates among the wealthy members of Western societies, excessive military zeal, and so forth—as aberrations caused by novel environments. But if these failures are mere accidents, then the successes—cases in which there is a crude, qualitative fit between behavior and inclusive fitness—may be mere accidents as well.

13.4 Evolutionary Psychology and Its Promise

The spectacular differences among human cultures have always been a source of skepticism about sociobiology. We have already argued that sociobiology is not linked to genetic determinism. Our adaptations may be dependent on features of the environment that we could change if we wished. Nevertheless, there is a strong tendency to believe that since adaptations are the products of cumulative selection, the development of an adaptation should be relatively stable, causing it to appear again and again despite cultural changes. For the environment is unlikely to have been constant throughout the period in which cumulative selection operated on us. Contrary to this expectation, human cultural life seems extraordinarily diverse. The way of life of the Australian Arunta seems very different from that of the New Zealand Maori. The differences in their facial features and the like may have (so this line of thought goes) genetic bases. But their linguistic and cultural differences do not, as the development of children moved from one culture to another makes clear. No one expects to find a "gene for hunting and gathering" in the Arunta that is less common in Maori populations. Furthermore, the pattern of genetic differences among humans is the exact opposite of their pattern of cultural differences. In genetics, it is at least arguable that the differences among the individuals in a single group swamp average differences between groups. In the cultural realm, we observe high within-group similarity and high between-group difference.

Within the social sciences it is common to suppose that our evolved "human nature" places only the broadest constraints on our cultural life (1.4). No doubt if we were asexual, or if we could photosynthesize our own food, our cultures would be very different. However, our evolutionary inheritance makes possible a wide range of cultural forms, probably far wider than has yet been exemplified in human history. Since every human group has a similar set of biological resources, the great differences between groups must be explained in terms of differing cultural resources. Difference explains difference. In this view, the job of evolutionary theory is rather limited: it should aim to explain the preconditions of culture. It should explain how hominids developed the ability to transmit culture and the plasticity to be shaped by that culture.

Evolutionary psychologists, sociobiology's latest defenders, fiercely resist this division of labor between evolutionary and cultural theory. First, they suspect that human cultural diversity is less profound than it may at first appear. Second, they argue that diversity itself may have an evolutionary explanation. Organisms are adapted to behave differently in different circumstances. In many species of wrasses, a female changes sex when she becomes the largest member of the group, yet we think of "her" sex determination mechanism as a single adaptation. Equally, a single mechanism of resource assessment might generate one behavior in the Australian desert and another in a London supermarket. This, of course, is an idea similar to Alexander's, discussed in the last section. Where Alexander talked of "learning rules," more recent theorists talk of "Darwinian algorithms."

The work of the linguist Noam Chomsky looms large in evolutionary psychologists' discussions of human diversity. If his theories are correct, the differences between human languages, while real, are not profound. There are many important features common to all human languages, even if they are not obvious at first glance. The class of humanly possible languages is quite tightly constrained by the nature of a domain-specific cognitive structure: the "language acquisition device." Moreover, that device contains conditional elements—"switches"—whose different settings explain many of the differences among languages. Language thus demonstrates both of the evolutionary psychologists' points: Diversity may be less than it appears, and diversity can be explained by a single mechanism, one that operates differently in different circumstances. We think that language is important for a third reason, too: it shows the inappropriateness of a nature/culture dichotomy. If language is a specific adaptation, then it evolved only because our ancestors were already a species with a culture. Moreover, an individual's acquisition of language depends on both the language acquisition device and the surrounding culture. Whether our context is developmental or evolutionary biology, it's wrong to think of language as exclusively a "cultural" or a "biological" phenomenon.

Evolutionary psychologists reject the metaphor of the human mind as a general-purpose computer programmed differently by different cultures. They replace this vision with an alternative, modular theory of mind. The mind is a cluster of evolved information-processing mechanisms. The main goal of evolutionary psychology is to characterize these *Darwinian algorithms*. For example, Buss (1994) and Symons (1979) think that there are Darwinian algorithms of sexual attraction that result in the tendency of human males to find attractive those females that bear the cultural marks of youth, and of women to find attractive those men that bear the cultural marks of high status. Cosmides and Tooby (1989) argue that specialist algorithms for regulating social exchange ensure that all human groups are aware of and have safeguards against the possibility of others defaulting on deals. The Darwinian algorithms are supposed to be mental *modules* in the sense of Jerry Fodor (1983): they are domain-specific, mandatory, opaque, and informationally encapsulated mechanisms. Darwinian algorithms are *domain-specific* because they deal with a specific class of situations in which the organism finds itself. The mate choice module is not used to choose food or clothing. They are *mandatory* because people do not choose to approach these problems in this specific way, as they would choose to use one algorithm rather than another to do a math problem. Rather, when a suitable problem presents itself, the appropriate module leaps into action. Darwinian algorithms are *opaque* because their internal processes are not consciously accessible. It takes scientific investigation to teach us what features of members of the opposite sex cause us to be attracted to them. Finally, Darwinian algorithms are *informationally encapsulated* because they do not make use of the information stored elsewhere in the cognitive system. Phenomena such as phobias are taken to represent a clash between the conclusions of a mental module and our conscious thought processes. The information that this particular spider is made of rubber cannot get into the module.

Evolutionary psychologists hope to identify the Darwinian algorithms by the strategy of *adaptive thinking* (10.7). Adaptive thinking infers the solution—the adaptation—from the problem—the ecological context in which the organism evolved. The first task of the evolutionary psychologist is therefore to identify the adaptive problems our ancestors confronted in their environments. In foraging for food, for example, our ancestors would have needed a good grip on the physical, social, and biological geography of a range that was likely to be extensive, and through which they would frequently shift.

The second task is to discover the stable correlations between those aspects of the environment humans are equipped to sense and those aspects they need to know about. We would expect natural selection to engineer into task-specific devices implicit knowledge of these correlations. If in the semi-arid environments in which humans lived for a long time, there was a stable correlation between a deeper green leaf color and an accessible underground water flow, the evolutionary psychologist would expect awareness of this correlation to be engineered into those mechanisms specialized for controlling movement through a complex and varied environment. If the nutritional value of food was reliably correlated with its sugar content, we would expect people both to be able to detect and to desire sugar's sweet taste. However, some adaptive problems may be recalcitrant. There may be no reasonably reliable environmental cue that can be used to solve them. In visual perception, abrupt transitions in light intensity on the retinal image covary with edges of objects in the environment. This covariation supports our seeing the world in three dimensions. Only cues like this make vision possible. Without them, the task of moving from two-dimensional information on the retinal image to a three-dimensional representation of the world would be intractable. No doubt being able to predict the weather four or five days in advance would have been an advantage of the first importance many times in human evolution. The "weather prediction module" has not evolved, we conjecture, because the environment does not provide suitable information to run one.

The third task of the evolutionary psychologist is to construct an information-processing design that could solve the adaptive problem using the available cues. Possible designs are then evaluated against one another using the techniques of optimality modeling described in section 10.6. This results in an adaptive hypothesis: the organism will use the most advantageous design to solve the problem. The fourth and final task is to experimentally test for the existence of the hypothesized mechanism, for many potentially useful adaptations will not actually be engineered into us. As Donald Symons says, "Although adaptive thinking is an important source of inspiration for the evolutionary psychologist, nature always gets the last word" (Symons 1992, 143–44).

13.5 Evolutionary Psychology and Its Problems

We agree with the central idea of evolutionary psychology, namely, that we should look for the effects of natural selection on the psychological mechanisms that explain our behaviors, rather than on those behaviors themselves. Moreover, we agree that it is very likely indeed that selection has been one of the forces that has transformed our cognitive system. That said, we think the standard formulation of evolutionary psychology suffers from two serious and linked problems. In sections 11.1 and 11.4, we discussed the idea that evolutionary change in a lineage is a response to the environment. The environment poses problems, and under the influence of selection, the lineage changes, becoming better adapted to that environment. This externalist picture fits some instances of adaptive change, but it fails to fit many others, including the ones of most concern to us here. Second, evolutionary psychology has made a somewhat premature commitment to the theory that sees the mind as an assemblage of special-purpose modules.

Evolutionary change that is driven by the social environment of a population should not be seen as an adaptive response of that population to its environment. For the social environment and the lineage change together. For example, one fashionable theory about cognitive evolution is the "Machiavellian intelligence hypothesis." According to this hypothesis, our mental capacities evolved in an "arms race" within human populations. Their evolution was driven, perhaps, by the hope of exploiting others, but certainly by the need to avoid exploitation by them (Byrne and Whiten 1988). If the selection pressures important in cognitive evolution derive from interactions within the group, then selective environment and adaptive response change together. There is no invariant environment to which the lineage is adapted.

The traditional oversimplified picture of the relation between environment and adaptation makes it easy to overlook the fact that evolutionary psychology has its own "grain problem." What are the problems that exist "out there" in the environment? Is the problem of mate choice a single problem or a mosaic of many distinct problems? These problems might include: When should I be unfaithful to my usual partner? When should I desert my old partner? When should I help my sibs find a partner? When and how should I punish infidelity? This grain problem in evolutionary psychology challenges the idea that adaptations are explained by the problem to which the adapted trait is a solution. If (but only if) there is a single cognitive device that guides an organism's behavior with respect to issues of mate choice, then mate choice is a single domain, and these are all different aspects of the same problem. It is not the existence of a single problem confronting the organism that explains the module, but the existence of the module that explains why we think of mate choice as a single problem.

Evolutionary psychologists have been very keen to reject the "general-purpose computer" model of the mind. But they should be cautious about accepting a modular theory of mind. For specialized mechanisms have a downside: they are vulnerable to exploitation in a malign world (Krebs and Dawkins 1984). If our minds are the result of an arms race, then they evolved in a hostile world, not merely an indifferent one. Evolutionary psychologists' adaptationist instincts should make them cautious about using language as their exemplar of an adapted psychological capacity. Game theoretic models of the evolution of language have a strong cooperative element. In fact, they are close to one end of a mathematical spectrum that runs from games of pure cooperation to games of pure conflict (the "zero-sum" games where my gain and your loss always cancel out to zero). Both parties in a linguistic interaction benefit from successfully communicating their intended meaning. Even if they have other exploitative agendas, neither will succeed unless the utterance is understood. A rigid, modular language acquisition device is unlikely to be exploited by other individuals to prevent someone from learning the language. A module for resource sharing, however, might well be manipulated to gain a better share of resources.

There are further reasons to doubt whether evolutionary theory predicts a modular mind. Perceptual systems exemplify a surprising truth about human mentality. The information processing tasks implicit in much human action are much more complex and difficult than one would intuitively expect. In many branches of cognitive psychology, this realization has generated a series of *poverty of the stimulus* arguments. These arguments attempt to show that we develop cognitive skills too fancy, and with too little information from the environment, for their development to be the result of general learning mechanisms. The outputs of the visual system are determinate and astonishingly reliable representations of what is seen, yet the stimuli to the perceptual mechanisms are typically fragmentary and equivocal. That does support the view that perceptual tasks could be carried out only by mental organs specifically adapted for those very tasks, like Chomsky's language acquisition device. Care is needed, however, in extrapolating from these examples. First, even superior performance in certain cognitive areas is not sufficient grounds for positing a Darwinian algorithm. We clearly have the potential to "automate" cognitive skills not subserved by purpose-built wetware. Chess, bridge, and other difficult cognitive games provide striking examples, as do the cognitive skills involved in car driving. These skills are domain-specific and widely spread through the population, but they cannot be based on a Darwinian algorithm.

Second, and very importantly, the poverty of the stimulus argument does not sustain some of the central hypotheses of evolutionary psychology. For example, Cosmides and Tooby argue that we have a module of social exchange (Cosmides 1989; Cosmides and Tooby 1989, 1992). But their reasoning is the inverse of poverty of the stimulus reasoning. We find a range of computationally trivial reasoning tasks extraordinarily difficult. One of the most famous demonstrations of this feature of our cognitive design is the Wason card selection task. Subjects are confronted with four cards, being able to see just one half of each card. They are asked which cards they need to examine fully in order to test the rule that a card with a circle on the left has a circle on the right. Logically, this task is trivial, for the rule is falsified only by a card with a circle on the left but none on the right. So cards with circles on the right, or blank on the left, are irrelevant. Yet subjects struggle to get it right. When the equivalent reasoning tasks are about social exchange, however, we do much better. So we can do an easy task in one domain with less difficulty than the same task in other domains. A poverty of the stimulus argument applies in the reverse situation, in which we can do a computationally complex task without much effort. For example, determining motion from changes in apparent shape and size of images on the retina is computationally extraordinarily complex, yet we do it easily.

The same contrast arises in mate selection. The Buss and Symons mating rule is not computationally complex. It is very simple: women find high status attractive, whereas men find youth attractive. There is no need for a specialized mechanism to operate this decision rule. It may, of course, be very difficult to determine whether someone is young or of high status, but that is not what the specialized mechanism has to do. There are many social interactions in which age and status judgments are important, and there is no evidence that our judgments of age or status for the purposes of mating ever conflict with the judgments we make for other purposes. So there seems no reason to suppose that assessing age and status is the work of the postulated mate choice module. All the module does is direct our sexual attention to those who have these properties, and that is a relatively simple task.

Finally, many important problems cannot be solved by modular mechanisms. Fodor (1983) has argued convincingly that the pragmatics of language cannot be handled by a specialist device. It is one thing to know what a sentence means; it's another to know the intentions that lie behind its utterance. The latter problem is not solvable by shortcuts from a restricted data

Figure 13.1 The Wason card selection test. Consider the following hypothesis: If there is a circle on the left, then there is a circle on the right. Which cards must you see in order to test this hypothesis? (Adapted from Wason 1968.)

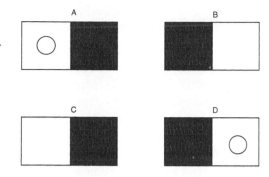

base—that is, by an encapsulated device. Everything the hearer knows is potentially relevant and potentially useful in decoding the speaker's intent. The same problem seems to arise in many of the domains of interest to evolutionary psychology. Could an encapsulated mechanism deliver reliable judgments about a prospective mate's status? A spouse's infidelity? The probability of a prospective partner's cheating? We need to be shown the equivalents in these domains of the reliable rules of thumb that our perceptual mechanisms exploit.

So evolutionary psychology has bought into an oversimplified view of the relationship between an evolving population and its environment, and has prematurely accepted a modular conception of the mind. These two problems are linked. We remarked above that hardwired mechanisms are vulnerable to deception in a malign world. The problems that confronted our ancestors did not stay the same, and the regularities in the world on which their solutions depended were apt to change. Traits are sometimes adaptations to an independent, impervious environment. But when evolution is driven by features of the social structure of the evolving species, evolution transforms the environment of the evolving organism. The evolution of language, of tool use, and of indirect reciprocity are not solutions to pre-existing problems posed to the organism. There are no stable problems in these domains to which natural selection can grind out a solution. The "adaptive problem" is always being transformed in an arms race. As we evolve to detect cheaters, these honesty-mimics evolve better and better imitations of a trustworthy and honest face. The heuristic recommended in evolutionary psychology is not just adaptationist, but sees adaptation as accommodation to the evolving lineage's environment (11.1). We suspect that cognitive adaptation often transforms the environment rather than being an accommodation to it. So there will be real troubles in store for a methodology of discovering the mechanisms of the mind that proceeds by first trying to discover the problems that it must solve, and then testing for the presence of the solutions. This methodology does not reflect the interactive character of social evolution.

This methodology is in even more trouble if D. S. Wilson and Sober (1994) are right in thinking that population structure (8.4, 8.5) has been very important in human evolution. That view is very plausible. Alexander (1987) devotes most of a book to arguing that competition between human groups has been an enormously significant factor in human evolution. For a very long time we have been one another's most deadly enemies. If so, this complicates the adaptationist heuristic in two ways. First, population structure is clearly not a stable background against which psychology changes. Alexander emphasizes that one of the chief effects of some cognitive changes is a change in group size. Second, as D. S. Wilson emphasizes, introducing population-structured evolution into the picture changes which adaptations it is sensible to expect. Many altruistic behaviors that would be selected against within a single population will evolve in a population divided into groups (Wilson 1997, 1998b).

In brief, then, we think that the development of an evolutionary psychology is the right aim for those who seek to apply evolutionary theory to human behavior. But both the objective and subjective obstacles to carrying out this program remain serious.

13.6 Memes and Cultural Evolution

So far we have considered the idea that evolutionary theory helps to explain human society by helping to explain the nature of individual humans. Since human cultures are products, though perhaps indirect ones, of biological evolution, they are best understood by understanding the processes that made them over deep time. An alternative to this idea is to treat the theory of biological evolution as an instance of a more general explanatory scheme. We can conceive of cultural change itself as an autonomous evolutionary process. If we consider the different aspects of human cultures, and the elements out of which they are composed, we see variation, differential fitness, and heritability. Ideas, fashions, inventions, and the like can spread through society. We can see these as replicators, variants competing with differential success. Ideas are replicators because they are potentially copied from human brain to human brain through indefinitely deep lineages. Moreover, they are active replicators, for ideas have effects that make them more or less likely to be copied. They are in competition, both in a general and a more specific way. They compete as a consequence of limits on the resource pool out of which new links are made. More specifically, some ideas are in direct conflict with one another: they compete, as it were, for the same slot in human brains. Rival political and religious ideas and rival sporting allegiances compete in this more direct sense. According to Dawkins, Dennett, Hull, and others, there is literally an evolutionary process operating on ideas. In Dawkins's language, ideas are *memes,* and meme lineages compete and grow differentially. The same generalizations that describe biological evolution describe the evolution of memes. An account of meme lineages, their phenotypic effects, and their environment is an account of human culture.

We are very skeptical about this way of applying evolutionary theory to the task of explaining features of human societies. Ideas may be copied, they may have effects that make their transmission to a new bearer more likely, and the success of one idea may be bad news for the prospects of others. So the world of memes may indeed show phenotypic variation, differential fitness, and heritability. Even so, we have three reasons for doubting that we will learn much about human society and culture from the theory of memes.

First, as we emphasized in section 2.2, biological evolution depends on cumulative selection, and that imposes extra and more demanding conditions on selective regimes. For instance, the mutation rate must be low, but not too low in comparison to the strength of selection. If the mutation rate is too high, the noise of random change will drown the signal of selective propagation. If it is too low, selection will use up all variation, and evolutionary change will grind to a halt. The power of natural selection to produce change over time depends on specific features of the biological world. It depends on the grubby details of biology. We see no reason to expect a parallel to these details in the selective environment of memes.

Second, we do not clearly see the explanatory target of meme theory. What is it supposed to explain? Perhaps we can redescribe various social processes in evolutionary language. There is a selective regime, in a sense, if we have a population of potentially persisting entities in which the persistence of one negatively affects that of the others, and in which persistence is not mere chance. So understood, the publishing industry is a selective regime. But what would that explain about the publishing industry? Natural selection is a *hidden hand* theory. It explains the appearance of conscious coordination and design without requiring a designer. But the social world—for instance, the world of publishing—is a world in which there are real intentions and real planning. Of course, market economics is a hidden hand theory too: indeed, it was the first hidden hand theory. It explains the coordination of production and consumption without requiring a planner to oversee the coordination. But it is not an evolutionary hidden hand theory: classic economics makes no reference to replication or inheritance mechanisms. So what features of the social world need explanation by a selection-like hidden hand theory? What features of, say, changes in fashion or in the publishing industry show the appearance of deliberate design without being the result of deliberate design? Perhaps in the specific domain of science there is such a feature: the growth over time of objective knowledge. But in society and culture more generally, we see no obvious candidate. So we do not see what treating ideas or social forms as memes is supposed to explain.

Third, the explanatory power of natural selection has been denied on the grounds that its central explanatory idea—"the survival of the fittest"—is a tautology, because "fittest" just means "the organisms that best survive" (4.2) This objection fails, for we do have an independent grip on the concept of fitness. It means "expected reproductive success," not "actual reproductive success." Our knowledge of an organism's morphology, behavior, and ecological circumstances tell us the success to expect. Ecological source laws explain fitness and its variation across different organisms in a population (11.1, 11.5). But, as Sober (1992) argues, a variant of the tautology objection seems much more damaging to the conception of memic evolution. With the possible exception of scientific ideas, we have no explanation of the nature of the fitness of ideas, not do we typically understand why they differ in fitness. We can call a tune "a meme with high replication potential" rather than "catchy" if we like. But without source laws, this adds nothing to our understanding of musical trends.

In the next chapter we explore these issues of human evolution further through the specific example of the evolution of emotions. There we discuss some ideas we think have real interest, though they are still very speculative.

Further Reading

13.1, 13.2   Wilson's views on human sociobiology are given in the final chapter of *Sociobiology* (Wilson 1975) and in *On Human Nature* (Wilson 1978). In later work with Lumsden (Lumsden and Wilson 1981, 1983) he attempts to integrate culture more deeply into his picture. There are a number of useful collections on these early versions of sociobiology. Caplan 1978 includes the early and vitriolic exchanges between Wilson and those that criticized his work on political grounds. Those critics included Lewontin and Gould, members of Wilson's own university. Montagu 1980 also collects some of these early reactions to sociobiology. Two book-length responses, wholly or largely focused on sociobiology, are Lewontin, Rose, and Kamin 1984 and Kitcher 1985. Of these, Kitcher's is very much the best (though written in a rather hectoring tone), discussing not just the early version of

Wilson's work, but also his joint work with Lumsden, and Alexander's program as well. Lewontin, Rose, and Kamin treat sociobiology as a species of biological determinism, misrepresenting the actual views of many of their targets, and in any case focusing on an inessential rather than an essential element of the program. Both the Wilson program and the Alexander program are discussed further in Sterelny 1992a. For a more nuanced but still rather adaptationist take on sex roles in humans and other primates, see Small 1993. The inference from biological premises to ethical and political conclusions is discussed in the final chapter of Kitcher 1985, section 10 of Sober 1994, and the final chapter of Sober 1993. Ruse (1986) is much more friendly than we are both to the original Wilson program in sociobiology and to the drawing of normative conclusions from that program. Section 6 of Hull and Ruse 1998 is a good, broad survey of many of the issues of this chapter and the next; section 8 is relevant to the more specific issues of genetic determinism and human variation.

13.3   Alexander sets out his program in *Darwinism and Human Affairs* (1979) and *The Biology of Moral Systems* (1987). Chagnon and Irons 1979 is a collection of anthropological papers exemplifying his program. Smith 1987 is a good overview. In 1990, the journal *Ethology and Sociobiology* devoted a special issue of volume 11 to the debate between defenders of an Alexander-style program and evolutionary psychology. The main argument centered on the significance of the current adaptive value of behavior. Defenders of Alexander's approach continue to think of it as significant, whereas evolutionary psychologists are skeptical. This whole issue is worth reading. Alexander's view continues to be quite important in anthropology. Two recent, representative examples are Chisholm 1994 and Smith and Smith 1994.

13.4, 13.5   Evolutionary psychology is the cutting edge of contemporary sociobiology, and is well served with literature. Crawford, Smith, and Krebs 1987 and Barkow, Cosmides, and Tooby 1992 are two important collections. Sterelny 1995 is an extensive critical review of the second of these. Symons 1979 might well be the first extended defense of this approach. Not surprisingly, sex has been followed up with papers in the two collections and in the journals; for a recent example, see Jones 1995. Much of the new material on sex is presented accessibly in Buss 1994. One of us (Sterelny) thinks that Frank 1988, which defends an evolutionary hypothesis on the role of the emotions, is probably the most plausible of all the current variants of evolutionary psychology. Evolutionary theories of emotion are the focus of the next chapter. Barkow (1989) defends a grand synthesis; Tooby and

Cosmides (1992) develop their own synthesis. In a series of papers, Cosmides and Tooby (1989, 1992; Cosmides 1989) develop their idea that social exchange depends on a specific cognitive specialization. Davies, Fetzer, and Forster (1995) reply. As we have noted, much of evolutionary psychology relies on the model of a domain-specific mechanism provided by contemporary linguistics. Pinker 1994 is a splendid introduction to, and exemplar of, the Chomskian world view: a world view that we think includes a rather simple-minded contrast between learned and innate capacities. Oyama (1985) provides a good dose of skepticism about this distinction. We think that R. Wright 1994, though readable, is a much too confident endorsement of current ideas. Hirschfeld and Gelman 1995 and Sperber, Premack, and Premack 1995 are two good recent collections on domain-specific cognition. A recent collection of responses to evolutionary psychology is Davies and Holcomb 1999. The link between adaptationism and externalism is discussed extensively by Godfrey-Smith (1996), who defends a very modest form of adaptationist externalism while discussing the evolution of very simple cognitive capacities. Sterelny (1997) responds.

13.6 The formal parallel between cultural evolution and genetic change has been well defined and defended by Boyd and Richerson (1985) and Cavalli-Sforza and Feldman (1981). Dawkins (1976) introduces and defends the idea of memes in *The Selfish Gene;* he backs off a little in *The Extended Phenotype* (1982). Dennett (1995) takes up the idea with great enthusiasm—indeed, we think excessive enthusiasm—in *Darwin's Dangerous Idea.* Hull (1988) defends a very cut-down version specifically in the context of science. Sterelny 1994 is partly a critique, partly a limited defense, of this restricted application of the central idea. Sober (1992) expresses his skepticism about the whole enterprise. Sperber (1996, chaps. 4 and 5) criticizes the meme theory on the interesting grounds that while in genetic replication correct copying is the norm and change the exception, in the transmission of ideas this is reversed. Thus his worry relates to ours about cumulative selection. Moreover, he points out that it is not at all obvious that ideas are replicated in anything like the sense that genes are. He points out that a child's version of a story—say, "Little Red Riding Hood"—is likely to be an amalgam—a composite—of many tellings of the story by parents, grandparents, and others. It has no specific ancestor. In contrast, each gene in the child has a specific and identifiable ancestor in one of her parents.

14

A Case Study: Evolutionary Theories of Emotion

14.1 Darwin on the Emotions

The Expression of the Emotions in Man and Animals began life as a chapter of Darwin's major work on human evolution, *The Descent of Man* (1871). Darwin found it impossible to compress his material on emotions into a single chapter, and it appeared as a separate book in the following year. So, unlike many topics in human evolution, the study of emotion has long rested on a mass of empirical detail about the products of evolution. It is an excellent case study in both the power and the limitations of the evolutionary approach to the human mind.

The emotions were a particularly controversial topic in Darwin's day because they were seen as uniquely human, even spiritual, traits. A well-known book of the period argued that the muscles of the human face served no practical purpose, but had been designed specifically to express the states of the human soul (Bell 1873). Other animals lacked these facial expressions because they lacked souls. The emotions thus provided not only evidence for the existence of a Creator, but evidence of the special position of humanity in His creation. Darwin undertook to show that human facial expressions were modifications of facial expressions seen in other primates. They had clear practical functions in earlier evolutionary phases and were now used to communicate with other members of the species:

> with mankind some expressions, such as the bristling of the hair under the influence of extreme terror, or the uncovering of the teeth under that of furious rage, can hardly be understood, except in the belief that man once existed in a much lower and animal-like condition. The community of certain expressions in distinct though allied species as in the movements of the same facial muscles during laughter by man and by various monkeys, is

rendered somewhat more intelligible, if we believe in their descent from a common progenitor. (Darwin 1965, 12)

Darwin's methods for studying facial expressions of emotion were extraordinarily advanced. The first experimental psychologists began work in the 1880s. In the 1860s, Darwin was already testing people with arrays of photographs to determine whether they could recognize emotions from facial expressions! The technique of electrically stimulating muscles to determine their anatomical action had recently been pioneered, and Darwin used these results to analyze facial expressions into particular muscle movements. This allowed him to determine whether the physiological basis of a human expression was the same as that of a monkey's, despite the huge differences in their faces. He succeeded in demonstrating many such homologies between humans and other primates. The least satisfactory of Darwin's methods were those he used to establish that the same facial expressions existed in different human cultures. Rather than asking for photographs, he sent questionnaires to traders and missionaries around the world asking them to look for particular facial expressions. The fact that Darwin described the expressions he hoped to find made it very likely that his correspondents would read the desired results into their observations. Darwin's failure to exploit the camera in this cross-cultural study is all the more surprising given the advances in ethnographic photography at the time. In fact, while Darwin's book was in preparation, his close associate T. H. Huxley was developing a uniform standard for ethnographic photographs in the hope of making them more useful (Spencer 1992). But despite this missed opportunity, Darwin correctly identified the majority of the facial expressions that are in fact found in all human cultures (14.3).

Darwin believed that the expression of emotion could be explained by three evolutionary principles. First, many expressions started as practical responses to the situation that elicits the corresponding emotion, and only later became signals. Surprised chimpanzees open their eyes wide and look in the direction of the surprising noise or movement. Chimpanzees expose their teeth to conspecifics whom they intend to attack as a display of fighting prowess. Humans display these behaviors too, and presumably inherited them from the common ancestor of chimps and humans. But humans rarely worry about the size of an opponent's teeth when picking a fight. So why do we bother to display them? One obvious explanation is that any behavior reliably correlated with an emotion acts as a signal. If such signals are valuable, then these behaviors will be retained as signals even when their other

Figure 14.1 Some homologous facial expressions in humans and chimpanzees. Anger, type 1 in humans *(top left)* is homologous with anger, type 1 in chimpanzees *(top right)*; anger, type 2 in humans *(bottom left)* is homologous with a fear-anger blend in chimpanzees *(bottom right)*. (From Chevalier-Skolnikoff 1973, 73, 76.)

functions become less relevant. Even in chimpanzees, the baring of the teeth is at least as much a signal of the intention to attack as it is a preparation for actually attacking.

Darwin's second way of explaining emotional expression was the *principle of antithesis*. It is nicely explained in his own words:

> When a dog approaches a strange dog or a man in a savage or hostile frame of mind he walks upright and very stiffly; his head is lightly raised, or not much lowered; the tail is held erect and quite rigid; the hairs bristle, especially along the neck and back; the pricked ears are directed forwards, and the eyes have a fixed stare. . . . These actions . . . follow from the dog's intention to attack his enemy. [But when he recognizes the "stranger" as his master] let it be observed how completely and instantaneously his whole bearing is reversed. Instead of walking upright, the body sinks downwards or even crouches, and is thrown into flexuous movements; his tail, instead of being held stiff and upright is lowered and wagged from side to side; his hair instantly becomes smooth; his ears are depressed and drawn backwards, but not closely to the head. From the drawing back of the ears, the eyelids become elongated, and the eyes no longer appear round and staring. . . . Not one of the movements, so clearly expressive of affection, are of the least direct service to the animal. They are explicable, as far as I can see, solely from their being in com-

Figure 14.2 The principle of antithesis. The behavior of an affectionate dog *(bottom)* is the opposite of that of a dog expressing the intent to attack *(top)*. (From Darwin 1965, 52–53.)

FIG. 5.—Dog approaching another dog with hostile intentions.   By Mr. Riviere.

FIG. 6.—The same in a humble and affectionate frame of mind.   By Mr. Riviere.

plete opposition or antithesis to the attitude expressive of anger. (Darwin 1965, 50–51)

These behaviors serve the function of clearly *not* being the behaviors associated with aggression. It is hard to imagine anything more important in a pack of wild dogs than to avoid giving the false impression that you want to fight!

Finally, Darwin considered many emotional behaviors such as trembling and sweating to be the "overflow" of "excess nerve energy." Like most of his contemporaries, he regarded the nervous system as akin to a hydraulic system. Once the mind is excited, this excitement must be released somewhere. This idea is no longer widely accepted. Many of Darwin's "overflows" have since been interpreted as the effects of physiological preparation for appropriate action. Walter D. Cannon argued in the 1920s that anger and fear release adrenaline, which readies the body for action in the famous "fight or flight" response (Cannon 1927).

14.2 Sociobiology and Evolutionary Psychology on the Emotions

Several sociobiologists have exploited general evolutionary models to develop hypotheses about the role of emotions as mechanisms of social control. In his famous paper *The Evolution of Reciprocal Altruism,* Robert Trivers (1971) suggested that a complex system for cheating and detecting cheating would arise in communities in which individuals exchanged favors. Guilt, anger, and vengefulness could all be part of this system. Guilt can play the role of the ropes that bound Ulysses to the mast as he sailed past the Sirens—it holds people to agreements they would otherwise break. The existence of anger and vengefulness also holds other people to agreements they would like to break. These suggestions were developed by Robert Frank into the commitment theory of emotion, which we discuss in section 14.4. These suggestions all turn on adaptations to human social life.

The evolutionary psychologists John Tooby and Leda Cosmides have predicted that emotions will turn out to be behavioral programs that are deployed in response to frequently recurring ecological situations that include, but are not restricted to, the social environment:

> Each emotion state—fear of predators, guilt, sexual jealousy, rage, grief, and so on—will correspond to an integrated mode of operation that functions as a solution designed to take advantage of the particular structure of the recurrent situation these emotions respond to. (Tooby and Cosmides 1990, 410)

They go on to list eight properties of environment and mechanisms that should characterize each emotion and seventeen classes of biological processes that should be partly governed by emotional state.

It is striking that sociobiologists and evolutionary psychologists use evolutionary theory to predict what will be discovered about the emotions. This is very different from Darwin's approach. He started by describing what had evolved and then tried to explain it. The modern authors start with the explanation and proceed (hopefully) to the discovery of the actual nature of human emotion. They use the heuristic of adaptive thinking for exploring the mind. As we suggested in sections 10.7 and 13.5, there are serious problems with this approach. Adaptive explanation is an inference from the current phenotype of an organism to the problems that organism faced in its evolutionary past. Obviously, that inference will be problematic if we do not have an accurate description of the current phenotype and its adaptive significance—of the solution that evolution actually produced. The inference from current adaptive importance to adaptation is problematic enough even when the adaptive and phenotypic claims on which it is based are uncontroversial (13.1). The inference is still more problematic when the nature of the phenotype and its adaptive importance are yet to be established.

For example, Roger Shepard (1992) gives an impressive account of the selective forces that shaped the human system of color perception. He argues that there are three types of color receptors, rather than two or four, because under natural lighting conditions the visual system must compensate for three kinds of change in background illumination if objects are to seem the same color at different times *(color constancy)*. Obviously, Shepard would have had difficulty constructing this explanation if we were still debating whether "eye-beams" go out and touch the surface of the object we are looking at! Shepard needed a very precise description of the phenotype of human color vision in order to develop his inference from utility to history. For all we know, our discussions of emotion may be in some respects as primitive as that ancient Greek theory of vision. The way in which emotions are classified displays considerable variation across cultures, and there is no agreed scientific taxonomy of emotions. The Japanese describe an emotion called *amae:* a deeply fulfilling and gratifying sense of childlike dependence on a person or institution. Is this the same as any of the emotions experienced by Europeans? Is my anger when I am shoved in a queue the same emotion as my anger at the atrocities in Bosnia? Is sadness a mood or an emotion? There are no generally agreed-upon answers to questions like these. We cannot even begin to probe the relation between the adaptive importance of emotions and their selective history when the nature of the phenotype itself remains so ill-defined.

The dependence of evolutionary theories of emotion on untested "folk theories" of emotion is evident in the work of Robert Plutchik. According to Plutchik, the emotions can be arranged in a circle, with each point on the circumference representing the "opposite" state to the emotion on the other side (Plutchik 1962, 1970, 1980a,b, 1984). Plutchik's evolutionary theory is that animal behaviors fit into four pairs of adaptive categories. They are to designed to protect/destroy, reproduce/deprive, incorporate/reject, or explore/orient. There are four pairs of emotions corresponding to these four pairs of adaptive functions. Plutchik is able to find four pairs of opposing segments on his emotion solid that correspond to these adaptive categories.

At first glance, Plutchik's "emotion solid" looks like an informal version of an optimality analysis of adaptation (10.6). He has asked what sort of emotion system would best fulfill the adaptive needs of the organism, and he has tested his model of emotion evolution against the emotion system we see in

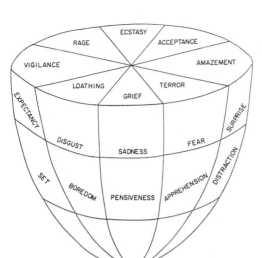

Figure 14.3   Plutchik's emotion solid. The four pairs of "polar opposites" are supposed to correspond to four fundamental pairs of adaptive functions. The vertical dimension represents intensity. (From Plutchik 1970, 10.)

human beings. The model passed the test by correctly predicting the human emotion system. This impression is, sadly, an illusion. First, let us look at how the predictions of the model were compared with reality. The emotion solid was generated by "semantic field-analysis" on a group of English emotion words. Competent English speakers were asked to make comparisons between words, rating them on a scale of similarity or difference. The positions of words in the circle reflect these judgments. In other words, the emotion solid is a model of what English speakers think about emotions. So the model is suspect in two ways. First, it is at best a model of how we perceive and classify emotions, but there is no reason to suppose that our common-sense judgments about the nature of emotions lock onto their most important features. Second, the model is derived from the judgments of a particular culture. In itself, the cultural restriction to speakers of English may not be a crippling problem. Different cultures have very different classifications of emotion, but perhaps there is some common core of truth to all of them. But now look at how Plutchik developed the evolutionary theory that supposedly explains this data. After the emotion solid was derived, Plutchik interpreted it in terms of functional categories that seemed to make some sort of evolutionary sense. The evolutionary theory was not independently derived, but constructed to fit the data. Hence no empirical challenge was ever offered to a theory built by gathering the untested intuitions of a single cultural group!

Not all evolutionary theories of emotion are this distantly connected with the real world. In the next section, we discuss a theory based on Darwin's own work that is as deeply grounded in empirical evidence as any theory of animal behavior. The results of this theory bear out some of the evolutionary psychologists' expectations, but not all of them.

14.3   The Modular Emotions

The notion of modularity was introduced in section 13.4. Modular cognitive systems can be compared to reflexes like the eye-blink response, except that they involve much more complex information processing. Like reflexes, modular cognitive systems are mandatory: they are not brought into operation by a conscious decision. They are opaque to consciousness: it is not possible to monitor their inner workings. They are informationally encapsulated: not all the information available to the organism is available to the module. And, most important in terms of evolutionary theory, they are domain-specific. They deal with one sort of cognitive process—say, constructing visual images, decoding speech sounds, determining the direction of sounds, or causing rapid emotional responses. Their domain specificity means that modules are adaptations to relatively sharply bounded features of the environment. At least some human emotions—the *modular emotions*—seem to have these characteristic features of modular cognition. They involve complex, coordinated, and automated responses that are not voluntary. Moreover, these responses often are not derailed by the agents' being aware of their inappropriateness to the situation in which they find themselves. So some emotions, at least, seem to be specialized, domain-specific cognitive and behavioral "programs" that are relatively independent of conscious voluntary control.

Darwin correctly identified the characteristic facial expressions produced by the modular emotions, but his work was neglected for most of the twentieth century. It was revived largely through the work of Paul Ekman and his collaborators. Other scientists had repeated Darwin's experiment in which emotions were reliably recognized from photographs of faces (Izard 1969). Ekman's special contribution was to conduct studies on isolated peoples who had had no previous exposure to European facial expressions of emotion. Ekman, Sorensen, and Friesen (1969) conducted studies on the recognition of emotions in two visually isolated, preliterate cultures in Borneo and New Guinea, using the experimental design that dates back to Darwin. Their most impressive work used a new version of this experimental design originally intended for working with children. The traditional version gives observers a list of emotions and asks them to choose one for each photograph. An illiterate observer has to hold the list in memory. Furthermore, the list must be translated into a language whose emotion vocabulary may be almost in-commensurable with the emotion vocabulary of English. In the new version of the design, both problems are avoided. Observers are shown three photographs at once and told a story involving something that would be an emotion stimulus in their culture. They are asked to indicate the person in the story.

Ekman and Friesen showed their photographs to observers from the Fore language group in New Guinea (Ekman and Friesen 1971). These people had seen no movies or magazines, and they neither spoke nor understood English or Pidgin (the official language of New Guinea). They had not lived in any European settlement or government town, and they had never worked for a European. Forty photographs were used in experiments with 189 adults and 130 children. High degrees of agreement were observed between the categories the experimenters intended the pictures to represent and the categories they were chosen as representing by the Fore. In one experiment, photographs of Europeans intended by the experimenters to represent sadness, anger, and surprise were shown to the New Guineans. They were told a story about a man whose child has just died and asked to indicate which photograph represented that man. Seventy-nine percent of adults and eighty-one percent of children selected the face intended to represent sadness. Europeans also seem to recognize the facial expressions of New Guineans. Ekman and Friesen asked the Foré people to act out the roles of the people described in the emotion stories by making appropriate facial expressions. Videotapes of nine New Guineans were shown to thirty-four U.S. college students. For the most part, the students correctly understood which emotion was intended.

Together with previous studies, these findings suggest that the same facial behaviors for a certain range of emotions can be found in all human cultures. This idea has been very well confirmed. Ekman and Friesen's New Guinean experiment was repeated by another team at a later date, and their results were confirmed. Ekman and his collaborators conducted a fascinating experiment showing that American and Japanese students displayed almost identical facial behavior while watching a stress-inducing film—unless an authority figure was observing and questioning them. In that case, the Japanese masked their facial muscle movements by imposing a polite smile on top of them (Ekman 1971). Finally, the ethologist Irenaus Eibl-Eibesfeldt (1973) confirmed several earlier findings that children born deaf and blind display the pan-cultural expressions under the same circumstances as normal children.

Results of this kind are often described as the discovery of *human universals*.

The term *universal* is ambiguous. If "universal" means "a trait that occurs in all or most cultures," then it is much clearer to call these traits *pan-cultural*. Another sense in which a trait might be "universal" is that every individual human being has the trait. It is much clearer to call these traits *monomorphic*. The contrast between polymorphic and monomorphic traits is standard in biology. *Polymorphic* traits are those that exist in several different forms in the same species. Eye color is polymorphic in humans. Monomorphic traits exist in the same form in every "normal" individual. Leg number is monomorphic in humans. None of the experiments described above were designed to show that facial expressions of emotion are monomorphic.

The New Guinean experiments concentrated on six facial expressions, to which Ekman gave the labels "surprise," "joy," "sadness," "fear," "anger," and "disgust." Later work has linked these expressions, at least tentatively, to specific activity patterns in the autonomic nervous system—the system involved in the "fight or flight" response. Ekman has proposed the theory that the modular emotions take the form of *affect programs:* complex, coordinated, and automated responses involving many different physiological systems. The elements of an affect program may include facial expressions, responses such as flinching and orienting to the stimulus, changes in the tone of voice, endocrine system changes and consequent changes in the level of hormones, autonomic nervous system changes, subjective emotional feelings, and such cognitive phenomena as direction of attention. The system that produces an affect program has the properties of a modular cognitive system. The programs occur without conscious decision whenever an appropriate stimulus is present. The programs are opaque to consciousness: they "happen" to a person, rather than being "done" by them. The affect program system is informationally encapsulated because it ignores much of what is known about the stimulus. After a severe electric shock some years ago, one of us (Griffiths) was unable to touch exposed electrical cables for some years, even when he could see that the other end of the cable was not connected. Similar "phobic" responses are familiar to most people. One study showed that people who were experimentally taught to associate a particular food with experiences of nausea continued to be disgusted by that food even when they understood that the food was not connected to the nausea they experienced during the experiment (Logue, Ophir, and Strauss 1986).

Evolutionary psychologists clearly regard Ekman's affect program theory as confirming their predictions about emotions (Tooby and Cosmides 1990). However, his affect programs do not fit their predictions very accurately. Tooby and Cosmides suggest that emotions are solutions to very specific evolutionary problems. They suggest that the emotion module is programmed to

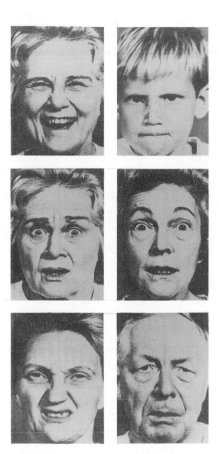

Figure 14.4   Some photographs used by Ekman and his collaborators in their cross-cultural research. *Clockwise from top left:* happiness, anger (closed-mouth type), surprise, sadness, disgust, fear. (From Ekman and Friesen 1975.)

respond to cues such as "looming approach of a large fanged animal" (for fear) and "seeing your mate have sex with another" (for a postulated emotion of sexual jealousy). But although *the output* of the module—the emotional response—is stereotyped and pan-cultural, *the input* to the module—the emotion stimulus—is very flexible and varies between cultures and individuals. The "stimulus appraisal mechanism" controlling the affect programs is not programmed to respond to specific stimulus situations like those Tooby and Cosmides describe. Newborn babies respond to loud sounds and loss of balance with fear, to prolonged restraint with rage, and to gentle forms of skin stimulation with pleasure. They respond consistently to very little else. Recent research has added only one important thing to this list of neonate responses: newborn babies are extremely responsive to human facial expressions (Meltzoff and Moore 1977; Izard 1978; Trevarthen 1984). The founder of behaviorism, J. B. Watson, used the fact that babies do not show the fear response to classic phobic stimuli such as snakes and spiders to support his view that nearly all human behavior results from individual conditioning (Watson 1930). In later life, fear is produced by any stimuli an individual has come to associate with danger, sadness by any stimuli associated with loss, and so forth. In this view, contrary to the predictions of the evolutionary psychologists, affect programs are designed to cope with quite general evolutionary problems, and the affect program system is designed to redefine those problems as the environment changes.

J. B. Watson's extreme environmentalism was mistaken, not because there are more inborn emotional responses, but because learning itself is a complex ability with an evolutionary history. The classic phobic stimuli must indeed be learned, but associations between these stimuli and fear may be easier to acquire and harder to lose than associations between fear and other stimuli, such as flowers and colored shapes (Ohman et al. 1976). The associations between animal-derived foods and disgust and between snakes and fear are "prepared associations" that the organism makes easily and discards with difficulty (Seligman and Hager 1972). The sources of information from which emotion stimuli are learned are also interesting. Associations are acquired through the child's own experience, but they are also acquired through observations of adult emotion (Klinnert et al. 1983). Children need not be hurt in the dark to learn to fear darkness; they need only witness fear of the dark in adults.

Many of the problems of evolutionary psychology discussed in the last chapter are brought out by the example of emotion stimuli. There is an obvious adaptive interpretation of the way in which emotion stimuli are learned, but it is not one that could have been thought out in advance. The affect program system may represent a subtle compromise between the need for flexibility in a changing world and the need to learn without many expensive trials. This particular compromise could not be predicted in advance by imaginatively reconstructing the evolutionary process. The example of emotion stimuli also exemplifies evolutionary psychology's "grain problem." Tooby and Cosmides implicitly treated "fear of predators" as one evolutionary problem. Evolution treated it as a part of a single, larger problem—"fear in general."

14.4 Beyond the Modular Emotions

Ekman's affect program system creates brief, highly stereotyped emotional reactions. But there are many emotions that are not like this. People can be guilty, envious, or jealous without displaying any stereotyped pattern of physiological effects. Affect programs are also strongly modularized, like the systems that construct sensory images of the world. These systems present finished products to the rest of the mind, but their inner workings are opaque. But many emotions seem intimately bound up in and influenced by conscious thought processes. The affect programs are also pan-cultural, whereas many emotions seem to be culture-specific. We have already mentioned the Japanese emotion *amae,* which involves a highly rewarding sense of dependence. Something about human development in Japan induces this feature of the psychological phenotype in a way that human development in Europe does not.

Some authors have argued that the more complex emotions are "blends" of the six or seven "basic" emotions. Everyone has the same basic building blocks, but they put them together in different ways, accounting for the variation across cultures. Ekman and others have shown that there can be blends of the various affect programs. If stimuli for two programs occur together, a blend of the two programs can be displayed. But we doubt that all emotional phenomena can be explained by gluing together these six basic programs. First, there is the problem already mentioned: many emotions do not have particular, stereotyped outputs. Any blend of two affect programs should contain such elements. Second, many emotions are sustained responses, not brief responses like the affect programs. Third, the differing situations that elicit, say, jealousy and moral indignation do not differ from each other merely in the proportions of danger, conspecific challenge, noxiousness, and potential loss that they involve. They have their own specific significances for the organism. Finally, merely blending together several reflexlike responses would not produce something more cognitively involved.

If evolutionary psychology is to explain the wider range of emotions, it will have to uncover other specialized cognitive mechanisms. These mechanisms, like Cosmides and Tooby's proposed specialized rules governing the exchange of resources (1992), should be domain-specific, but able to affect conscious decision making and long-term planning. Frank (1988) proposes just such mechanisms. In his *commitment model* of emotion, emotions are motivations that conflict with the cool, rational calculation of immediate rewards. Loyalty leads people to keep an agreement, even when this brings them no advantage and they have no recourse if the other party fails to do so. Resentful or vengeful people often cause further harm to their own interests in order to avenge themselves on someone who has injured them. Tolstoy's character Anna Karenina committed suicide because of the psychological pain she knew this would cause the lover who had abandoned her. A strong emotional response to perceived exploitation (the "sense of fairness") may lead people to refuse to participate in an arrangement if they believe that the other party is exploiting them, even if there is still an absolute gain to be had by participating in the transaction. The models used in modern economics and decision theory assume that people make choices that maximize their individual economic welfare. These models do not predict the kinds of "emotional" decisions just described.

All these "irrational" behaviors have something in common: if an individual were known to be committed to them in advance, that individual would be treated differently by other agents. This different treatment can have advantages. If a person is known to be loyal, one can make mutually advantageous but unenforceable agreements with them. If a person is known to be vengeful, then it is unwise to wrong them. A person with a sense of fairness must be offered a fair deal because they will turn down anything less. So Frank proposes that many emotions, including anger, contempt, disgust, envy, shame, and guilt, may be powerful and spontaneous motivations designed to enforce commitment to powerful strategies that would otherwise be disrupted by calculations of immediate self-interest.

Several experiments support the existence of Frank's "sense of fairness." In one experiment, pairs of players were asked to divide $10.00 between them. Player one was given the chance to propose a division of the money, player two the chance to accept or reject the proposed division. If player two rejected the division, the money was lost to both. Since the players never met again, there was no question of player two establishing a precedent, so to reject any deal at all was to throw money away. Nevertheless, player two frequently rejected unfair offers. The average division in cases in which an agreement was reached was only 61/39 in player one's favor (Guth, Schmittberger, and Schwarze 1982). A later experiment showed that the average person in a similar game was happy to lose up to $2.59 rather than accept an unfair bargain (Kahnemann, Knetsch, and Thaler 1983). A follow-up experiment cast light on the motivation of the "fair-minded" players. Players were offered the choice of dividing $12.00 on a 50/50 basis with a player known to have proposed an unfair bargain to a third party or $10.00 on a 50/50 basis with a player known to have proposed a fair bargain. 74% of players chose to sacrifice the extra dollar to punish the unfair player. This suggests that the financial sacrifice in the original experiments was made in order to punish player one.

The evolutionary psychologists like to contrast their view of the mind with what they call the "standard social science model," which says that all mental activity is controlled by general-purpose mechanisms. Such mechanisms are not designed to solve any specific adaptive problem. They are supposed to have been selected for their ability to cope with a changing environment. Evolutionary psychology views the mind very differently, as a bundle of specialized adaptations. Frank's proposal involves a typical contrast between specialized and general-purpose mechanisms. He suggests that the emotions are special mechanisms designed to solve the "commitment problem." Traditional decision theoretic models assume that this problem is solved with a general-purpose cognitive mechanism. The two theories thus give very different accounts of human motivation. In the traditional vision, people have general goals such as pleasure, wealth, and social status. They use general principles of rationality and their background knowledge to derive instrumental goals from these general goals. A person might form the desire for a new job because it will increase their purchasing power and give them a higher social status. So in this model, the desires a person has in some specific situation represent means to more general ends. In contrast to this, Frank proposes that certain situations call forth special, irruptive motivations that are not derived from more general ends. People are deflected from the project of maximizing their overall payoffs by emotion. Loyalty to a sacked friend, for example, can prevent their taking the new job. Frank proposes that his theory gives a better explanation of actual human behavior. The experiments described above are meant to show that there are motivations that cannot be derived from general goals. The traditional model must argue that they can be so derived. Perhaps the desire for good relations with the sacked friend is derived from a general goal of social integration. If so, then no special cause of this motivation is needed.

Frank's theory depends on the ability of individuals to communicate their emotions, and hence their behavioral dispositions, to one another. Verbal threats and promises are insufficient because they do not guarantee future behavior. The best way to tell whether someone is loyal or vengeful would be to study their past behavior, but this is often impractical. The connection between facial behavior and emotion might help to bridge the gap. Expressions of affection or threat, although perhaps not as stereotyped and reliable as the expressions of the affect programs, might be more reliable than verbal threats and promises. Some of Frank's experiments suggest that people can gain information about who will cheat in forthcoming games through unrelated social interactions.

In evolutionary terms, however, signals of intention are inherently evolutionarily unstable. An individual who can signal the intention of cooperating without following it up with action will reap the rewards of the cooperative agreement, or deter exploitation, without incurring the costs. This mimic strategy will spread until the signal is no longer reliably associated with the behavior. As this happens, it will be less and less advantageous to respond to the signal. Eventually the whole signaling system will break down. Some game theoretic models predict a cycle in which new signals continually evolve as old ones become discredited. Robert Trivers suggested that an elaborate and unstable system of signaling, cheating, and detecting cheating should result, and that emotions like guilt and anger might be part of that system (Trivers 1971). Frank himself has constructed an evolutionary game theoretic model in which cheaters coexist with honest signalers at an equilibrium ratio. If it is possible to detect cheaters, then cheaters will not drive out honest signalers. But if there is a cost involved in detecting cheaters, then cheaters will not themselves be eliminated. The proportion of cheaters will merely be driven down to the point at which the chance of meeting a cheater is too small to make it worth paying the cost of checking. Those who rely on signals will tolerate a reduction in the reliability of those signals rather than pay the cost of detecting cheaters or the cost of ignoring honest signals.

14.5 Emotion, Evolution, and Evolved Psychology

Several of the concerns we expressed about evolutionary psychology in section 13.5 have emerged in a more concrete form in the study of emotion. One of these is the "grain problem"—the problem of separating out individual evolutionary challenges. A solution to this problem is essential if the mind is to be considered as a collection of special-purpose cognitive adaptations. Unfortunately, the "grain" of the evolution of psychological mechanisms is determined at least in part by the pre-existing cognitive capacities of the organism: its ability to separate out the neurological mechanisms that are relevant to one problem from those relevant to another. This ability is not something that can be predicted without a detailed knowledge of how the brain develops.

The problems of adaptive thinking presented themselves in a very vivid form in the case of emotion. Evolutionary psychology has suggested that adaptive thinking can act as a heuristic guiding the search for psychological mechanisms. In the case of emotion, this seems very optimistic. Emotion theory has clearly made the most progress when it has taken the results of physiological or psychological investigations of emotion and looked for evolutionary explanations of those discoveries. Many emotions seem to solve evolutionary problems that are themselves the products of human psychological and social evolution. The "problem" is best identified by finding the solution.

Emotion, like other topics in human evolution, may hold particular perils for adaptive thinking. The construction of evolutionary "explanations" for claims that we would very much like to be true can obscure the fact that there is little hard evidence that they *are* true. Frank suggests that the emotion of love is an adaptation that allows people to make a real commitment to their partners. Partners who believe they are loved will share their resources freely, because they believe they will not be abandoned if they become poor or ill. "Love is not love which changeth when it alteration finds," as Shakespeare puts it. Sadly for Shakespeare's reputation as an evolutionary psychologist, Melvin Konner (1982) points out that societies have only rarely established pair bonds on the basis of romantic love, and suggests that if this emotion had any biological function, it would be to facilitate adultery and abandonment rather than continuation of the pair bond. The "fact" that there is an emotion called love that holds partners together should be established before it is explained.

The pattern of explanation that Darwin established for emotion, in which behaviors with one function are retained to perform another, demonstrates the importance of a historical, phylogenetic perspective on the evolutionary process. The "self-propelling" nature of human social evolution suggests the same conclusion. Adaptation occurs in a specific historical context, and genuine "adaptive thinking" is the reconstruction of successive phases of evolutionary history. In our view, there have been various promising starts and half-starts on the project of developing an evolutionary understanding of human nature and human society. But the practical and theoretical problems that infest this project are far from being overcome.

Further Reading

14.1   The significance of Darwin's *The Expression of Emotions In Man and Animals* is discussed at length by Ghiselin (1981), Richards (1987), and Griffiths (1997). For a good recent review of the communication of emotion in an evolutionary and comparative context, see Hauser 1996. For a lively read on the evolution of communication in general, with many applications to human communication, see Zahavi and Zahavi 1997.

14.2   The early sociobiologists' discussions of emotions are reviewed by Weinrich (1980). See Tooby and Cosmides 1990 for one version of an evolutionary psychology of the emotions. Both views are analyzed by Griffiths (1997).

14.3   Ekman presents his theory in Ekman 1972, and much important research in the Darwinian mold is outlined in the collection Ekman 1973. The view that emotions are modular is defended at length by Rozin (1976). Another important exploration of the evolutionary perspective on emotion is by McNaughton (1989).

14.4, 14.5   Frank's *Passion within Reason* (1988) is probably the best book yet on the possible evolutionary significance of a wider range of emotions. It is extremely accessible. The issue of commitment is discussed in an evolutionary perspective, but from a point of view more sympathetic to standard ideas about rational decision, by Skyrms (1996). Antonio Damasio's *Descartes' Error* (1994) is another very readable attempt to extend a biological perspective beyond the affect program responses. Both Frank and Damasio are analyzed by Griffiths (1997).

VI

Concluding Thoughts

15

What Is Life?

15.1 Defining Life

Definitions explain the meaning of a term by relating the defined term to other expressions in the language. For example, a definition of *acid* specifies the necessary and sufficient conditions that all, and only, acids share. More generally, definitions relate items in a language to other items in that language. Some of these other terms, in turn, may have their meanings explained through definitions. But at some point the chain of definitions must end. Some concepts must be understood without the help of other verbal formulae. So in semantics and psychology, it is now realized that our capacity to use concepts and refer to kinds need not depend on a grasp, implicit or explicit, on the necessary and sufficient conditions of membership of those categories. Humans have been able to use terms for chemical and physical kinds *(iron, liquid, salt, planet)* long before they understood the nature of those kinds. Though natural kinds may have essences, those essences are discovered not through the construction of definitions at the beginning of inquiry, but, if we are lucky, as the culmination of inquiry.

So biologists do not need a definition of *life* to help them recognize what they are talking about. But definitions are often useful. When categories overlap, or are easily confused with one another, the precision induced by definition is important, for definitions enable us to notice important distinctions that are easily overlooked. Confined as we are to the surface of a near-spherical globe, we can easily overlook the distinction between mass and weight, which is the interaction of mass and a gravitational field. So definitions that made this distinction explicit were important in the development of physics. As we saw in part 2, *gene* has been used to name very different kinds in biology; making these distinctions explicit avoids confusion. Similarly, different concepts of the organism may be important, and hence it is

important that the distinctions among them be explicitly marked. So definition is sometimes an important tool in theoretical advance.

However, we doubt that biology is currently impeded by biologists using
life for distinct though related kinds. For example, we do not see how a definition of *life* is likely to help us with odd and hard-to-classify cases: prions,
viruses, social insect colonies, or the much less plausible idea that the earth
itself is a living system. The adequacy of the definition is settled by our view
of the case, not vice versa. Consider for a moment the *Gaia hypothesis,* the
idea (in one of its forms) that the earth itself, or perhaps just the biosphere, is
a living organism. We see no useful role for a definition of life in evaluating
this metaphor. In some very important ways, the earth is obviously unlike an
organism. It is not the result of evolution through competition within an
ancestral population of proto-Gaias. Nor does the biosphere result from a
developmental cycle. The biosphere we have now will not produce a world-
seed that grows into the biosphere of the earth at some later stage.

If we emphasize the typical histories of living things, then Gaia is not
lifelike. But so what? Defenders of the Gaia idea emphasize the interconnections and reciprocal causal influences of living things with one another and
the abiotic environment. These reciprocal interactions, they suggest, act like
stabilizing or homeostatic mechanisms. There are both conceptual and empirical problems in evaluating this claim. As Kirchner (1991, 41) points out,
in some respects, clearly life has not been homeostatic. Life, after all, radically
altered the composition of the earth's atmosphere. So without an exact specification of the particular homeostatic mechanisms under consideration, the
idea that the biosphere is a connected set of self-sustaining homeostatic
mechanisms is too vague to evaluate. But even if it is made precise, the issue
of whether the biosphere is alive is irrelevant. We do not need to detour
through that question to evaluate the various Gaia hypotheses about the extent to which living systems and their environment change one another, the
extent and ways in which these interactions are stabilized, or the extent to
which these mutual changes make the earth more life-friendly.

So defining life is not a prerequisite for determining the scope of biology.
The revival of interest in definitions of life has a different source: an interest
in *universal biology.* All living systems on earth share many important properties. They are cells or are built from cells. Proteins play an essential role in
the metabolism of all living things, and nucleic acids play an essential role in
the process through which life gives rise to life. Replication and reproduction
results in populations in competition, and natural selection on variation
within those populations produces adaptation, sometimes complex adaptation. For all living things live in regimes in which natural selection is at work.
But are these and other universal features of life on earth characteristic only

of life as we find it here and now? Or are some of these features truly universal: features of life anywhere, any time? Those interested in universal biology seek a characterization not just of life *as it happens to be,* but of life *as it must be.* For them, a definition of life is a specification of life's real essence (Bedau 1996; Langton 1996; Ray 1996).

It is worth pausing for a moment to remind ourselves just how ambitious this project really is. Biologists have always been interested in general principles. We have discussed plenty of candidates from ecology and evolutionary biology. It has been tough enough to find principles that are true of all life here and now. We have argued that adaptive and ecological hypotheses are best seen as hypotheses about particular clades, particular branches in the tree of life, not life as a whole. If that is right, then what price really universal biology: generalizations true not just of our life-world, but of any life-world?

Despite the ambition of the project, a number of biologists have explored the distinction between the specific features of life on earth and those features that life necessarily has. Gould, Kauffman, Goodwin, and Dawkins have deeply contrasting ideas on evolution, but they share this interest. We considered in chapter 12 both Gould's idea that the array of complex adaptations evolution on earth has produced is contingent, and his idea that the complexity of life tends to drift upward over time as a matter of statistical rather than evolutionary necessity. Gould's main emphasis is on the contingency of life's actual history. In contrast, Dawkins argues against a "historical accident" view of life's most central mechanisms. The most central features of both developmental biology and genetics are, he claims, features of universal biology. He argues against the possibility of *Lamarckian evolution,* at least if we understand Lamarckian evolution to involve the inheritance of only *adaptive* changes by the next generation. An organism's phenotype can certainly change its germ line genotype. For example, an organism may expose itself to mutagens in the environment, or act in ways that lower the efficiency of its DNA proofreading mechanisms. But that is not yet Lamarckian, for those changes in the stream of influence from parent to offspring do not make the offspring more likely to resemble the parent in this respect. A rat with a taste for nesting in nuclear reactors is unlikely to produce offspring with their DNA altered in such a way as to induce in them the same preference. Dawkins concedes that it is possible, though difficult, to imagine mechanisms in which the acquisition of a novel phenotypic trait changes the replicators responsible for the phenotype of the next generation in ways that make that novel phenotype reappear. In his view, it is much harder to imagine mechanisms that are sensitive to the distinction between adaptive and other novelties, and which make only adaptive changes more likely to reappear in the next generation (Dawkins 1986).

We are skeptical about Gould's ideas on contingency. We are also very wary of plausibility arguments for impossibility claims—"arguments from personal incredulity," as Dawkins himself has called them in a different context. Dawkins, after all, thinks that memes are replicators (13.6). Memes—if memes are taken to be the information content of ideas—do change, and sometimes adaptively, during the time they are in a particular interactor. If someone using a stone tool of a standard pattern discovers that grinding its edge on sandstone gives it a sharper cutting surface, that is a change in a specific meme token. It is a mutation, and one likely to be passed on because it is adaptive. So if interaction between phenotype and environment can improve a meme that is carried and transmitted, it is not obvious why Dawkins thinks that no similar mechanism could work with other replicators. Admittedly, if memes are replicators at all, they are late-model replicators. They are replicators that emerge deep into the history of a life-world. So perhaps the mechanisms that permit their evolution to be in this sense "Lamarckian" depend on a rich history of prior evolutionary change. But we do not see why this must be so. After all, the fidelity of genetic replication, and the sequestering of the germ line genes in many species, is itself the product of much evolution.

Despite our skepticism about these particular claims, we agree that there is a very good question lurking behind the idea of a universal biology. We seek not just an account of actual biology in all its diversity, but also an explanation of why that diversity is not greater still. However, we see two problems in asking for an explanation of the limits on life's diversity.

First, we should not conceptualize this question by contrasting chance with necessity. Consider, for example, David Raup's representation of possible and actual shell shapes. He shows that, to a first approximation, shell form can be represented as the outcome of only three different growth parameters. In light of this understanding, actual shells occupy a rather small region of the space of possible shells (for an elegant discussion, see Dawkins 1996, chap. 6). Why? Is this restriction a consequence of function, of subtle constraints on development, or of historical contingency? These are clearly difficult but interesting questions. But it is surely unlikely that most of the unoccupied region is literally impossible to occupy. It is equally unlikely that the occupied region is occupied through nothing but historical chance. Similarly, there are no species with three sexes, and that is no accident. As the literature on the evolution of sex makes clear, sex has a cost, and that cost would increase with the number of sexes. But should we infer that the evolution of three sexes is impossible? That would surely be rash: we can conceive of a developmental biology that might work with three sexes. Nuclear DNA has two parents, so we could have three if mitochondrial DNA came from a third. But an evolutionary trajectory leading to three sexes would be both available to a lineage and favored by selection only in very extraordinary circumstances. So, as Dennett (1995) has noted, contrasting historical accident with necessity is likely to be the wrong way of posing this problem. It is probably too crude a distinction to get at the questions that really interest us. Instead, we need some notion of a phenomenon's *improbability*. Bats evolved; no marsupial equivalent did. Is there some reason why a flying marsupial is less likely than a flying placental? Difficult though this question is to answer, it is surely a better question than asking whether a flying marsupial is impossible.

A second problem is the difficulty of testing conjectures about universal biology. This problem of testing is one of the fuels of the developing but over-hyped field of *artificial life*. One of the repeated themes of A-life literature is the "$N = 1$" problem, the problem of distinguishing between accidental and essential features of life with a sample size of one.

> Ideally, the science of biology should embrace all forms of life. However, in practice, it has been restricted to the study of a single instance of life, life on earth. Because biology is based on a sample size of one, we cannot know what features of life are peculiar to earth, and what features are general, characteristic of all life. (Ray 1996, 111)

One aim of A-life is to increase N, and in doing so, generate a definition of life that tells us which features of life are essential to life in and of itself. Just as "strong AI" claims that some computing systems housed in current or near-current computers are not mere simulations of thought, but instances of it, the defenders of "strong A-life" argue that some computer models of lifelike interactions are not simulations of life, but instances of it. They are alive.

The defenders of strong AI argue that a cognitive system is any system organized in the right way. Whether a system thinks is independent of its physical constitution. The essence of mind is form, organization, or function: some abstract property. Because the essential features of having a mental life are not tied to a specific physical implementation, thinking is *substrate-neutral*. Mental properties are functional properties, not physical ones. Strong A-life models itself on this line of argument. Being alive is substrate-neutral. Life is a feature of form, not matter. A living system is any system with the right organization or structure.

> Life is a property of form, not matter, a result of the organization of matter rather than something that inheres in the matter itself. Neither nucleotides nor amino acids nor any other carbon-chain molecule is alive—yet put them together in the right way, and the dynamic behavior that

> emerges out of their interactions is what we call life. It is effects, not things, upon which life is based—life is a kind of behavior, not a kind of stuff—and as such it is constituted of simpler behaviors, not simpler stuff. (Langton 1996, 53)

and therefore,

> it is possible to abstract the logical form of a machine from its physical hardware, it is natural to ask whether it is possible to abstract the logical form of an organism from its biochemical wetware. (Langton 1996, 55)

So in this view, the data structures in, for example, Thomas Ray's famous Tierra program are alive, not merely illustrations of life.

We see no merit at all in these claims. First, the form/matter distinction, the distinction on which the whole idea rests, is an untenable dichotomy. There is no single level of function or organization resting on a single level of matter. Rather, there is a cascade of increasingly or decreasingly abstract descriptions of any one system. In philosophy of psychology, the original home of the function/realization distinction, "two-levelism" has been powerfully criticized by William Lycan (1990). In David Marr's famous description of the structure of psychological theories (1980), there are at least two functional levels alone. The highest level describes the task that the psychological system accomplishes. In the case of vision, Marr claims that the task is to interpret the world in terms of moving, three-dimensional colored objects using patterns of stimulation of the retina as data. An intermediate level might describe how the system processes information in order to accomplish this task. It details the algorithms by which retinal patterns are transformed into representations of the world. The lowest level describes how these computational processes are physically implemented in the brain. Many authors have argued for a number of separate algorithmic or computational levels of description between the superficial level of task description and anything resembling a direct description of brain structure. Lycan has pointed out that much of what passes for a description of the "physical realization" of the mind is really a description of function. Synapses, the connections between brain cells, come in radically different forms, but for most purposes we can abstract away from this detail and describe them by their function: transferring excitation from one cell to the next.

Exactly the same multilevel picture applies to biological systems. For some purposes, a highly abstract, purely informational description of the genome may be appropriate. For others, we want to know in great physical detail the structure of the DNA molecule; for instance, in explaining its coiling properties. Other needs will call for intermediate degrees of detail. There is a whole language of genetics—of introns and exons, of crossing-over, of gene duplication and gene repair—that is functional in abstracting away from the intricate details of molecular mechanisms (some of which are still, indeed, not known), but which is not wholly abstract. This is certainly not a language of form as opposed to matter. So the substrate neutrality thesis rests on a false dichotomy.

Moreover, we think that the idea that simulations are instances of life is an unnecessary hostage to fortune, for the importance of A-life models does not depend on the claim that they create life. The $N = 1$ problem is indeed a serious obstacle to the testing of conjectures in universal biology. But the $N = 1$ problem has been exaggerated, and in any case, the testing problem is not solved by deeming computer simulations to be alive. Of course exobiology would be great if we could do it; a genuinely independent life-world could scarcely fail to tell us much of importance about what is robust about biological process and what is not. But the problem of universal biology can be attacked here and now by the construction of distinct theories that have different implications for evolutionary, developmental, and ecological possibilities, and which can be tested by their application to the huge and varied experiment we actually have available. We do not have a wonderful array of theories that are well confirmed and empirically equivalent with respect to life on earth, but with different implications about how life might have been. $N = 1$ may begin to bite if and when we have to decide between empirically well-confirmed and locally equivalent theories: theories that make the same predictions about life here—predictions that are confirmed—but which make different predictions about what life might be like elsewhere. But we are yet to be indulged with such choices.

Evolutionary simulations will have an important role to play in constructing these theories of life's robust properties. Such models could test conditions under which particular developmental, genetic, ecological, or evolutionary phenomena would arise. Under what circumstances could a third sex evolve? Under what circumstances could variation be directed rather than random? Well-calibrated models that showed the evolution of exotic phenomena not observed in the natural world would be very suggestive indeed. But they can play that role as representations of biological processes, not manifestations of them.

In running simulations, we are trying to find out what those models predict, when those predictions are inaccessible to analytic techniques. The great virtue of these simulations is that one can play with various parameters and thus get a feel for which outcomes are robust under fine-scale changes in the model, and which are not. Thus, for example, Nilsson's model of the evolution of eyes is impressive because the parameters are chosen conservatively, and yet eyes evolve, by geological standards, with great speed (Nilsson and Pelger 1994; see also Dawkins 1996, chap. 5). Simulations are important, and we will consider their message further in section 15.3. But nothing of what these models tell us depends on thinking of them as actually alive. Indeed, we think that the view that these programs are instances of life rather than representations of it trivializes the real questions that motivate universal biology. Consider, again, three sexes. We would like to know whether there are circumstances that would effectively select for three sexes. It is likely that only evolutionary modeling will advance our grip on this problem. But to do so, such models must be well calibrated. Their assumptions must be realistic. Suppose we were to accept that the data structures manipulated in a Tierra-like program were themselves alive. Suppose, further, that we accept that sex is defined not by the physical exchange of nucleic acids, but abstractly and functionally, as the A-life program urges. Sex, in this abstract conception, is information exchange. So any information exchange between token data structures before they are replicated is sex. There is no doubt that it is possible to develop models with three-way exchange of information between data structures. Hence, by this A-life definition, we could have life with three sexes. But this would be a trivial solution to the problem; it is too cheap. Unless the model faithfully represented the constraints on physically embodied living things—for example, constraints on development—it would not tell us what we wanted to know about the possibility of three sexes. If it did faithfully represent those constraints, we could learn what we wanted to know. But nothing would be added by insisting that the model manifests as well as represents life.

15.2 Universal Biology

So we interpret the project of defining life—investigating the extent to which features of the tree of life are historically specific to life here and now—as the program of universal biology. Until quite recently, the issue of universal biology was enmeshed with the issue of biological laws. Scientists and philosophers of science have often taken the main aim of scientific investigation to be the discovery of "laws of nature," such as Newton's laws of motion and of gravitational attraction. Ernest Rutherford, the famous New Zealand physicist who discovered that atoms are mostly empty space, thought that the discovery of such laws was an essential feature of science. Newton's "laws" turned out not to be laws after all, but nonetheless they were the central exemplar of scientific discovery for two hundred years. Scientists record many particular facts: about the charge of particles, the structure of particular compounds, or the age and composition of a particular star. But their main task, in this conception, is to discover the universal principles that particular facts instantiate. Particular domains of science are characterized by distinctive laws, the laws that organize all the innumerable singular facts in each domain. For example, the laws of chemistry might be the general principles that specify the array of possible molecular structures while ruling out others as impossible.

One way of asking questions about biology's status as a science is to ask whether there are any *distinctly biological* laws of nature. To see what such laws might look like, consider von Baer's laws of embryology. In 1828 Karl Ernst von Baer suggested the following generalizations about development:

1. In development, generalized features appear before specialized ones.

2. Within major taxonomic groups, the embryos of different species resemble one another more in early development than they do in late development.

3. The embryos of higher species are like the embryos, but not the adults, of lower species.

4. The embryos of different taxonomic groups diverge progressively and do not recapitulate different levels of adult organization.

Suppose these or other generalizations turn out to hold true. A further question then arises: Are these generalizations reducible, in one of the senses we distinguished in section 6.1, to more general principles? That is, can they be incorporated within chemistry or physics as special cases of more general chemical or physical principles? The status of biology as a good and autonomous science has sometimes been tied to the existence of biological laws and their relation to the laws of more general disciplines. Biology, in this view, is an autonomous science in good repute only if biologists have discovered laws—and moreover, laws that are not just special cases of more fundamental principles. Physics-oriented philosophers of science such as J. J. C. Smart (1963) have suspected that biology is not in this sense a real science, but instead a technical discipline like civil engineering. In this view, biology merely explores the consequences of the operation of general physical and chemical principles in particular contexts. We saw in chapters 6 and 7 that the principles of Mendelian genetics are probably not reducible in any simple way to those of biochemistry. So if those principles counted as laws, they would form the subject matter of an autonomous discipline. However, as we shall see, they probably do not count as laws by the classic criteria of lawlikeness.

Laws of nature have two features. First, they are exceptionless universal generalizations. The generalization

> On earth, all organisms have a particular genetic code in which four distinct bases specify twenty amino acids and a stop signal

is not a law of nature because it is spatiotemporally restricted, and it is not quite exceptionless (Dyer and Obar 1994, 73–74). Mendel's laws are not exceptionless either. Second, in a sense that no one has ever succeeded in making properly clear, laws of nature hold *necessarily*. Their truth is no accident. Consider the contrast between "No dense object 20 kilometers in diameter consists of chemically pure gold" and "No dense object 20 kilometers in diameter consists of chemically pure plutonium." The first statement may well be true. Quite likely, no large planetoid of chemically pure gold has ever formed. But if true, its truth is accidental. There is nothing about the way the universe works that debars such a possibility. The truth of the second statement, however, is no accident, as a lump of plutonium that large would be above critical mass and would blow apart. So while there could be a gold planetoid, there could not be a plutonium one. Hence only the second generalization is an application of a law of nature. So even if the "genetic code" were universal (perhaps because life here on earth is all the life there happens to have been), a specification of the codon/amino acid pairing is not a law of nature unless this pairing is the only pairing there could be, which it is not.

It is now widely accepted that in this sense, there are no biological laws of nature. Rosenberg argues that this follows from the fact that biological kinds are functionally rather than structurally identified. There are, he says, no interesting true generalizations about marine animals because of the great physical and structural heterogeneity of those animals (1994, 33–34). He takes biology to contrast with physics and chemistry in this respect. But it's very far from clear that he is right. To the contrary, the picture of physics and chemistry as scientific domains in which myriads of particular facts are organized by exceptionless laws, laws that it is our aim is to discover, may be wrong. It is arguable that this picture depends on an oversimplified view of those disciplines (Cartwright 1983, 1989). If so, then law-hunting is the wrong aim for universal biology. Universal biology will not consist of a set of exceptionless generalizations. And earthly biology cannot be segmented into a universal part—generalizations true of all life everywhere—and historically contingent, spatiotemporally restricted generalizations that happen to be true of life here and now.

Nonetheless, we agree that there is an itch to be scratched. We see two different routes by which universal biology can be pursued, one focusing on pattern and the other on mechanism. First, we can think of universal biology as a set of hypotheses—speculations might be a better term—about robust patterns in the history of life, here and now and in such other life-worlds as there may be. We have already seen one example of such a hypothesis. In section 12.1 we discussed Gould's claim that an increase in mean complexity over time depends on the fact that life starts close to the point of minimum complexity. If he is right, this is a pattern we would expect to see in most life-worlds. Dawkins has floated a much more ambitious set of ideas about robust patterns of complexity, outlining a series of complexity thresholds through which he expects all or most life histories to pass. Dawkins defends a "replicator-first" view of the origin of life, so for him, the first of these thresholds is the formation of a replicating molecule. The second he calls the "phenotype threshold," which is passed when replicators begin to increase not by virtue of their intrinsic chemical properties, but through phenotypic effects on their environment. A third critical threshold, in his view, is passed when replicators and their phenotypes become linked in teams; we might think of this as the invention of something like an organism (Dawkins 1995, 151–155).

These are very large scale hypotheses about robust evolutionary patterns. Many much more particular hypotheses have also been proposed. For example, Dennett (1995) discusses "forced moves" and "good tricks" in design space: adaptations we might reasonably expect to find in an independent experiment in life. If organisms that move and explore their environment evolve in a world, then vision will be a "good trick." It is clearly not inevitable: organisms need to be big enough to support eyes, and some kind of light-sensitive pigment must be available. But we certainly would not be surprised to find vision in an independent life-world. These specific pattern hypotheses are likely to be conditional rather than categorical. If avoiding Muller's ratchet explains the existence of sex, then we should not expect to find the equivalent of sex in a life-world unless creatures in it have segregated into distinct species and some of those species have small population sizes. For it is in such populations that mutations accumulate.

An alternative approach to universal biology focuses on mechanisms. We noted above Rosenberg's skepticism about biological laws, even earth-bound ones. We agree with his views on biological generalizations, but are not convinced that biology is distinctive in its lack of laws. Rosenberg himself thinks otherwise. He thinks that the physical sciences, dealing with a simpler and more structurally uniform domain, can still hope to discover simple universal principles that underpin and explain "the buzzing, blooming confusion of nature" (Rosenberg 1994, 33, in turn borrowing from William James). Rosenberg takes this to mark an important difference between biology and other sciences. His pessimism about biology—his insistence that we can conceive of it only as a kind of useful instrument—seems to overlook the possibility that realist biology can be pursued not by seeking exceptionless general laws, but by discovering recurrent causal mechanisms.

Most obviously, natural selection itself will be a distinctive and critical mechanism operating in any living world, for the complex adaptive mechanisms distinctive of life can arise only by cumulative natural selection. We have emphasized that cumulative natural selection depends on more than variation, heritability, and differential fitness (2.2). So we cannot rule out the existence of semi-life-worlds, worlds in which replicators of a sort exist and interact with their environment in ways that enable them to gather the resources to replicate, but in which replication is so inaccurate, and the direction of selection so variable, that no complex structures have ever evolved. But these are precisely worlds in which we would be pushed to decide whether there was life or not.

Natural selection might not be the only universal or near-universal mechanism. It might turn out that the chemistry of life is inevitably carbon-based, so some biochemical mechanisms might be universal. At a larger scale, there might be universal aspects of development. Consider, for example, gastrulation, the first major reorganization of a developing animal embryo. In this process, the hollow ball of cells formed by the initial divisions of a fertilized egg folds to form a cup with an inner lining, beginning the more obvious process of cell differentiation. There are probably no important universal and exceptionless generalizations about gastrulation. Nonetheless, it is a conservative and conserved process. Gastrulation takes place in quite similar ways across animal life. It is a very important developmental mechanism, even if it is expressed somewhat differently in many developmental processes (Buss 1987, chap. 2). We doubt that gastrulation as such is likely to be a feature of a really universal biology, if such is to be had. But some developmental mechanisms might be. We might risk a modest wager that some form of developmental entrenchment—early aspects of development are increasingly difficult to change—will be a robust feature of life. So despite their antiquity, von Baer's laws may be part of a future exobiology!

Other claims about necessary mechanisms strike us as more suspect. Both Dawkins and Maynard Smith have floated the idea that the information transmitted from generation to generation must be digitally coded (Dawkins 1995; Maynard Smith 1996). They argue that if natural selection is to build complex organisms, the fundamental mechanisms of replication must be digital. Complex living structures arise only through long histories of gradual change under cumulative selection, and that, in turn, requires high-fidelity replication. *Digital codes* can be replicated many times with high fidelity, for they are inherently far less ambiguous than *analog codes*. Analog codes, on the other hand, are impossible to replicate many times without critical degradation. A document photocopy chain a hundred links long will have an unreadable blur at the hundredth link. Send the same document through a hundred-link e-mail chain, however, and the first and the hundredth will probably be identical.

> An analog genetic system could be imagined. But we have already seen what happens to analog information when it is recopied over successive generations. It is Chinese Whispers. Boosted telephone systems, recopied tapes, photocopies of photocopies—analog systems are so vulnerable to cumulative degradation. Genes . . . can self-copy for ten million generations, and scarcely degrade at all. Darwinism works only because—apart from discrete mutations which natural selection either weeds out or preserves—the copying process is perfect. Only a digital genetic system is capable of sustaining Darwinism over eons of geological time. (Dawkins 1995, 19)

We discussed in chapter 5 our general worries about the idea of genes and genotypes as codes. Let us set these aside. We are still unconvinced of the digital encoding hypothesis. The fidelity of replication depends not only on the ease with which distinct characters in the code can be recognized for what they are, but on error correction systems as well. So even if analog replication has a higher error risk, if it is supported by good error detection and correction mechanisms, long chains of high-fidelity analog representations are possible. Thus if at each link in the photocopy chain, thousands of slightly varying copies are made, and only the best is retained for copying into the next link, then a long high-fidelity series is possible. Actual biology shows that this is no idle possibility. For, as Dawkins himself points out, a fertilized cell is no mere package of DNA. Cell differentiation in the early embryo depends on a series of chemical gradients in the fertile egg: from top to bottom, from front to back, and, often, from left to right. It is these gradients that cause different cells in the early embryo to differentiate from one another. If genes are digitally coded information, then chemical gradients are analog instructions telling cells where they are. Yet this information, this gradient, is reconstructed with high fidelity generation by generation.

Despite our skepticism about this hypothesis, we suspect that somewhat more convincing arguments can be mounted for a universal biology of mechanism than of pattern, just because the basic mechanisms of life are more directly constrained by the physical and chemical basis of life. Thus the fact that energy is never converted with perfect efficiency has implications in ecology for the structure of food chains and communities. Big fierce animals are rare, for they can never harvest more than a smallish fraction of the energy potentially available to the primary producers (Colinvaux 1980).

15.3 Simulation and Emergence

Universal biology has been most consistently pursued in the field of artificial life, most importantly in the work of Stuart Kauffman (1993, 1995a,b). Much of this literature itself, and even more of the philosophical reflection on it, has focused on the issues of *emergence* and *self-organization*. The contribution this work makes to universal biology is the claim that there are both very general patterns and very general constraints that emerge out of the complexity of the organization of life. As we shall see, these constraints are often read as constraints on selection.

The idea of self-organization is the idea that living systems are inherently organized; organization arises spontaneously in the system itself rather than having to be imposed from the outside through the mechanism of selection. We shall see the importance of this idea in Kauffman's work shortly. This discussion of emergence links an empirical idea to a conceptual one. The central empirical idea defining emergence is that surprisingly complex system-level behavior can arise out of locally interacting simple units. Complexly behaving systems require neither complex parts nor central direction. The elements in A-life models are often quite simple units whose interactions are all governed by local rules—indeed, relatively simple local rules. But the behavior of the system as a whole is often adaptively complex. Some social insect colonies may provide natural examples of the phenomenon in question. Simply interacting simple creatures nonetheless produce complex, adaptive, and patterned behavior. So a good many of the most striking examples of A-life models can be seen as undercutting the idea that fancy systems must be built of fancy components. They show that complex system-level behavior may arise out of interacting simple components.

The conceptual idea is methodological. Since the interaction of the components determines system-level behavior, we will not get much of a handle on what the system will be like by studying the components in isolation. Understanding emergence as an empirical phenomenon will require new models of scientific explanation (see Burian and Richardson 1996; Clark

1996, 1997; Hendriks-Jansen 1996). We noted in section 7.3 that some ideas in developmental biology are thought to support a similar message.

An example might make these abstract points clearer. One good example is Reynolds's model of flocking behavior. He calls his simulated creatures "boids," and the rules they follow are very simple. Each acts

> to maintain a minimum distance from other objects in the environment, including other boids,
>
> to match velocities with boids in its neighborhood, and
>
> to move toward the perceived center of mass of the boids in its neighborhood. (Langton 1996, 66)

Despite the simplicity of these rules, boids simulate flocking rather well. Boids flow naturally around obstacles, and they show the illusion of coordination that we see in schools of fish and flocks of birds. So this example shows how creatures following very simple, locally cued behavioral rules could form flocks whose global behavior appears coordinated.

So some of these A–life simulations are very suggestive. But what, exactly, do they show? What is their evidential status? This question is particularly important in thinking about Kauffman's work, for many see him as developing a picture of life that underplays the role of natural selection. His work is often presented both as showing restrictions on the power of natural selection and as showing that we do not need to invoke selection to explain order. Order arises "naturally."

Kauffman's work exemplifies the idea that complex macroscopic organization can derive from the interactions of simple systems under local control. For example, Kauffman argues against a "replicator-first" version of the origin of life. He claims that two constraints make replicator-first views implausible. If the simplest bacteria are any guide, even first-generation replicators would have to be quite long, for short sequences would not exert phenotypic power over their propensity to be copied. Yet, despite the lack of evolved catalysts and evolved error-correcting machinery, these first longer sequences—the first sequences with phenotypic power—would have had to be replicated accurately enough to avoid an error catastrophe that would destroy the biological properties of their copies. Kauffman doubts that this is possible (1993, 287–291; 1995a, 41–43). He defends instead a "metabolism-first" or a "cell-first" view of life's origins. When enough biochemicals are confined in a single system (and he suggests ways in which this might happen), the chance becomes quite high that there will be sufficient catalytic links between the individual constituents for the "soup" as a whole to become "autocatalytic," sustaining itself without there being any element dedicated to replication. The properties of life emerge spontaneously at some threshold of complexity of the system as a whole without that system containing any element that plays a distinct role in its maintenance or replication. Life, in Kauffman's view, is an *emergent phenomenon:* it arises from relatively simple locally interacting constituents. It is a property of an ensemble, not of any special element within the ensemble.

We find these ideas on the origin of life interesting and suggestive, but Kauffman is probably best known for his ideas on evolution. These ideas are generated from very simple, very abstract models in which just two elements vary. N is the size of a population whose units vary in fitness. In these models, N is often thought of as a population of genes. K measures the "connectedness" between members of N. The more other units each unit of N interacts with, the greater is K. So K measures the extent to which the fate of each unit is determined locally: as K goes up, local control goes down. If we think of N as the genes in a genotype, K might measure the number of genes that determine whether a given gene is switched on or off. K's role can be modified by a third parameter, P, which modulates K. P measures the sensitivity of our target gene to its promoters and repressors. If P is high (near 1.0), the target gene's action is insensitive to its environment; for example, it will remain on unless all its inputs are telling it to turn off. If P is low (near 0.5), the target gene is sensitive to all its inputs, and high values of K will have a profound effect.

Kauffman derives some striking and lifelike general results from these models. Dennett (1995) suggests that we think of selection as an engine that, granted order of a certain kind—order with variation—generates design. One way to think of Kauffman's results—a way he often suggests—is to see them as showing that selection has rather more order to work with than we might have thought. At the beginning of life, selection would not have to build cells all the way up from amino acid biochemistry. Instead, richer and more complex structures would automatically arise and become available for selection. Given the size of gene populations in cells, selection would not have to build the whole array of differentiated cells from single-celled prototypes that varied only slightly from one another. For if K were low, but not too low, different gene activity patterns would automatically generate an array of cell types. In this sense, Kauffman thinks his models yield "order for free," not as a replacement for selection in the explanation of organic differentiation and adaptive design, but as a richer input to that process. This is at once a constraint on selection, for fewer apparently possible biological structures are really possible, and a boost to selection, for it makes it easier for selection to reach some regions of design space.

Even so, the most general and important result in the *NK* models is that connectedness damps down the effect of selection. Kauffman agrees that selection is central to the history of life. But he argues that it is effective only in certain adaptive landscapes. Selection, recall, can take a population from one phenotype to another only if the intermediate phenotypes are of intermediate fitness (2.2). Consider a rat population whose body weight averages about 1 kilogram, living in an environment in which rats would be better adapted if they weighed 2 kilos. If a 1.5 kilo rat is less fit than either a 1 kilo or a 2 kilo rat (too fat to run; too small to fight), then selection alone cannot edge the phenotype to 2 kilos, even if the 1 kilo average is disastrously less fit than the ideal 2 kilo rat. If (holding other aspects constant) there is a steady increase in fitness as weight approaches 2 kilos, the phenotype fitness landscape is *smooth*. If, instead of a smooth upward curve, when we plot weight against fitness we see a jagged curve with many rises and dips between 1 and 2 kilos, the fitness landscape is *rugged,* and there are many local optima.

How effective will selection be in these different fitness landscapes? This in part depends on a third factor. We have spoken so far of the fitness of *phenotypes*—in particular, of body weight. A further condition for effective selection is a reasonably systematic relationship between genotype and phenotype. Let's call the genotype of a rat that weighs 0.98 kilo **R. R*** is the genotype of a 1 kilo rat, and 1 kilo is the local optimum. If you are a rat in the range 0.8 to 1.2 kilos, you are best off being exactly 1 kilo. But can selection push a population of 0.98 kilo rats (with genotype **R**) to a population of 1 kilo rats (with genotype **R***)? Only if genotypes that are similar to **R** have a similar fitness, presumably because they have a similar phenotype with respect to body weight and other traits relevant to rat survival and reproduction. If a small variation in **R** (say, a change in one gene) produces a distinctly different phenotype, and hence a genotype of distinctly different fitness, then the fitness landscape is *uncorrelated*. Selection is ineffective if the fitness landscape is uncorrelated. If it is rugged but correlated, selection can at least push populations to local optima. If it is smooth and correlated, we can get to a global optimum. But in uncorrelated landscapes, selection takes us nowhere.

Recall that in Kauffman's models, *K* measures the connectedness of genes in a genome. *K* measures the number of genes that determine, say, whether a given gene is turned on. Now, according to Kauffman's models, if *K* gets too high, we should expect uncorrelated fitness landscapes. (Values above

3 or 4 are high unless a gene is relatively unresponsive to the genes connected to it—unless P is high.) So evolution under natural selection is possible only in a rather abstractly defined class of environments in which the linkage of the components is not too tight, and in which the fitness landscape is not too rugged.

What are we to make of these results? Some of them are genuinely striking. They seem to accord well with what we know of development. For instance, the models predict that early ontogeny should be more fixed than it is later because of the entrenchment of early mechanisms. Furthermore, Kauffman also argues that his models predict the pattern of the Cambrian radiation. We should expect evolutionary histories to be characterized by a "Cambrian explosion" pattern, with most diversity generated early and relatively less originating later. Kauffman's reasons depend in part on the broad developmental considerations we discussed in chapters 10 and 12, but also on the idea that many relatively low fitness peaks will be unoccupied early in an evolutionary radiation. As these peaks become occupied, it becomes harder to find a higher peak. Imagine, for example, a previously unoccupied region being penetrated for the first time by plant-eating insects. At first, many different varieties will find ways of making a living. But as time goes by, fewer and fewer changes will result in organisms whose lifestyles have not been pre-empted.

> The Cambrian explosion is like the earliest stages of the technological evolution of an entirely new invention, such as the bicycle. Recall the funny early forms: big front wheels, little back ones; little front wheels, big back ones. A flurry of forms branched out . . . giving rise to major and minor variants. Soon after a major innovation, discovery of profoundly different variations is easy. Later innovation is limited to modest improvements on increasingly optimised designs. (Kauffman 1995a, 13–14)

So a lot happens fast, then not much happens at all.

These models are very clearly suggestive, but we remain cautious. Note, for example, that Kauffman's "Cambrian pattern" is not the pattern of the actual Cambrian radiation. As we discussed in sections 12.2 and 12.3, the critical claim about the Cambrian is about *morphological* diversity, not *adaptive* diversity. Gould claims that the Cambrian saw morphological diversity at its maximum. He thinks that even after the Permian mass extinction, there was no comparable invention of new body organizations. In some views, Gould's picture reflects taxonomic practice rather than biological reality. But no one claims that the adaptation-building engine switched off after the Cambrian.

Many major adaptive complexes postdate that era. Yet Kauffman's claim is, as we have seen, one about adaptive evolution and adaptive complexes.

More generally, the very abstractness of these models makes their connection with real biological phenomena difficult to evaluate. A mean-spirited approach would be to argue that these models are like the "proofs" nineteenth-century scientists are alleged to have produced showing that bumblebees cannot fly. Consider, for example, the idea that genotypes are self-organized. Given their degree of interconnection, it is unlikely that selection could prevent mutation and other disruptions from "spreading genotypes more evenly over the fitness landscape" (Burian and Richardson 1996, 157–58). Yet genotypes are not just ordered and complex: they are very considerably differentiated from one another, and this in many ways must be the result of selection. The differences between primate genotypes may be partly due to drift, but surely many are the result of selection. So we *already know* that selection can change genotypes, despite their apparently high connectedness. That knowledge cuts across the model result that as the connectedness of a system goes up, and the number of elements in that system goes up, selection becomes increasingly ineffective. Kauffman's investigations into universal biology have discovered a "constraint on selection" that shows that most actual biology is impossible, and that much actual evolution has not happened.

But there is a more generous way of thinking about these models. They lead us to ask how evolution under natural selection dodges the apparent constraints that would seem to make it impossible. Is the number of effective units (the size of N) smaller than it would seem? Is effective connectivity less than it seems? It is often thought that if natural selection is to be effective, the phenotypes of organisms must be modular, with some traits able to vary independently of one another. So perhaps we should see these models as offering a hint that genotypes, too, are more modular than they seem. More generally, we should treat these models as "how possibly" explanations. Adaptationism's critics have often made the point that we should not conflate "how-possibly" explanations with "how-actually" explanations, and that point is well taken. Even so, how-possibly explanations are important. First, even in those areas in which we think we have approximately the right how-actually story, an expansion of the space of possible explanations is often useful, for competing explanations suggest critical tests. How-possibly explanations are still more important when we deal with puzzling phenomena. A how-possibly explanation of, say, the evolution of human language would be useful because we have no good grip on what intermediate forms of language might be like and why they were adaptive. Language poses a *trajectory problem*. The same is true of the origin of life.

When it comes to universal biology, then, we are left shivering on the brink, nervous virgins wondering about sex. There is no denying the fascination of the problems posed. We would love to know which features of the tree of life are robust. Would relatively small, chemically possible changes in the DNA-RNA-protein transcription machinery preclude the evolution of sex? Would relatively small changes in mitochondrial inheritance make a third sex possible? Yet though speculating on these questions is fun, and simulation imposes some discipline on our speculation, we suspect that they remain empirically recalcitrant.

Further Reading

15.1 Putnam's work is primarily responsible for the insight that definition plays a relatively minor role in our grasp of concepts (see especially Putnam 1975, chapters 11 and 12). For an introduction to this view of our concepts and the way they relate to the world, see Devitt and Sterelny 1986. Griffiths (1997) applies these ideas specifically to concepts in biology. We rather doubt that the Gaia hypothesis has been worth all the ink spilled in its elaboration. But readers who think otherwise might find Joseph 1990, a very friendly overview and history of the hypothesis, enjoyable. Schneider and Boston 1991 is a well-balanced collection on the subject.

For a general introduction to A-life, see Emmeneche 1994. Chris Langton is a central figure in the development of A-life. He has edited a series of collections on A-life for the Sante Fe Institute for Studies in the Sciences of Complexity. These are published as *Artificial Life 1* to *Artificial Life N,* with a rapidly growing N. These volumes are very variable in content. Boden 1996 is a very useful anthology, partly but not wholly drawn from this series. Langton 1995 is also a good anthology, though much less philosophically oriented. For a fine critique of the form/matter dichotomy on which the strong A-life program rests, see Lycan 1990. Sober (1996) surveys this program in his typically lucid and sensible way. See also Sterelny 1997a.

15.2 Ernst Mayr has campaigned long and hard in defense of the idea that biology is both a good and an autonomous science (see the first two essays in Mayr 1988, and more recently, Mayr 1996). The relationship between biology and other sciences is central to the work of both Rosenberg (1985, 1994) and Dupré (1993). In very different ways, they both end up with the view that biology has a different character than physics and chemistry, though only

Rosenberg reads this as an indication that biology has a different status as a science. In the supplement to volume 64, number 4 (1997), *Philosophy of Science* has published an important symposium on laws in biology, with papers by Beatty, Brandon, Sober, and Mitchell. Weinert 1995 is a recent good collection on the general issue of laws of nature.

15.3 Kauffman's magnum opus (1993), as we have noted before, is very difficult. Kauffman 1995a is much more readable, though somewhat infested by musings on the meaning of life. In Kauffman 1995b, he gives a good short introduction to his views. There are good introductory discussions of his work in Emmeneche 1994, Depew and Weber 1995, Burian and Richardson 1996, and Weber and Depew 1996. There is a briefer introduction in Dennett 1995.

Final Thoughts

Dobzhansky remarked that "nothing in biology makes sense except in the light of evolution." This famous slogan is sometimes interpreted to mean that natural selection has a profound and pervasive effect in every nook and cranny of the biological world. Indeed it does: it is no part of our vision to underplay the significance of selection. To the contrary: it is part of our message that we should not see selective and historical explanation in evolutionary biology as alternatives to each other. But equally, we can and should take Dobzhansky's slogan to mean that *history matters*. Nothing in biology makes any sense except in the context of its place in phylogeny, its context in the great tree of life. As we see it, time and time again it turns out that reconstructing that tree is critical to understanding the living world. Species, the basic units of life's variety, are the most fundamental segments of that tree. Interpreting and testing adaptationist hypotheses, reconstructing the relationship between ecology and evolutionary history, displaying and explaining the pattern of evolutionary history at the very largest scale, all depend on reconstructing that tree. The argument of this book reveals, we think, a critical foundational role for systematics in all these fields. So cladism, the view that both sets discovering phylogenetic history as the goal of systematics and develops techniques for carrying that goal through, turns out to be of fundamental importance. Many evolutionary hypotheses cannot be interpreted, let alone effectively tested, without a reasonable estimate of a group's phylogeny. Gould's hypothesis about the upward drift of complexity, for example, can be neither interpreted nor tested except against the background of a phylogenetic hypothesis about the lineage in question. So one general conclusion that emerges from our argument is the centrality of phylogenetic hypotheses to evolutionary debates.

Like most other authors in this field, we have made extensive use of the replicator/interactor distinction. This distinction is often associated with the rhetoric of reductionism, which suggests that only the genes really matter in evolution, although the distinction itself makes no such commitment. We are sympathetic to the replicator/interactor distinction as a way of conceptualizing the two fundamental mechanisms that together generate evolutionary history, though Griffiths, in particular, thinks that there is another fully adequate formulation of evolutionary theory that is not centered on this distinction. But we have no sympathy with a "reductionist" reading of the distinction. That is, we have no sympathy with an interpretation that downplays the importance of interaction or interactors, nor with one that prejudges the cast of replicators and interactors. We have suggested expanding the usual list of replicators, for we think that nongenetic inheritance is more pervasive and more important than previously thought. Moreover, we regard the cast of interactors as open, too. Both group and species selection (taking groups and species as interactors) raise, as we have shown, difficult empirical and conceptual questions. But in our view, there is a persuasive—though very far from decisive—case in favor of both. More importantly, as we see it, there is no general conceptual or theoretical argument against the possibility of selection on high-level interactors. Suggestions about group and species selection therefore need to be looked at empirically, on a case-by-case basis.

So a second general conclusion is that the macroscopic levels of organization biologists study—organisms, groups, species, clades—are most unlikely to be epiphenomena of processes at lower levels of biological organization. They may sometimes be so: Vrba points out that differential species survival is often a mere effect of differential success at the level of individual organisms. But there is no reason to suppose, as a general operating principle, that this *must* be so. That would conflate the uncontroversial "no miracles" version of the reductive demand with the much more ambitious and controversial demand that all theories about macroscopic processes be incorporated by theories of the microlevel. We think that the replicator/interactor conception of evolution is best combined with a recognition of the reality and importance of the nested levels of organization of the living world—that is, of its hierarchical organizations both in ecology and evolutionary history.

We began this book by suggesting that philosophy of biology has a role to play in the debates arising out of evolutionary theory and related branches of biology because those debates have both a conceptual and an empirical element. We hope that the long march through fifteen chapters will have convinced the hardy survivor of two things: first, that this assessment is right. The debates about the nature and importance of selection, about species and their role in evolution, about Gould's disparity hypothesis, and the like are not simple empirical disputes that could be settled by appropriate observation or experiment. But neither are they "merely semantic," easily settled by agreeing on a convention of how to use terms such as "species," "disparity," or "reduction." Second, we hope we have shown that this combination of conceptual and empirical elements does not make these problems *utterly intractable;* difficult, yes, but not impossible. For we think this book shows that the community of biologists and philosophers of biology has been making real progress on these issues. We think the discussions of, say, the units of selection, reduction, species, disparity, and the like all show that we are collectively generating real progress, not just endlessly chasing our own tail.

So we do claim progress. But we also agree that many problems remain open and difficult. As we mention in chapter 5, the right picture of the relation between developmental and evolutionary biology remains very much a matter of debate. So too does the whole question of genetic information: whether there is some important and distinct sense in which the genes (and other replicators) carry information about an interactor's phenotype. As we noted in the discussion of group selection, there are many unresolved issues in the proper understanding of organisms and superorganisms. Philosophy of ecology has barely got going. And there are many other examples. So philosophers of biology need not fear running out of business, at least in the short run.

Enough, already.

Glossary

actual sequence explanation   an explanation that characterizes events in fine detail, so that substituting other similar events would make the explanation invalid. See *robust process explanation*.

adaptation   a feature of an organism whose presence today can be explained by the fact that it served some useful purpose in previous generations and hence helped some organisms to reproduce more than others. A cat's claws, for example, are adaptations for catching prey.

adaptive   contributing positively to the current fitness of the organism that possesses it.

adaptive radiation   a process by which, if the members of a species find themselves in vacant territory (by being the first to reach an island, by surviving an extinction event, or by invading a new type of habitat), their descendants, over evolutionary time, often diversify into many new species. Some of these will make their living in ways very different from the founding species.

allele   an alternative form of a *gene*. Genes are located at particular regions of a *chromosome* known as loci (singular: *locus*). In a given population and at a given locus there may be several alleles.

amino acids   the building blocks of *polypeptides* and hence of *proteins*. One amino acid corresponds to one *codon* in the genetic code.

analog code   a system in which a range of continuous values of a variable in the receiver represent continuous values of a variable in the sender

analogous traits   see *homologous traits*

arms race   a competitive ecological interaction between two species as a result of which each becomes better adapted to cope with the presence of the other.

arthropods   a phylum of metazoans with a segmented body and an exoskeleton divided into jointed units. Insects and crustaceans are living examples.

Bauplan (plural: **Baupläne**)   the fundamental body plan of a group of related species; the basic layout manifested in various forms in these organisms. Also known as *type*. The persistence of body plans over long periods of evolution and through many episodes of speciation is known as the *unity of type*.

biogeography   the study of the distribution of plants and animals across the globe.

biological determinism   the view that important features of human psychological or social organization are in some way "fixed" by human biology. Different accounts of what and how characteristics are fixed generate different variants of biological determinism.

biota   the totality of living things in a region.

chromosome   a long DNA molecule that is wound around supporting, structural proteins; found in *eukaryotes*.

clade   a group of species and their common ancestor; hence a segment of the tree of life (derived from the Greek word for "branch"). See also *monophyletic*.

clone   a sequence of identical copies of some biological entity. Since no two biological entities are strictly identical—alike in every respect—describing a sequence as a clone implies a judgment about important similarities. Usually, genetic identity is the identity in question, and so clones of organisms arise only through asexual reproduction.

codon   a group of three *nucleotides* in the genetic code. One codon specifies one amino acid, the unit from which proteins are built. Some codons signal the beginning and end of a gene rather than specifying an amino acid.

conspecific   an organism of the same species as the one under discussion.

crossing-over   a process that occurs during *meiosis* in which homologous chromosomes cross over and recombine, so that a part of each chromosome is exchanged with the other, before the chromosomes split into different *gametes*. Hence a gamete can have a sequence at a locus different from that of either parent.

cumulative selection   selection acting repeatedly on a population. A new adaptation will evolve only as the result of many generations of selection preserving the favored feature and, in partnership with the mechanisms of variation, gradually enhancing it. Cumulative selection is much more powerful than single-step selection.

derived trait   see *primitive trait*

digital code   a system in which discrete states of the receiver represent discrete states of the sender.

diploid   having two versions of each chromosome; if the organism is sexual, one of these comes from each parent.

disparity   the variety or range of biological forms manifested at a particular time. Disparity is often conceptualized through spatial metaphors. Gould, for example, thinks of "morphospace" as the array of all possible ways organisms could be physically organized. Disparity would then be the chunk of morphospace colonized at a particular time. It is controversial whether disparity can be measured in any objective way.

dispersal   the distribution of plants and animals as a result of their own movement or that of their *gametes.*

diversity   the number of species extant at a particular time.

dominant   the relationship of one *allele, A,* to another, *a,* at the same locus when the *heterozygote Aa* has the same *phenotype* as the *homozygote AA.* The other allele, *a,* is said to be *recessive* if the phenotype of the homozygote *aa* is distinct from the *Aa* phenotype. Because the *Aa* phenotype can differ from both *AA* and *aa,* the dominant/recessive distinction is not exhaustive.

eliminativism   the idea that the processes and entities mentioned by a theory can be shown not to exist by refuting that theory

epistasis   an interaction among genes in which the effect of an allele at one locus depends on which alleles are present at some other locus.

epistemology   the theory of knowledge and its nature. For example, an epistemological, or epistemic, question about biology is a question about how that biological fact is known.

ethology   the study of animal behavior under its normal ecological conditions (as opposed to unusual laboratory conditions) and from an evolutionary perspective

eugenics   the improvement of human fitness through selective breeding

eukaryotes   organisms built from complex cells that have a discrete nucleus and much other cellular machinery, typically including *mitochondria* and (in plants) chloroplasts.

eusociality   a form of group life in extended families in which some members of the group become specialized for reproductive functions, and other members of the group give up reproduction entirely. These nonreproducing animals live as members of a sterile worker caste (or castes), and are often physically very different from one another and the reproducing animals despite not being genetically distinct from them.

exaptation   an *adaptive* trait whose current adaptiveness is not due to the same effect on fitness by virtue of which it was initially favored by natural selection

exobiology   the biology of life on other planets.

exon   see *reading sequence*

fitness   a measure of the ability of a gene, organism, or other biological unit to reproduce itself.

frequency-dependent selection   selection in which the *fitness* of a trait depends on the proportion of other, competing phenotypes in the population as a whole.

function   the purpose of a biological trait; what it is for.

Gaia hypothesis   the idea that the *biota* of the entire earth constitutes a single super-organism.

game theory   the mathematical study of competitive interaction. In evolutionary "games," the players are organisms of alternative design, and their "payoffs" are increases in their *fitness.*

gametes   sex cells. Gametes are *haploid,* having half the usual chromosome complement, and fuse in sexual reproduction to form a *diploid* cell. Sperm, ova, and pollen are all gametes.

gene   a unit of heredity. Genes have no uncontroversial definition; however, almost everyone accepts that they are, or include, DNA sequences of some kind. *Reading sequences* are commonly regarded as the paradigm genes.

gene pool   the totality of genes in a breeding population or species.

genome   all the genes of one organism.

genotype   a synonym for *genome;* sometimes used more narrowly as a specification of all the genes an organism has at a specific locus or set of loci.

germ line   the cell lineages in a multicellular organism that can potentially give rise to sex cells or *gametes*. See *somatic line.*

grade   a type of biological organization. The same grade can potentially be found in several different *clades.*

group selection   *natural selection* operating on groups of organisms

haplodiploid genetic system   a genetic system in which females grow from fertilized eggs and have genes from both parents and are thus *diploid,* but males grow from unfertilized eggs and are *haploid.* Males have no father and have no sons. They have a random selection of one from each of their mother's pairs of chromosomes (perhaps modified by *crossing-over*), and they transmit all their genes to their daughters. Ants, bees and wasps use this system.

haploid   having a single set of chromosomes.

heritability   a measure of the probability that an offspring will share a trait possessed by its parent.

heterozygous   see *homozygous*

homeostasis   the process by which an organism maintains physiological variables, such as temperature and salinity, within acceptable limits. Sometimes used to refer to feedback processes more generally.

homologous traits traits that taxa have in common through inheritance from a common ancestor. These contrast with *analogous* or *homoplastic* traits, qualitatively similar traits that have evolved independently.

homoplastic traits see *homologous traits*

homozygous having two *alleles* at a *locus* on a *chromosome* that are identical. When the alleles are different, the organism is *heterozygous*.

inclusive fitness see *kin selection*.

independent assortment, law of one of Mendel's laws, which states that the probability of a gamete receiving a particular allele at one locus is independent of which allele it receives at another locus. It is false when the two loci are on the same chromosome (gene *linkage*).

interactor a theoretical unit consisting of structures built as the result of the influence of a *replicator* (such as a gene) or replicators in development and acting so as to assist the reproduction of the replicator(s) responsible for its production. See also *phenotype*.

intron see *reading sequence*

kin selection the process of ensuring the presence of copies of one's distinctive genes in the next generation by helping one's relatives to breed, since they are likely to share those genes. A trait is kin-selected if it evolves by causing organisms to assist their relatives.

Lamarckian evolution a theory proposing that characters acquired during the lifetime of an organism can be passed on to that organism's offspring.

lineage a sequence of ancestors and descendants; parents and offspring.

linkage the tendency of two genes to be inherited together (thus violating the law of *independent assortment*). Genes on the same *chromosome* are linked, and the closer together their loci, the tighter their linkage; that is, the higher the probability that if one is inherited, the other will be. The closer genes are, the less likely they are to be split apart by *crossing-over*.

locus (plural: **loci**) the position on a *chromosome* occupied by a particular *gene*.

meiosis the type of cell division that gives rise to *gametes,* in which cells divide to form new cells with half the number of *chromosomes*. It contrasts with *mitosis,* cell division in which the daughter cells have the same chromosome number as the mother cell.

meiotic drive gene an allele that has an effect through which it has a greater than 50% chance of making it to a gamete when these are formed through *meiosis* (violating the law of *segregation*). Such alleles may spread even if they have adverse effects on an organism's fitness.

memes   postulated units of cultural inheritance, intended to be analogous to *genes*.

Mendelian genetics   the discipline that studies heredity by performing breeding experiments and observing the effect of crossing organisms with different characters on the characters manifested in their offspring. Also known as classic genetics or transmission genetics. Contrasts with *molecular genetics,* which studies the physical nature of the units of heredity.

messenger RNA (mRNA)   the RNA sequence *transcribed* from a gene and later *translated* into protein. See *reading sequence.*

Metazoa   the animals; a kingdom of life characterized by multicellularity, cells organized into tissues, an alimentary canal, and a nervous system.

mitochondria   organelles within *eukaryote* cells that have their own DNA, reproduce themselves by splitting, and are inherited in the cytoplasm of the egg (hence only from the mother). They play a critical role in the production of energy in the cell.

mitosis   see *meiosis*

molecular genetics   see *Mendelian genetics*

monomorphic   see *polymorphic*

monophyletic   containing an ancestral species and all, and only, its descendant species. More controversially, a single species is called monophyletic if it contains all, and only, the organisms descended from a single event of speciation or hybridization.

morph   one of several different phenotypes found in a single population.

mRNA   see *messenger RNA*

natural kind   a category postulated to correspond to some real distinction in the subject matter being classified, rather than being an arbitrary way of classifying.

natural selection   the process by which some traits come to predominate in a population, by virtue of superior *fitness,* while others decline in frequency.

niche   the ecological role played by a species in an ecosystem. The same species in different ecosystems can play different ecological roles and hence occupy different niches.

nucleotide   one the chemical bases from which DNA is composed: adenine, thymine, guanine or cytosine; or of which RNA is composed (in RNA, uracil replaces thymine). See *codon*

ontology   in philosophy, the study of what broad categories exist. Whether God exists, or whether minds are entities distinct from brains, are ontological questions. The ontology of a theory is the claims the theory makes about what exists.

phenotype   the manifested morphology, physiology, and behavior of an organism; contrasts with the *genotype,* the total collection of genes that an organism carries.

pleiotropic   having effects on more than one trait.

polygenic   affected by more than one gene.

polymorphic   having more than one form simultaneously in a population. Contrasts with *monomorphic* traits, which are the same in each individual in a population.

polypeptide   a molecule made up of many *amino acids* joined by peptide bonds. One or more polypeptide chains makes up a *protein.*

primitive trait   a trait that a species inherits from an ancestor in an unmodified form. In contrast, a *derived trait* is a trait that has appeared for the first time in the species or group of species in question. Primitive traits need not be simple. For example, if we are considering a group of cave-dwelling creatures, the primitive trait may be the possession of functional eyes, and the derived trait vestigial, nonfunctioning ones.

process explanation   see *robust process explanation*

prokaryotes   single-celled organisms without a nucleus or mitochondria. The bacteria are one main subdivision of the prokaryotes.

proteins   a class of very large molecules made up of *polypeptide* chains of *amino acids* that are central to the chemistry of life.

reading sequence   a sequence of DNA that is *transcribed* into messenger RNA (mRNA), which is in turn *translated* into protein. Some sections of the mRNA, called *introns,* are cut out and discarded before translation, leaving only the intervening *exons.* Hence not all the DNA in a reading sequence contributes to the final protein product.

realism   in philosophy, the view that a particular entity exists and exists independently of humans, their conceptions, and their observations—that it exists objectively. For example, realism about species is the view that species exist independently of their recognition and classification by humans.

recessive   see *dominant*

reduction   (1) the controversial idea that later, superior theories can explain and treat as special cases earlier, inferior theories of the same subject matter. (2) the controversial idea that the processes and entities of a "higher-level" theory, such as psychology or biology, can be explained in terms of the processes and entities of a "lower-level" or "more fundamental" science, such as physics. (3) The noncontroversial idea that theories of higher-level processes must not rely on causal mechanisms that are inexplicable or mysterious from the perspective of more fundamental lower-level theories.

replicator   a theoretical unit of heredity and selection; an entity that makes copies of itself and may cause the existence of an *interactor* or *vehicle.*

robust process explanation   an explanation that characterizes events in very broad terms, so that the explanation remains valid when other, similar events are substituted for those that actually occurred. See *actual sequence explanation.*

segregation, law of   one of Mendel's laws, which states that the two alleles are separated in the formation of the gametes (sex cells), with each gamete receiving only one allele. See *meiotic drive gene*.

segregation distorter gene   see *meiotic drive gene*

selection   see *natural selection*

sociobiology   the study of the evolution of the social behavior of animals. This term is sometimes used more narrowly to refer to the theory of the evolution of human social behavior.

somatic line   that part of the organism consisting of cell lineages that are unable to give rise to sex cells *(gametes)*. See *germ line*.

species   the smallest *taxa* mentioned in a system of biological classification. Species are sometimes thought of as the units within which a single evolutionary process unfolds.

taxon (plural: **taxa**)   a group of organisms recognized in biological systematics (taxonomy). Taxa are traditionally organized into a hierarchy of species, genera, families, orders, classes, and phyla.

transcription   the production from a *reading sequence* of DNA of a matching sequence of *messenger RNA (mRNA)*

translation   the production from a transcript of *mRNA* of a *protein* (a chain of *amino acids*)

type   see *Bauplan*

unity of type   see *Bauplan*

vehicle   a slightly more controversial term for an *interactor,* carrying the implication that the *replicator* is the core unit of the evolutionary process.

vicariance   the distribution of plants and animals as a result of geological processes. Contrasts with *dispersal*.

viviparous   giving birth to live young, rather than laying eggs.

Weissmanism   the doctrine that there is a distinct *germ line*. True mainly in animals.

References

Agar, N. 1996. Teleology and genes. *Biology and Philosophy* 11:289–300.

Alberts, B., D. Bray, J. Lewis, M. Raff, K. Roberts, and J. D. Watson. 1994. *Molecular Biology of the Cell.* 2d ed. New York and London: Garland.

Alberts, B., D. Bray, J. Lewis, M. Raff, K. Roberts, and J. D. Watson. 1996. *Molecular Biology of the Cell.* 3d ed. New York and London: Garland.

Alexander, R. 1979. *Darwinism and Human Affairs.* Seattle: Washington University Press.

———. 1987. *The Biology of Moral Systems.* New York: De Gruyter.

———. 1990. Darwinian algorithms: The adaptive study of learning and development. *Ethology and Sociobiology* 11:241–303.

Allee, W. C. 1951. *Cooperation among Animals: With Human Implications.* New York: Henry Schuman.

Allen, C., M. Bekoff, and G. Lauder, eds. 1998. *Nature's Purposes: Analyses of Function and Design in Biology.* Cambridge, Mass: MIT Press.

Amundson, R. 1996. Historical development of the concept of adaptation. In *Adaptation,* edited by M. R. Rose and G. V. Lauder. San Diego: Academic Press.

———. 1998. Two conceptions of constraint: Adaptationism and the challenge from developmental biology. In *Philosophy of Biology,* edited by D. Hull and M. Ruse. Oxford: Oxford University Press. First published in *Philosophy of Science* (1994) 61: 556–78.

Amundson, R., and G. V. Lauder. 1994. Function without purpose: The uses of causal role function in evolutionary biology. *Biology and Philosophy* 9:443–70.

Archibald, J. D. 1996. *Dinosaur Extinction and the End of an Era.* New York: Columbia University Press.

Arms, K., and P. S. Camp. 1987. *Biology.* 3d ed. Philadelphia: Saunders.

Aviles, L. 1986. Sex ratio bias and possible group selection in the social spider *Anelosimus eximius. American Naturalist* 128:1–12.

Bailey, W. J., W. J. J. L. Slighthorn, and M. Goodman. 1992. Rejection of the "flying primate" hypothesis by phylogenetic evidence from the e-globin gene. *Science* 256: 86–89.

Baird, J. 1990. The fifth day of creation. *BioEssays* 12:303–6.

Baker, R. R., and M. Bellis. 1995. *Human Sperm Competition*. London: Chapman and Hall.

Barkow, J. H. 1989. *Darwin, Sex and Status: Biological Approaches to Mind and Culture*. Toronto: University of Toronto Press.

Barkow, J. H., L. Cosmides, and J. Tooby, eds. 1992. *The Adapted Mind: Evolutionary Psychology and the Generation of Culture*. Oxford: Oxford University Press.

Barr, T. C. 1968. Cave ecology and the evolution of troglodytes. In *Evolutionary Biology*, vol. 2, edited by T. Dobzhansky, M. Hecht, and W. Steere, 35–102. Amsterdam: North Holland.

Barton, R. A., and R. I. Dunbar. 1997. Evolution of the social brain. In *Machiavellian Intelligence II: Extensions and Evaluations,* edited by A. Whiten and R. Byrne, 240–63. Cambridge: Cambridge University Press.

Bateson, P. 1976. Specificity and the origins of behavior. *Advances in the Study of Behavior* 6:1–20.

———. 1983. Genes, environment and the development of behaviour. In *Animal Behaviour: Genes, Development and Learning,* edited by P. Slater and T. Halliday. Oxford: Blackwell.

———. 1991. Are there principles of behavioural development? In *The Development and Integration of Behaviour,* ed. P. Bateson, 19–31. Cambridge: Cambridge University Press.

Beatty, J. 1985. Speaking of species: Darwin's strategy. In *The Darwinian Heritage,* edited by D. Kohn. Princeton, NJ: Princeton University Press.

Bedau, M. 1996. The nature of life. In *The Philosophy of Artificial Life,* edited by M. Boden, 332–60. Oxford: Oxford University Press.

Belew, R. K., and M. Mitchell, eds. 1996. *Adaptive Individuals in Evolving Populations*. Sante Fe Institute Studies in the Sciences of Complexity. Santa Fe, NM.: Santa Fe Institute.

Bell, C. 1873. *Expression: Its Anatomy and Philosophy*. Reprint of 1844 edition. New York: Wells.

Bell, G. 1982. *The Masterpiece of Nature*. London: Croom Helm.

Bell, M. A. 1997. Origin of metazoan phyla: Cambrian explosion or Proterozoic slow burn? *Trends in Ecology and Evolution* 12:1–2.

Beukeboom, L. B., and J. H. Werren. 1993. Transmission and expression of the parasitic paternal sex ratio (PSR) chromosome. *Heredity* 70:437–43.

Bigelow, J., and R. Pargetter. 1987. Functions. *Journal of Philosophy* 54:181–96.

Blackstone, E. 1995. A units-of-evolution perspective on the endosymbiont theory of the origin of the mitochondron. *Evolution* 49:785–96.

Boden, M., ed. 1996. *The Philosophy of Artificial Life*. Oxford: Oxford University Press.

Bonner, J. T. 1993. *Life Cycles: Reflections of an Evolutionary Biologist*. Princeton, NJ: Princeton University Press.

Bourke, A., and N. Franks. 1995. *Social Evolution in Ants*. Princeton, NJ: Princeton University Press.

Bowler, P. J. 1989. *Evolution: The History of An Idea*. Rev. ed. Berkeley: University of California Press.

Boyd, R., P. Gasper, and J. D. Trout, eds. 1991. *The Philosophy of Science*. Cambridge, MA: MIT Press.

Boyd, R., and P. Richerson. 1985. *Culture and the Evolutionary Process*. Chicago: University of Chicago Press.

Brandon, R. 1982. Levels of selection. In *Proceedings of the Philosophy of Science Association*, vol. I, edited by P. Asquith and T. Nickles, 315–22. East Lansing, MI: Philosophy of Science Association.

———. 1988. The levels of selection: A hierarchy of interactors. In *The Role of Behavior in Evolution*, edited by H. Plotkin, 51–71. Cambridge, MA: MIT Press.

———. 1990. *Adaptation and Environment*. Cambridge, MA: MIT Press.

———, ed. 1996. *Concepts and Methods in Evolutionary Biology*. Cambridge: Cambridge University Press.

Brandon, R., and J. Antonovics. 1996. The coevolution of organism and environment. In *Concepts and Methods in Evolutionary Biology*, edited by R. N. Brandon, 161–78. Cambridge: Cambridge University Press.

Brandon, R., and R. Burian, eds. 1984. *Genes, Organisms and Populations*. Cambridge, MA: MIT Press.

Brandon, R., and M. D. Rausher. 1996. Testing adaptationism: A comment on Orzack and Sober. *American Naturalist* 148 : 189–201.

Briggs, D., D. Erwin, and F. Collier. 1994. *The Fossils of the Burgess Shale*. Washington and London: Smithsonian Institution Press.

Briggs, D. E. G., R. A. Fortey, and M. A. Wills. 1992a. Morphological disparity in the Cambrian. *Science* 256 : 1670–73.

Briggs, D. E. G., R. A. Fortey, and M. A. Wills. 1992b. Reply to Foote and Gould and Lee. *Science* 258 : 1817–18.

Brook, J. D., M. E. McCurrach, H. G. Harley, A. J. Buckler, D. Church, H. Aburatani, K. Hunter, V. P. Stanton, J. P. Thirion, T. Hudson, R. Sohn, B. Zemelman, R. G. Snell, S. A. Rundle, S. Crow, J. Davies, P. Shelbourne, J. Buxton, C. Jones, V. Juvonen, K. Johnston, P. S. Harper, D. J. Shaw, and D. E. Housman. 1992. Molecular basis of myotonic dystrophy: Expansion of a trinucleotide (CTG) repeat at the $3'$ end of a transcript encoding a protein kinase family member. *Cell* 68 : 799–808.

Brooks, D., and D. McLennan. 1991. *Phylogeny, Ecology and Behavior: A Research Program in Comparative Biology*. Chicago: University of Chicago Press.

Brown, J. R. 1987. *Helping and Communal Breeding in Birds: Ecology and Evolution*. Princeton, NJ: Princeton University Press.

Brusca, G. J., and R. C. Brusca. 1990. *Invertebrates*. Sunderland, MA: Sinauer Associates.

Budiansky, S. 1995. *Nature's Keepers: The New Science of Nature Management*. New York: Free Press.

Buller, D. J., ed. In press. *Function, Selection and Design*. New York: SUNY Press.

Burian, R. M. 1983. Adaptation. In *Dimensions of Darwinism*, edited by M. Grene, 287–314. Cambridge: Cambridge University Press.

———. 1992. Adaptation: Historical perspectives. In *Keywords in Evolutionary Biology*, edited by E. F. Keller and E. Lloyd, 7–12. Cambridge, MA: Harvard University Press.

————. 1997. On conflicts between genetic and developmental viewpoints—and their attempted resolution in molecular biology. In *Structures and Norms in Science,* edited by M. L. Dalla Chiara, 243–64. Dordrecht: Kluwer.

Burian, R. M., and R. C. Richardson. 1996. Form and order in evolutionary biology. In *The Philosophy of Artificial Life,* edited by M. Boden, 146–72. Oxford: Oxford University Press.

Buss, D. M. 1994. *The Evolution of Desire: Strategies of Human Mating.* New York: Basic Books.

Buss, L. 1985. The uniqueness of the individual revisited. In *Population Biology and Evolution of Clonal Organisms,* edited by J. Jackson, L. Buss, and R. Cook, 467–505. New Haven, CT: Yale University Press.

————. 1987. *The Evolution of Individuality.* Princeton, NJ: Princeton University Press.

Butlin, R. K. 1987a. Speciation by reinforcement. *Trends in Ecology and Evolution* 2(1): 8–13.

————. 1987b. Species, speciation and reinforcement. *American Naturalist* 130(3): 461–64.

Byerly, H. C., and R. E. Michod. 1991. Fitness and evolutionary explanation. *Biology and Philosophy* 6:1–22.

Byrne, R., and A. Whiten, eds. 1988. *Machiavellian Intelligence.* Oxford: Oxford University Press.

Cannon, W. D. 1927. The James-Lange theory of emotions: A critical examination and an alternative theory. *American Journal of Psychology* 39:106–24.

Caplan, A., ed. 1978. *The Sociobiology Debate.* New York: Harper and Row.

Cartwright, N. 1983. *How the Laws of Physics Lie.* Oxford: Clarendon Press.

————. 1989. *Nature's Capacities and Their Measurement.* Oxford: Clarendon Press.

Cavalli-Sforza, L., and M. Feldman. 1981. *Cultural Transmission and Evolution: A Quantitative Approach.* Princeton, NJ: Princeton University Press.

Cavalli-Sforza, L. L., P. Menozzi, and A. Piazza. 1994. *The History and Geography of Human Genes.* Princeton, NJ: Princeton University Press.

Chadarevian, S. de. 1998. Of worms and programmes: *Caenorhabditis elegans* and the study of development. *Studies in History and Philosophy of Biological and Biomedical Sciences* 29(1):81–105.

Chagnon, N., and W. Irons, eds. 1979. *Evolutionary Biology and Human Social Behavior: An Anthropological Perspective.* North Scituate, MA: Duxbury Press.

Chaloner, W. G., and A. Hallam, eds. 1989. *Evolution and Extinction.* Cambridge: Cambridge University Press.

Chevalier-Skolnikoff, S. 1973. Facial expression of emotion in non-human primates. In *Darwin and Facial Expression: A Century of Research in Review,* edited by P. Ekman, 11–89. New York and London: Academic Press.

Chisholm, J. 1994. Death, hope and sex: Life history theory and the development of reproductive strategies. *Current Anthropology* 34:1–24.

Chomsky, Noam. 1980. *Rules and Representations.* New York: Columbia University Press.

Churchland, P. 1986. *Neurophilosophy: Towards a Unified Science of Mind-Brain.* Cambridge, MA: MIT Press.

Churchland, P. M. 1989. *A Neurocomputational Perspective: The Nature of Mind and the Structure of Science.* Cambridge, MA: MIT Press.

Claridge, M. F., H. A. Dawah, and M. R. Wilson, eds. 1997. *Species: The Units of Biodiversity.* New York: Chapman and Hall.

Clark, A. 1996. Happy couplings: Emergence and explanatory interlock. In *The Philosophy of Artificial Life,* edited by M. Boden, 262–81. Oxford: Oxford University Press.

———. 1997. *Being There: Putting Brain, Body, and World Together Again.* Cambridge, MA: MIT Press.

Coddington, J. 1988. Cladistic tests of adaptational hypotheses. *Cladistics* 4:3–22.

Colinvaux, P. 1980. *Why Big Fierce Animals Are Rare.* London: Penguin.

Colwell, R. K. 1981. Group selection implicated in the evolution of female biased sex ratios. *Nature* 290:401–4.

———. 1992. Niche: A bifurcation in the conceptual lineage of the term. In *Keywords in Evolutionary Biology,* edited by E. F. Keller and E. Lloyd, 241–248. Cambridge, MA: Harvard University Press.

Conner, E. F., and D. Simberloff. 1986. Competition, scientific method and null models in ecology. *American Scientist* 74:153–62.

Conway Morris, S. 1989. Burgess Shale faunas and the Cambrian explosion. *Science* 246:339–46.

———. 1993. The fossil record and the early evolution of the Metazoa. *Nature* 361:219–25.

———. 1998. *The Crucible of Creation: The Burgess Shale and the Rise of Animals.* Oxford: Oxford University Press.

Cooper, A., and R. Fortey. 1998. Evolutionary explosions and the phylogenetic fuse. *Trends in Ecology and Evolution* 13(4):151–55.

Cooper, G. 1990. The explanatory tools of theoretical population biology. In *Proceedings of the Philosophy of Science Association,* vol. 1, edited by A. Fine, M. Forbes, and L. Wessells, 165–78. East Lansing, MI: Philosophy of Science Association.

———. In press. *The Science of the Struggle for Existence.* Cambridge: Cambridge University Press.

Cosmides, L. 1989. The logic of social exchange: Has natural selection shaped how humans reason? Studies with the Wason Selection Task. *Cognition* 31:187–276.

Cosmides, L., and J. Tooby. 1989. Evolutionary theory and the generation of culture. Part II. Case study: A computational theory of social exchange. *Ethology and Sociobiology* 10:51–97.

———. 1992. Cognitive adaptations for social exchange. In *The Adapted Mind,* edited by J. H. Barkow, L. Cosmides, and J. Tooby, 163–227. Oxford: Oxford University Press.

Cowie, F. 1998. *What's Within? Nativism Reconsidered.* New York: Oxford University Press.

Cracraft, J. 1983. Cladistic analysis and vicariance biogeography. *American Scientist* 71:273–81.

Crawford, C., M. Smith, and D. Krebs, eds. 1987. *Sociobiology and Psychology: Ideas, Issues and Applications.* New York: Lawrence Erlbaum Associates.

Cronin, H. 1991. *The Ant and the Peacock; Altruism and Sexual Selection from Darwin to Today.* Cambridge: Cambridge University Press.

Cronquist, A. 1987. A botanical critique of cladism. *Botanical Review* 53 : 1–52.

Culp, S., and P. Kitcher. 1989. Theory structure and theory change in contemporary molecular biology. *British Journal for the Philosophy of Science* 40 : 459–83.

Cummins, R. 1994. Functional analysis. In *Conceptual Issues in Evolutionary Biology,* 2d ed., edited by E. Sober. Cambridge, MA: MIT Press. First published in *Journal of Philosophy* (1973) 72 : 741–64.

Damasio, A. R. 1994. *Descartes' Error: Emotion, Reason and the Human Brain.* New York: Grosset/Putnam.

Damuth, J. 1985. Selection among "species": A formulation in terms of natural functional units. *Evolution* 39(5):1132–46.

Damuth, J., and L. Heisler. 1988. Alternative formulations of multilevel selection. *Biology and Philosophy* 3 : 407–30.

Darwin, C. 1860. *Journal of Researches into the Geology and Natural History of the Countries visited during the Voyage of H.M.S. Beagle Round the World under the Command of Capt. Fitzroy R.N.* Rev. ed. London: John Murray.

———. 1871. *The Descent of Man and Selection in Relation to Sex.* London: John Murray.

———. 1964. *On the Origin of Species: A Facsimile of the First Edition.* 1859. Cambridge, MA: Harvard University Press.

———. 1965. *The Expression of the Emotions in Man and Animals.* 1872. Chicago: University of Chicago Press.

Davies, P. S., and H. Holcomb, eds. 1999. *The Evolution of Minds: Psychological and Philosophical Perspectives.* Dordrecht: Kluwer.

Davies, S., J. Fetzer, and T. Forster. 1995. Logical reasoning and domain specificity: A critique of the social exchange theory of reasoning. *Biology and Philosophy* 10(1):1–38.

Dawkins, R. 1976. *The Selfish Gene.* Oxford: Oxford University Press.

———. 1982. *The Extended Phenotype.* Oxford: Oxford University Press.

———. 1986. *The Blind Watchmaker.* New York: Norton.

———. 1989a. The evolution of evolvability. In *Artificial Life VI: Proceedings, Santa Fe Institute Studies in the Sciences of Complexity,* ed. C. Langton. Reading, MA: Addison-Wesley.

———. 1989b. *The Selfish Gene.* New ed. Oxford: Oxford University Press.

———. 1990. Parasites, desiderata lists and the paradox of the organism. *Parasitology* 100: S63–S73.

———. 1992. Progress. In *Keywords in Evolutionary Biology,* edited by E. F. Keller and E. Lloyd, 263–72. Cambridge, MA: Harvard University Press.

———. 1994. Burying the vehicle. *Behavioral and Brain Sciences* 17 : 617.

———. 1995. *River out of Eden: A Darwinian View of Life.* New York: Basic Books.

———. 1996. *Climbing Mount Improbable.* New York: W. W. Norton.

———. 1997. Human chauvinism: A review of S. J. Gould's *Full House. Evolution* 51: 1015–20.

Del Moral, R., and L. C. Bliss. 1993. Mechanisms of primary succession: Insights resulting from the eruption of Mount St Helens. *Advances in Ecological Research* 24 : 1–66.

Dennett, D. C. 1983. Intentional systems in cognitive ethology: The "Panglossian paradigm" defended. *Behavioral and Brain Sciences* 6:343–90.

―――. 1987. *The Intentional Stance.* Cambridge, MA: MIT Press.

―――. 1991. Real patterns. *Journal of Philosophy* 87:27–51.

―――. 1995. *Darwin's Dangerous Idea.* New York: Simon and Schuster.

Depew, D., and B. H. Weber. 1995. *Darwinism Evolving: Systems Dynamics and the Genealogy of Natural Selection.* Cambridge, MA: MIT Press.

de Queiroz, K. 1986. Systematics and the Darwinian revolution. *Philosophy of Science* 55: 238–59.

de Queiroz, K., and D. A. Good. 1997. Phenetic clustering in biology: A critique. *Quarterly Review of Biology* 72(1):3–30.

Desmond, A. 1982. *Archetypes and Ancestors: Palaeontology in Victorian London 1850–1875.* London: Blond and Briggs.

Devitt, M., and K. Sterelny. 1986. *Language and Reality: An Introduction to Philosophy of Language.* Oxford: Blackwell.

de Waal, F. 1982. *Chimpanzee Politics: Power and Sex Amongst the Apes.* New York: Harper and Row.

Diamond, J. 1991. The Saltshaker's Curse. *Natural History* 1991(10):20–26.

―――. 1992. *The Third Chimpanzee: The Evolution and Future of the Human Animal.* New York: Harper Collins.

Dobzhansky, T. 1951. *Genetics and the Origin of Species.* New York: Columbia University Press.

Donoghue, M. J., and P. D. Cantino. 1988. Paraphyly, ancestors, and the goals of taxonomy: A botanical defense of cladism. *Botanical Review* 54:107–28.

Dretske, F. 1981. *Knowledge and the Flow of Information.* Oxford: Blackwell.

―――. 1983. Precis of *Knowledge and the Flow of Information. Behavioural and Brain Science* 6:55–90.

Duellman, W. E., and E. R. Pianka. 1990. Biogeography of nocturnal insectivores: Historical events and ecological filters. *Annual Review of Ecology and Systematics* 21:57–68.

Dugatkin, L. 1997. *Cooperation among Animals: An Evolutionary Perspective.* Oxford: Oxford University Press.

Dugatkin, L. A., and H. K. Reeve. 1994. Behavioral ecology and levels of selection: Dissolving the group selection controversy. *Advances in the Study of Behavior* 23:101–33.

Dunbar, R. I. 1996. *Grooming, Gossip and the Evolution of Language.* London: Faber and Faber.

―――. In press. The social brain hypothesis. *Evolutionary Anthropology.*

Dupré, J., ed. 1987. *The Latest on the Best: Essays on Optimality and Evolution.* Cambridge, MA: MIT Press.

―――. 1993. *The Disorder of Things.* Boston: Harvard University Press.

Dyer, B. D., and R. A. Obar. 1994. *Tracing the History of the Eukaryotic Cell: The Enigmatic Smile.* New York: Columbia University Press.

Eberhard, W. 1990. Evolution in bacterial plasmids and levels of selection. *Quarterly Review of Biology* 65(1):3–22.

Egerton, F. N. 1973. Changing concepts of the balance of nature. *Quarterly Review of Biology* 48:322–50.

Eggleton, P., and R. Vane-Wright, eds. 1994. *Phylogenetics and Ecology.* London: Academic Press, for the Linnean Society of London.

Ehrlich, P. 1988. *The Machinery of Nature.* London: Paladin.

Ehrlich, P., and P. Raven. 1969. The differentiation of populations. *Science* 165:1228–32.

Eibl-Eibesfeldt, I. 1973. Expressive behaviour of the deaf and blind born. In *Social Communication and Movement,* edited by M. von Cranach and I. Vine, 163–94. London and New York: Academic Press.

Ekman, P. 1971. Universals and cultural differences in facial expressions of emotion. In *Nebraska Symposium on Motivation 4,* edited by J. K. Cole, 207–83. Lincoln, Nebraska, University of Nebraska Press.

———. 1972. *Emotion in the Human Face.* New York: Pergamon Press.

———, ed. 1973. *Darwin and Facial Expression: A Century of Research in Review.* New York and London: Academic Press.

Ekman, P., and W. V. Friesen. 1971. Constants across cultures in facial displays of emotion. *Journal of Personality and Social Psychology* 17:124–29.

———. 1975. *Unmasking the Face: A Guide to Recognising Emotions from Facial Expressions.* Englewood Cliffs, NJ: Prentice-Hall.

Ekman, P., E. R. Sorensen, and W. V. Friesen. 1969. Pan-cultural elements in facial displays of emotion. *Science* 164:86–88.

Eldredge, N. 1985a. *Time Frames: The Rethinking of Darwinian Evolution and the Theory of Punctuated Equilibria.* New York: Simon and Schuster.

———. 1985b. *Unfinished Synthesis: Biological Hierarchies and Modern Evolutionary Thought.* Oxford: Oxford University Press.

———. 1989. *Macroevolutionary Dynamics: Species, Niches and Adaptive Peaks.* New York: McGraw-Hill.

———. 1995. *Reinventing Darwin.* New York: John Wiley and Son.

Eldredge, N., and S. J. Gould. 1972. Punctuated equilibria: An alternative to phyletic gradualism. In *Models in Paleobiology,* edited by T. J. Schopf, 82–115. San Francisco: Freeman, Cooper.

Ellstrand, N. 1983. Why are juveniles smaller than their parents? *Evolution* 37:1091–94.

Elton, C. S. 1927. *Animal Ecology.* London: Sidgwick and Jackson.

Emmeneche, C. 1994. *The Garden in the Machine: The Emerging Science of Artificial Life.* Princeton, NJ: Princeton University Press.

Epp, C. 1997. Definition of a gene. *Nature* 389:537.

Ereshefsky, M., ed. 1992. *The Units of Evolution: Essays on the Nature of Species.* Cambridge, MA: MIT Press.

Erwin, D. H. 1993a. *The Great Paleozoic Crisis: Life and Death in the Permian.* New York: Columbia University Press.

———. 1993b. The origin of metazoan development: A paleobiological perspective. *Biological Journal of the Linnean Society* 50:255–74.

Fagerstrom, T., D. A. Briscoe, and P. Sunnucks. 1998. Evolution of mitotic cell-lineages in multicellular organisms. *Trends in Ecology and Evolution* 13(3):117–21.

Falk, D. 1990. Brain evolution in *Homo:* The "radiator" theory. *Behavioral and Brain Sciences* 13:331–81.

Falk, R. 1984. The gene in search of an identity. *Human Genetics* 68:195–204.

———. 1986. What is a gene? *Studies in the History and Philosophy of Science* 17:133–73.

Feyerabend, P. K. 1962. Explanation, reduction and empiricism. In *Minnesota Studies in the Philosophy of Science,* III, *Scientific Explanation, Space and Time,* edited by H. Feigl and G. Maxwell, 28–97. Minneapolis, University of Minnesota Press.

Fodor, J. A. 1974. Special sciences. *Synthese* 28:77–115.

———. 1975. *The Language of Thought.* New York: Thomas Y Crowell.

———. 1983. *The Modularity of Mind.* Cambridge, MA: MIT Press.

Fogle, T. 1990. Are genes units of inheritance? *Biology and Philosophy* 5:349–72.

Fong, D. W., T. C. Kane, and D. C. Culver. 1995. Vestigialization and loss of nonfunctional characters. *Annual Review of Ecology and Systematics* 26:249–68.

Foote, M. 1993. Discordance and concordance between morphological and taxonomic diversity. *Paleobiology* 19:185–204.

Foote, M., and S. J. Gould. 1992. Cambrian and Recent morphological disparity. *Science* 258:1816.

Ford Doolittle, W. F., and J. Brown. 1995. Tempo, mode, the progenote and the universal Root. In *Tempo and Mode in Evolution: Genetics and Paleontology 50 Years after Simpson,* edited by W. M. Fitch and F. A. Ayala, 3–24. Washington, DC: National Academy of Sciences.

Fox-Keller, E. 1995. *Refiguring Life: Metaphors of Twentieth-Century Biology.* New York: Columbia University Press.

Frank, P. 1988. *Passions Within Reason: The Strategic Role of the Emotions.* New York: W. W. Norton.

Futuyma, D. 1998. *Evolutionary Biology.* 3d ed. Sunderland, MA: Sinauer Associates.

Gatens, M. 1991. *Feminism and Philosophy.* Cambridge and Oxford: Polity Press.

Ghiselin, M. T. 1974a. *The Economy of Nature and the Evolution of Sex.* Berkeley: University of California Press.

———. 1974b. A radical solution to the species problem. *Systematic Zoology* 23:536–44.

———. 1981. Categories, life and thinking. *Behavioral and Brain Sciences* 4:269–313.

Gilbert, S. F., J. M. Opitz, and R. A. Raff. 1996. Resynthesising evolutionary and developmental biology. *Developmental Biology* 173:357–72.

Gilinsky, N. L. 1986. Species selection as a causal process. *Evolutionary Biology* 20:249–73.

Glen, W., ed. 1994. *The Mass Extinction Debate: How Science Works in a Crisis.* Stanford, CA: Stanford University Press.

Godfray, H. C. J., and J. H. Werren. 1996. Recent developments in sex ratio studies. *Trends in Ecology and Evolution* 11:59–63.

Godfrey-Smith, P. 1989. Misinformation. *Canadian Journal of Philosophy* 19(5):533–50.

———. 1992. Indication and adaptation. *Synthese* 92:283–312.

———. 1993. Functions: Consensus without unity. *Pacific Philosophical Quarterly* 74: 196–208.

———. 1994a. A continuum of semantic optimism. In *Mental Representation,* edited by S. P. Stich and T. A. Warfield, 259–77. Oxford: Blackwell.

———. 1994b. A modern history theory of functions. *Nous* 28:344–62.

———. 1996. *Complexity and the Function of Mind in Nature.* Cambridge: Cambridge University Press.

———. In press-a. Genes and codes: Lessons from the philosophy of mind? In *Biology Meets Psychology: Constraints, Conjectures, Connections,* edited by V. Hardcastle. Cambridge, MA: MIT Press.

———. In press-b. Review of D. C. Dennett's *Darwin's Dangerous Idea. Philosophy of Science.*

———. In press-c. Three kinds of adaptationism. In *Optimality and Adaptationism,* edited by S. Orzack and E. Sober. Cambridge: Cambridge University Press.

Godfrey-Smith, P., and R. C. Lewontin. 1993. The dimensions of selection. *Philosophy of Science* 60:373–95.

Goldberg, S. 1973. *Male Dominance: The Inevitability of Patriarchy.* New York: Morrow.

Goldschmidt, T. 1996. *Darwin's Dream Pond: Drama in Lake Victoria.* Boston: MIT Press.

Golley, F. B. 1993. *A History of the Ecosystem Concept in Ecology: More Than the Sum of the Parts.* New Haven, CT: Yale University Press.

Goodwin, B. C. 1994. *How The Leopard Changed Its Spots: The Evolution of Complexity.* New York: Charles Scribner and Sons.

Goodwin, B. C., and P. Saunders. 1989. *Theoretical Biology: Epigenetic and Evolutionary Order from Complex Systems.* Edinburgh: Edinburgh University Press.

Gottlieb, G. 1981. Roles of early experience in species-specific perceptual development. In *Development of Perception,* edited by R. N. Aslin, J. R. Alberts, and M. P. Petersen. New York: Academic Press.

Gould, S. J. 1977. *Ontogeny and Phylogeny.* Cambridge, MA: Harvard University Press.

———. 1980a. The episodic nature of evolutionary change. In *The Panda's Thumb.* New York: W. W. Norton.

———. 1980b. Is a new and general theory of evolution emerging? *Paleobiology* 6: 119–30.

———. 1980c. The return of the hopeful monster. In *The Panda's Thumb.* London: Penguin.

———. 1983a. The hardening of the Modern Synthesis. In *Dimensions of Darwinism,* edited by M. Grene, 71–93. Cambridge: Cambridge University Press.

———. 1983b. The meaning of punctuated equilibrium and its role in validating a hierarchial approach to macroevolution. *Scientia* 1:135–57.

———. 1985. The paradox of the first tier: An agenda for paleobiology. *Paleobiology* 11: 2–12.

———. 1989. *Wonderful Life: The Burgess Shale and the Nature of History.* New York: W. W. Norton.

————. 1990. Speciation and sorting as the source of evolutionary trends, or "Things are seldom what they seem." In *Evolutionary Trends,* edited by K. J. McNamara. Tucson: University of Arizona Press.

————. 1991. The disparity of the Burgess Shale arthropod fauna and the limits of cladistic analysis: Why must we strive to quantify morphospace? *Paleobiology* 17:411–23.

————. 1993a. *Eight Little Piggies: Reflections in Natural History.* New York: W. W. Norton.

————. 1993b. Reply to Ridley. *Paleobiology* 19:522–23.

————. 1994. Common pathways of illumination. *Natural History* 103(12):10–20.

————. 1995. A task for paleobiology at the threshold of majority. *Paleobiology* 21:1–14.

————. 1996a. *Full House: The Spread of Excellence from Plato to Darwin.* New York: Harmony Press.

————. 1996b. In the mind of the beholder. In *Dinosaur in a Haystack,* 93–107. London: Jonathan Cape.

————. 1996c. Microcosmos. *Natural History* 105:20–23, 66–68.

————. 1996d. Of tongue worms, velvet worms and water bears. In *Dinosaur in a Haystack,* 108–20. London: Jonathan Cape.

————. 1996e. Triumph of the root-heads. *Natural History* 105:10–17.

————. 1997. Self-help for a hedgehog stuck on a molehill. *Evolution* 51(3):1020–24.

Gould, S. J., and N. Eldredge. 1993. Punctuated equilibrium comes of age. *Nature* 366: 223–27.

Gould, S. J., and R. C. Lewontin. 1978. The spandrels of San Marco and the Panglossian paradigm: A critique of the adaptationist programme. *Proceedings of the Royal Society of London* B 205:581–98. (Reprinted in *Conceptual Issues in Evolutionary Biology,* 2d ed., edited by E. Sober. Cambridge, MA: MIT Press.)

Gould, S. J., and E. S. Vrba. 1982. Exaptation: A missing term in the science of form. *Paleobiology* 8:4–15.

Grafen, A. 1982. How not to measure inclusive fitness. *Nature* 298: 425–26.

Grant, B. 1996. *Indonesia.* Melbourne: Melbourne University Press.

Grantham, T. 1995. Hierarchical approaches to macroevolution: Recent work on species selection and the effect hypothesis. *Annual Review of Ecology and Systematics* 26: 301–21.

Gray, R. D. 1987. Faith and foraging: A critique of the "paradigm argument from design." In *Foraging Behavior,* edited by A. C. Kamil, J. R. Krebs, and H. R. Pulliam, 69–140. New York: Plenum Press.

————. 1992. Death of the gene: Developmental systems strike back. In *Trees of Life: Essays in the Philosophy of Biology,* edited by P. E. Griffiths, 165–209. Dordrecht: Kluwer.

————. 1997. In the belly of the monster: Feminism, developmental systems and evolutionary explanations. In *Evolutionary Biology and Feminism,* edited by P. A. Gowaty, 385–413. New York: Chapman and Hall.

Gray, R. D., and J. L. Craig. 1991. Theory really matters: Hidden assumptions in the concept of habitat requirements. *Acta XX Congressus Internationalis Ornithologici* 20: 2553–60.

Grbic, M., P. J. Ode, and M. R. Strand. 1992. Sibling rivalry and brood sex ratios in polyembryonic wasps. *Nature* 360:254–56.

Grehan, J. R. 1991. Panbiogeography 1981–91: Development of an earth/life synthesis. *Progress in Physical Geography* 15(4):331–63.

Griesemer, J. R. 1992a. The informational gene and the substantial body: On the generalization of evolutionary theory by abstraction. Unpublished manuscript.

———. 1992b. Niche: Historical perspectives. In *Keywords in Evolutionary Biology,* edited by E. F. Keller and E. Lloyd, 231–40. Cambridge, MA: Harvard University Press.

Griesemer, J. R., and W. C. Wimsatt. 1989. Picturing Weismannism: A case study of conceptual evolution. In *What the Philosophy of Biology Is,* edited by M. Ruse, 75–137. Dordrecht: Kluwer.

Griffiths, P. E. 1992. Adaptive explanation and the concept of a vestige. In *Trees of Life: Essays in the Philosophy of Biology,* ed. P. E. Griffiths, 111–31. Dordrecht: Kluwer.

———. 1994. Cladistic classification and functional explanation. *Philosophy of Science* 61: 206–27.

———. 1996a. Darwinism, process structuralism and natural kinds. *Philosophy of Science* 63(suppl.) (3):S1–S9.

———. 1996b. The historical turn in the study of adaptation. *British Journal for the Philosophy of Science* 47:511–32.

———. 1997. *What Emotions Really Are: The Problem of Psychological Categories.* Chicago: University of Chicago Press.

Griffiths, P. E., and R. . D. Gray. 1994. Developmental systems and evolutionary explanation. *Journal of Philosophy* 91: 277–304.

———. 1997. Replicator II: Judgement Day. *Biology and Philosophy* 12(4):471–92.

Griffiths, P. E., and E. Neumann-Held. In press. The many faces of the gene. *BioScience.*

Grosberg, R. K., and R. R. Strathmann. 1998. One cell, two cell, red cell, blue cell: The persistence of a unicellular stage in multicellular life histories. *Trends in Ecology and Evolution* 13(3):112–16.

Guth, W. T., R. Schmittberger, and B. Schwarze. 1982. An experimental analysis of ultimatum bargaining. *Journal of Economic Behavior and Organization* 3:367–88.

Haig, D. 1992. Genomic imprinting and the theory of parent-offspring conflict. *Seminars in Developmental Biology* 3:153–60.

Haig, D., and A. Grafen. 1991. Genetic scrambling as a defense against meiotic drive. *Journal of Theoretical Biology* 153:531–58.

Hamilton, W. D. 1964a. The genetical evolution of social behaviour, I. *Journal of Theoretical Biology* 7:1–16.

———. 1964b. The genetical evolution of social behaviour, II. *Journal of Theoretical Biology* 7:17–52.

———. 1980. Sex versus non-sex versus parasite. *Oikos* 35:282–90.

―――. 1988. Sex and disease. In *The Evolution of Sex,* edited by R. Bellig and G. Stevens, 65–95. New York: Harper and Row.

―――. 1996. *Narrow Roads of Gene Land: The Collected Papers of W. D. Hamilton.* Volume 1, *Evolution of Social Behaviour.* New York: W. H. Freeman.

Hardy, I. W. 1995. Protagonists of polyembryony. *Trends in Ecology and Evolution* 10(3): 179–80.

Harvey, P., and M. Pagel. 1991. *The Comparative Method in Evolutionary Biology.* Oxford: Oxford University Press.

Hauser, M. D. 1996. *The Evolution of Communication.* Cambridge, MA: MIT Press.

Heinrich, B. 1990. *Ravens in Winter.* London: Barrie and Jenkins.

Hendriks-Jansen, H. 1996. In praise of interactive emergence, or why explanation doesn't have to wait for implementations. In *The Philosophy of Artificial Life,* edited by M. Boden, 282–302. Oxford: Oxford University Press.

Hennig, W. 1966. *Phylogenetic Systematics.* Urbana: University of Illinois Press.

Herbers, J. M., and R. J. Stuart. 1996. Multiple queens in ant nests: Impact on genetic structure and inclusive fitness. *American Naturalist* 147: 161–87.

Hirschfeld, L., and S. Gelman, eds. 1995. *Mapping the Mind: Domain Specificity in Cognition and Culture.* Cambridge: Cambridge University Press.

Hoffman, A. 1989. *Arguments on Evolution.* Oxford: Oxford University Press.

Hölldobler, B., and E. O. Wilson. 1990. *The Ants.* Cambridge, MA: Harvard University Press.

―――. 1994. *Journey to the Ants.* Cambridge, MA: Harvard University Press.

Horan, B. 1989. Functional explanations in sociobiology. *Biology and Philosophy* 4: 131–58.

Hull, D. 1974. *Philosophy of Biological Science.* Englewood Cliffs, NJ: Prentice-Hall.

―――. 1978. A matter of individuality. *Philosophy of Science* 45: 335–60.

―――. 1981. Units of evolution: A metaphysical essay. In *The Philosophy of Evolution,* edited by R. Jensen and R. Harre. Brighton, Harvester.

―――. 1986. On human nature. In *Proceedings of the Philosophy of Science Association,* vol. 2, edited by A. Fine and P. Machamer, 3–13. East Lansing, MI: Philosophy of Science Association.

―――. 1988. *Science as a Process.* Chicago: University of Chicago Press.

―――. 1997. The ideal species concept—and why we can't get it. In *Species: The Units of Biodiversity,* edited by M. F. Claridge, H. A. Dawah, and M. R. Wilson. New York: Chapman and Hall.

―――. In press. Explanatory pluralism. In *Species: New Interdisciplinary Essays,* edited by R. A. Wilson. Cambridge, MA: MIT Press.

Hull, D., and M. Ruse, eds. 1998. *Philosophy of Biology.* Oxford: Oxford University Press.

Humphries, C. J., and J. A. Chappill. 1988. Systematics as science: A response to Cronquist. *Botanical Review* 54: 129–44.

Hurka, T. 1993. *Perfectionism.* New York: Oxford University Press.

Hurst, L., A. Atlan, and B. O. Bengtsson. 1996. Genetic conflicts. *Quarterly Review of Biology* 71: 317–64.

Hutchinson, G. E. 1959. Homage to Santa Rosalia; or, why are there so many kinds of animals? *American Naturalist* 93 : 145–59.

———. 1965. *The Ecological Theater and the Evolutionary Play.* New Haven, CT: Yale University Press.

———. 1975. Variations on a theme by Robert MacArthur. In *Ecology and Evolution of Communities,* edited by M. Cody and J. Diamond. Cambridge, MA: Harvard University Press.

———. 1978. *Introduction to Population Ecology.* New Haven, CT: Yale University Press.

———. 1991. Concluding remarks: Cold Spring Harbor Symposium 1957. In *Foundations of Ecology,* edited by J. Brown and L. E. Real. Chicago: University of Chicago Press.

Immelmann, K. 1975. Ecological significance of imprinting and early learning. *Annual Review of Ecology and Systematics* 6 : 15–37.

Izard, C. 1969. The emotions and emotion constructs in personality and culture. In *Handbook of Modern Personality Theory,* edited by R. B. Cattell and R. M. Dreger. Washington and London: Hemisphere Publishing.

———. 1978. On the development of emotions and emotion-cognition relationship in infancy. In *The Development of Affect,* edited by M. Lewis and L. Rosenblum. New York: Plenum Press.

Jablonka, E., and M. J. Lamb. 1995. *Epigenetic Inheritance and Evolution: The Lamarckian Dimension.* Oxford: Oxford University Press.

Jablonka, E., and E. Szathmary. 1995. The evolution of information storage and heredity. *Trends in Ecology and Evolution* 10(5) : 206–11.

Jablonski, D. 1986a. Background and mass extinctions: The alternation of macroevolutionary regimes. *Science* 231 : 129–33.

———. 1986b. Evolutionary consequences of mass extinctions. In *Patterns and Processes in the History of Life,* edited by D. Raup and D. Jablonski, 313–30. Berlin: Springer-Verlag.

———. 1987. Heritability at the species level: Analysis of geographic ranges of Cretaceous mollusks. *Science* 238 : 360–63.

———. 1991. Extinctions: A paleontological perspective. *Science* 253 : 754–57.

Jackson, F., and P. Pettit. 1992. In defence of explanatory ecumenicalism. *Economics and Philosophy* 8 : 1–21.

Jamieson, I. G. 1989. Levels of analysis or analysis at the same level? *Animal Behavior* 37 : 696–97.

Jamieson, I. G., and J. Craig. 1993. Inbreeding in the marshes. *Natural History* 102 : 50–56.

Janis, C. 1994. The sabretooth's repeat performances. *Natural History* 103(4) : 78–83.

Janzen, D. H. 1977. What are dandelions and aphids? *American Naturalist* 111 : 586–89.

Johnston, T. 1987. The persistence of dichotomies in the study of behavioural development. *Developmental Review* 7 : 149–82.

Jones, D. 1995. Sexual selection, physical attractiveness and facial neoteny. *Current Anthropology* 36 : 723–48.

Jones, D. L. 1987. *Encyclopedia of Ferns: An Introduction to Ferns, Their Structure, Biology, Economic Importance and Propagation*. Melbourne: Lothian.

Joseph, L. E. 1990. *Gaia: The Growth of an Idea*. New York: St. Martins Press.

Judson, H. L. 1997. *The Eighth Day of Creation*. 2d ed. New York: Simon and Schuster.

Judson, O., and B. B. Normark. 1996. Ancient asexual scandals. *Trends in Ecology and Evolution* 11(2): 41–46.

Kahnemann, D., J. Knetsch, and R. Thaler. 1983. Fairness and the assumptions of economics. *Journal of Business* 59: s285–s300.

Kauffman, S. A. 1993. *The Origins of Order: Self-Organisation and Selection in Evolution*. New York: Oxford University Press.

———. 1995a. *At Home in the Universe*. New York: Oxford University Press.

———. 1995b. "What is Life?" Was Schrodinger Right? In *What is Life? The Next Fifty Years: Speculations on the Future of Biology*, edited by M. P. Murphy and L. A. J. O'Neill, 83–114. Cambridge: Cambridge University Press.

Keller, E. F., and E. A. Lloyd, eds. 1992. *Keywords in Evolutionary Biology*. Cambridge, MA: Harvard University Press.

Keller, L., and K. G. Ross. 1993. Phenotypic plasticity and cultural transmission in the fire ant, *Solenopsis invicta*. *Behavioural Ecology and Sociobiology* 33: 121–29.

Kennedy, M. R., H. G. Spencer, and R. D. Gray. 1996. Hop, step, and gape: Do the social displays of the Pelecaniformes reflect phylogeny? *Animal Behaviour* 51: 273–91.

Kevles, D. J. 1986. *In the Name of Eugenics: Genetics and the Uses of Human Heredity*. Berkeley: University of California Press.

Kingsland, S. 1985. *Modeling Nature: Episodes in the History of Population Ecology*. Chicago: University of Chicago Press.

Kirchner, J. W. 1991. The Gaia hypotheses: Are they testable? Are they useful? In *Scientists on Gaia*, edited by S. Schneider and P. J. Boston, 38–46. Cambridge, MA: MIT Press.

Kitcher, P. 1984. 1953 and All That: A tale of two sciences. *Philosophical Review* 93: 335–73.

———. 1985. *Vaulting Ambition: Sociobiology and the Quest for Human Nature*. Cambridge, MA: MIT Press.

———. 1989. Some puzzles about species. In *What the Philosophy of Biology Is*, edited by M. Ruse, 183–208. Dordrecht: Kluwer.

———. 1994. Four ways of biologicizing ethics. In *Conceptual Issues in Evolutionary Biology*, edited by E. Sober, 439–50. Cambridge, MA: MIT Press.

———. In press. Battling the undead: How (and how not) to resist genetic determinism. In *Thinking about Evolution: Historical, Philosophical and Political Perspectives (Festschrift for Richard Lewontin)*, edited by R. Singh, K. Krimbas, D. Paul, and J. Beatty. Cambridge: Cambridge University Press.

Klinnert, M. D., J. J. Campos, J. F. Sorce, R. N. Emde, and M. Svejda. 1983. Emotions as behavior regulators: Social referencing in infants. In *Emotion: Theory, Research and Experience*, vol. 2, *Emotions in Early Development*, edited by R. Plutchik and H. Kellerman, 57–86. New York: Academic Press.

Knoll, A. H. 1994. Proterozoic and Early Cambrian protists: Evidence for accelerating evolutionary tempo. *Proceedings of the National Academy of Sciences USA* 91:6743–50.

———. 1996. Daughter of time. *Paleobiology* 22:1–7.

Kohler, R. E. 1994. *Lords of the Fly: Drosophilia Genetics and the Experimental Life*. Chicago: University of Chicago Press.

Kondrashov, A. S. 1993. Classification of hypotheses on the advantage of amphimixis. *Journal of Heredity* 84:372–87.

Konner, M. 1982. *The Tangled Wing: Biological Constraints on the Human Spirit*. London: William Heinemann.

Kornet, D. J. 1993. Permanent splits as speciation events: A formal reconstruction of the internodal species concept. *Journal of Theoretical Biology* 164:407–35.

Krebs, J., and R. Dawkins. 1984. Animal signals, mind-reading and manipulation. In *Behavioural Ecology: An Evolutionary Approach,* edited by J. R. Krebs and N. B. Davies. Oxford: Blackwell Scientific.

Krebs, J. R., and N. R. Davies. 1981. *An Introduction to Behavioural Ecology.* Oxford: Blackwell Scientific.

———, eds. 1984. *Behavioural Ecology: An Evolutionary Approach.* 2d ed. Oxford: Blackwell Scientific.

Kuhn, T. 1970. *The Structure of Scientific Revolutions.* Chicago: University of Chicago Press.

Lack, D. 1966. *Population Studies of Birds.* Oxford: Oxford University Press.

———. 1971. *Ecological Isolation in Birds.* Oxford and Edinburgh: Blackwell Scientific Publications.

Lakatos, I. 1970. The methodology of scientific research programmes. In *Criticism and the Growth of Knowledge,* edited by I. Lakatos and A. Musgrave, 91–96. Cambridge: Cambridge University Press.

Lambert, D. M., and H. G. Spencer, eds. 1995. *Speciation and the Recognition Concept.* Baltimore and London: Johns Hopkins University Press.

Langton, C., ed. 1995. *Artificial Life: An Overview.* Cambridge, MA: MIT Press.

———. 1996. Artificial life. In *The Philosophy of Artificial Life,* edited by M. Boden, 39–94. Oxford: Oxford University Press.

Lauder, G. V., M. L. Armand, and M. Rose. 1993. Adaptations and history. *Trends in Ecology and Evolution* 8:294–97.

Leakey, R., and R. Lewin. 1996. *The Sixth Extinction: Biodiversity and Its Survival.* London: Weidenfeld and Nicolson.

Legrand, E. K. 1997. An adaptationist view of apoptosis. *Quarterly Review of Biology* 72(2): 135–47.

Le Grand, H. E. 1988. *Drifting Continents and Shifting Theories.* Cambridge, MA: MIT Press.

Lehrman, D. S. 1953. Critique of Konrad Lorenz's theory of instinctive behaviour. *Quarterly Review of Biology* 28(4):337–63.

———. 1970. Semantic and conceptual issues in the nature-nurture problem. In *Development and Evolution of Behaviour,* edited by L. R. Aronson, E. Tobach, D. S. Lehrman and J. S. Rosenblatt, 17–52. San Francisco: W. H. Freeman.

Leibold, M. A. 1995. The niche concept revisited: Mechanistic models and community context. *Ecology* 76:1371–82.

Levins, R., and R. Lewontin. 1985. *The Dialectical Biologist.* Cambridge, MA: Harvard University Press.

Lewontin, R. C. 1957. The adaptations of populations to varying environments. *Cold Spring Harbor Symposium on Quantitative Biology* 22:395–408.

———. 1966. Is nature probable or capricious? *Bioscience* 16:25–27.

———. 1972. The apportionment of human diversity. In *Evolutionary Biology,* vol. 6, edited by T. Dobzhansky, M. K. Hecht, and W. C. Steere, 381–98. New York: Appleton Century Crofts.

———. 1974. The analysis of variance and the analysis of causes. *American Journal of Human Genetics* 26:400–411.

———. 1979. Sociobiology as an adaptationist program. *Behavioral Science* 24:5–14.

———. 1982a. *Human Diversity.* New York: Scientific American Press.

———. 1982b. Organism and environment. In *Learning, Development and Culture,* edited by H. C. Plotkin, 151–70. New York: Wiley.

———. 1983. The organism as the subject and object of evolution. *Scientia* 118:65–82.

———. 1985a. Adaptation. In *The Dialectical Biologist,* edited by R. Levins and R. C. Lewontin. Cambridge, MA: Harvard University Press.

———. 1985b. The organism as subject and object of evolution. In *The Dialectical Biologist,* edited by R. Levins and R. C. Lewontin Cambridge, MA: Harvard University Press.

———. 1987. The shape of optimality. In *The Latest on the Best,* edited by J. Dupré, 151–59. Cambridge, MA: MIT Press.

Lewontin, R. C. 1991. *Biology as Ideology: The Doctrine of DNA.* New York: Harper.

Lewontin, R. C., S. Rose, and L. J. Kamin. 1984. *Not in Our Genes.* London: Penguin.

Lickliter, R., and T. Berry. 1990. The phylogeny fallacy. *Developmental Review* 10:348–64.

Lloyd, E. 1993. Unit of selection. In *Keywords in Evolutionary Biology,* edited by E. Keller and E. Lloyd, 334–40. Cambridge, MA: Harvard University Press.

Lloyd, E., and S. J. Gould. 1993. Species selection on variability. *Proceedings of the National Academy of Sciences USA* 90:595–99.

Logue, A. W., W. I. Ophir, and K. E. Strauss. 1986. Acquisition of taste aversion in humans. *Behavior Research and Therapy* 19:319–33.

Lorenz, K. 1965. *Evolution and The Modification of Behaviour.* Chicago: University of Chicago Press.

———. 1966. *On Aggression.* London: Methuen.

Lumsden, C., and E. O. Wilson. 1981. *Genes, Mind and Culture.* Cambridge, MA: Harvard University Press.

———. 1983. *Promethean Fire.* Cambridge, MA: Harvard University Press.

Lycan, W. G. 1990. The continuity of levels of nature. In *Mind and Cognition,* edited by W. G. Lycan, 77–96. Oxford: Blackwell.

MacArthur, R. H. 1958. Population ecology of some warblers of northeastern coniferous forests. *Ecology* 30:599–619.

———. 1972. *Geographical Ecology: Patterns in the Distribution of Species.* New York: Harper and Row.

Maclaurin, J. 1998. Reinventing molecular Weissmanism. *Biology and Philosophy* 13(1): 37–59.

MacLeod, N. 1996. K / T redux. *Paleobiology* 22: 311–17.

Mann, S., N. H. C. Sparks, and R. G. Board. 1990. Magnetotactic bacteria: microbiology, biomineralization, paleomagnetism, and biotechnology. *Advances in Microbial Physiology* 31: 125–81.

Margulis, L., and K. Schwartz. 1988. *Five Kingdoms: An Illustrated Guide to the Phyla of Life on Earth*. New York: W. H. Freeman.

Marr, D. 1980. *Vision*. New York: Freeman.

May, R. M. 1984. An overview: Real and apparent patterns in community structure. In *Ecological Communities: Conceptual Issues and the Evidence,* edited by D. R. Strong, D. Simberloff, L. G. Abele, and A. B. Thistle, 3–16. Princeton, NJ: Princeton University Press.

May, R. M., and J. Seger. 1986. Ideas in ecology. *American Scientist* 74: 256–67.

Mayden, R. L. 1997. A hierarchy of species: The denouement in the saga of the species problem. In *Species: The Units of Biodiversity,* edited by M. F. Claridge, H. A. Dawah, and M. R. Wilson. New York: Chapman and Hall.

Maynard Smith, J. 1964. Group selection and kin selection. *Nature* 201: 1145–47.

———. 1976. Group selection. *Quarterly Review of Biology* 51: 277–83.

———. 1978. *The Evolution of Sex*. Cambridge: Cambridge University Press.

———. 1982. *Evolution and the Theory of Games*. Cambridge: Cambridge University Press.

———. 1984. Optimization theory in evolution. In *Conceptual Issues in Evolutionary Biology,* edited by E. Sober. Cambridge, MA: MIT Press. First published in *Annual Review of Ecology and Systematics* 9: 31–56.

———. 1987. How to model evolution. In *The Latest on the Best,* edited by J. Dupré. Cambridge, MA: MIT Press.

———. 1988. The evolution of sex. In *The Evolution of Sex,* edited by R. Bellig and G. Stevens, 3–19. New York: Harper and Row.

———. 1989a. *Did Darwin Get It Right?* New York: Chapman and Hall.

———. 1989b. *Evolutionary Genetics*. Oxford: Oxford University Press.

———. 1989c. Weismann and modern biology. In *Oxford Surveys in Evolutionary Biology,* vol. 6, edited by P. H. a. L. Partridge, 1–13. Oxford: Oxford University Press.

———. 1993. *The Theory of Evolution*. 2d ed. Cambridge: Cambridge University Press.

———. 1996. Evolution—natural and artificial. In *The Philosophy of Artificial Life,* edited by M. Boden, 173–78. Oxford: Oxford University Press.

Maynard Smith, J., and E. Szathmary. 1995. *The Major Transitions in Evolution*. New York: W. H. Freeman.

Mayr, E. 1961. Cause and effect in biology. *Science* 134: 1501–1606.

———. 1976a. *Evolution and the Diversity of Life*. Cambridge, MA: Harvard University Press.

———. 1976b. Typological versus Population thinking. In *Evolution and The Diversity of Life,* edited by E. Mayr, 26–30. Cambridge, MA: Harvard University Press.

——. 1982a. *The Growth of Biological Thought: Diversity, Evolution and Inheritance*. Cambridge: Cambridge University Press.

——. 1982b *Systematics and the Origin of Species*. Reprint of 1942 edition. New York: Columbia University Press.

——. 1988. *Towards a New Philosophy of Biology*. Cambridge, MA: Harvard University Press.

——. 1991. *One Long Argument: Charles Darwin and the Genesis of Modern Evolutionary Thought*. London: Penguin.

——. 1996. The autonomy of biology: The position of biology among the sciences. *Quarterly Review of Biology* 71 : 97–106.

——. 1997. The objects of selection. *Proceedings of the National Academy of Sciences USA* 94 : 2091–94.

McEvey, S., ed. 1993. *Evolution and the Recognition Concept of Species: The Collected Writing of Hugh E. H. Paterson*. Baltimore: Johns Hopkins University Press.

McKitrick, M. 1993. Phylogenetic constraint in evolution: Has it any explanatory power? *Annual Review of Ecology and Systematics* 24 : 307–30.

McMahon, T. A., and J. T. Bonner. 1983. *On Size and Life*. New York: Scientific American Books.

McMenamin, M., and D. McMenamin. 1990. *The Emergence of Animals: The Cambrian Breakthrough*. New York: Columbia University Press.

——. 1994. *Hypersea: Life on Land*. New York: Columbia University Press.

McNaughton, N. 1989. *Biology and Emotion*. Cambridge: Cambridge University Press.

McShea, D. W. 1991. Complexity and evolution: What everybody knows. *Biology and Philosophy* 6 : 303–24.

——. 1992. A metric for the study of evolutionary trends in the complexity of serial structures. *Biological Journal of the Linnean Society of London* 43 : 39–55.

——. 1993. Arguments, tests, and the Burgess Shale: A commentary on the debate. *Paleobiology* 19 : 399–402.

——. 1994. Mechanisms of large scale evolutionary trends. *Evolution* 48 : 1747–63.

——. 1996a. Complexity and homoplasy. In *Homoplasy: The Recurrence of Similarity in Evolution*, edited by M. J. Sanderson and L. Hufford, 207–25. San Diego: Academic Press.

——. 1996b. Metazoan complexity and evolution: Is there a trend? *Evolution* 50 : 477–92.

Mealey, L. 1995. The sociobiology of sociopathy: An integrated evolutionary model. *Behavioral and Brain Sciences* 18 : 523–99.

Meltzoff, A. N., and M. K. Moore. 1977. Imitation of facial and manual gestures by neonates. *Science* 198 : 75–78.

Meyer, A. 1993. Phylogenetic relationships and evolutionary processes in East African cichlid fishes. *Trends in Ecology and Evolution* 8(8) : 279–84.

Michod, R. E. 1995. *Eros and Evolution*. Reading, MA: Addison-Wesley.

Mikkleson, G. M. In press. Methods and metaphors in community ecology: The problem of defining stability. *Perspectives on Science*.

Millikan, R. 1989a. Biosemantics. *Journal of Philosophy* 86:281–97.

———. 1989b. In defense of proper functions. *Philosophy of Science* 56:288–302.

Mills, S., and J. Beatty. 1994. The propensity interpretation of fitness. In *Conceptual Issues in Evolutionary Biology,* 2d ed, edited by E. Sober. Cambridge, MA: MIT Press. First published in *Philosophy of Science* (1979) 46:263–88.

Minelli, A. 1993. *Biological Systematics: The State of the Art.* London: Chapman and Hall.

Mishler, B. D., and R. N. Brandon. 1987. Individuality, pluralism and the phylogenetic species concept. *Biology and Philosophy* 2:397–414.

Mitchell, S. 1992. On pluralism and competition in evolutionary explanation. *American Zoologist* 32:135–44.

Mitchell, S., and R. Page. 1992. Idiosyncratic paradigms and the revival of the super-organism. Bielefeld Research Group on Biological Foundations of Human Culture, University of Bielefeld.

Moeller, A. 1993. A Fungus infecting domestic flies manipulates sexual behavior of its host. *Behavioral Ecology and Sociobiology* 33(6):403–7.

Moksnes, A., E. Roskaft, A. T. Braa, L. Korsnes, H. M. Lampe, and H. C. Pedersen. 1990. Behavioral responses of potential hosts towards artificial cuckoo eggs and dummies. *Behaviour* 116:64–89.

Montagu, A., ed. 1980. *Sociobiology Examined.* Oxford: Oxford University Press.

Moore, J. A. 1993. *Science as a Way of Knowing: The Foundations of Modern Biology.* Cambridge, MA: Harvard University Press.

Morgan, E. 1982. *The Aquatic Ape.* London: Souvenir Press.

Morgan, N., and P. Baumann. 1994. Phylogenetics of cytoplasmically inherited micro-organisms of arthropods. *Trends in Ecology and Evolution* 9:15–20.

Moritz, R. F., and E. E. Southwick. 1992. *Bees as Superorganisms: An Evolutionary Reality.* Berlin: Springer-Verlag.

Moss, L. 1992. A kernel of truth? On the reality of the genetic program. In *Proceedings of the Philosophy of Science Association,* vol. 1, edited by D. Hull, M. Forbes, and K. Okruhlik, 335–48. East Lansing, MI: Philosophy of Science Association.

Mueller, G. B., and G. P. Wagner. 1996. Homology, *Hox* genes and developmental integration. *American Zoologist* 36:4–13.

Nagel, E. 1961. *The Structure of Science: Problems in the Logic of Scientific Discovery.* London: Routledge and Kegan Paul.

Neander, K. 1991. Functions as selected effects: The conceptual analyst's defence. *Philosophy of Science* 58:168–84.

———. 1995. Misrepresenting and malfunctioning. *Philosophical Studies* 79(2):109–41.

Nelson, G. J., and N. I. Platnick. 1981. *Systematics and Biogeography: Cladistics and Vicariance.* New York: Columbia University Press.

Neumann-Held, E. M. 1998. The gene is dead—long live the gene: Conceptualising the gene the constructionist way. In *Sociobiology and Bioeconomics: The Theory of Evolution in Biological and Economic Theory,* edited by P. Koslowski, 105–137. Berlin: Springer-Verlag.

Nijhout, H. F. 1990. Metaphors and the role of genes in development. *Bioessays* 12(9):441–46.

Nijhout, H. F., and S. M. Paulsen. 1997. Developmental models and polygenetic characters. *American Naturalist* 149(2):394–405.

Niklas, K. 1997. *The Evolutionary Biology of Plants.* Chicago: University of Chicago Press.

Nilsson, D.-E., and S. Pelger. 1994. A pessimistic estimate of the time required for an eye to evolve. *Proceedings of the Royal Society of London, B* 256:53–58.

Nitecki, M. H., ed. 1988. *Evolutionary Progress.* Chicago: University of Chicago Press.

Nunney, L. 1989. The maintenance of sex by group selection. *Evolution* 43:245–57.

Odling-Smee, F. J., K. N. Laland, and M. Feldman. 1996. Niche construction. *American Naturalist* 147:641–48.

O'Hara, R. J. 1993. Systematic generalization, historical fate and the species problem. *Systematic Biology* 42:231–46.

———. 1994. Evolutionary history and the species problem. *American Zoologist* 34:12–22.

Ohman, A., M. Fredrikson, K. Hugdahl, and P. Rimmo. 1976. Premise of equipotentiality in human classical conditioning. *Journal of Experimental Psychology* 105:313–37.

Ohno, S. 1996. The notion of the Cambrian pananimalia genome. *Proceedings of the National Academy of Sciences USA* 93:8475–78.

Olby, R. 1985. *Origins of Mendelism.* 2d ed. Chicago: University of Chicago Press.

O'Neill, R. V., D. L. DeAngelis, J. B. Waide, and T. F. H. Allen. 1986. *A Hierarchical Concept of Ecosystems.* Princeton, NJ: Princeton University Press.

Oosterzee, P. v. 1997. *Where Worlds Collide: The Wallace Line.* Melbourne: Reed Books.

Orzack, S. H., and E. Sober. 1994. Optimality models and the test of adaptationism. *American Naturalist* 143:361–80.

———. 1996. How to formulate and test adaptationism. *American Naturalist* 148:202–10.

———, eds. In press. *Optimality and Adaptationism.* Cambridge: Cambridge University Press.

Oster, G. F., J. D. Murray, and P. Maini. 1985. A model for chondrogenic condensations in the developing limb. *Journal of Embryology and Experimental Morphology* 89:93–112.

Otte, D., and J. A. Endler, eds. 1989. *Speciation and Its Consequences.* Sunderland, MA: Sinauer Associates.

Owen, D. 1980. *Camouflage and Mimicry.* Chicago: University of Chicago Press.

Oyama, S. 1985. *The Ontogeny of Information.* Cambridge: Cambridge University Press.

———. 1995. The accidental chordate: Contingency in developmental systems. *South Atlantic Quarterly* 94:509–26.

———. In press. *Evolution's Eye: Biology, Culture, and Developmental Systems.* Durham, NC: Duke University Press.

Padisak, J. 1994. Identification of relevant time-scales in non-equilibrium community dynamics: Conclusions from phytoplankton surveys. *New Zealand Journal of Ecology* 18:169–76.

Page, R. D. M. 1989. New Zealand and the new biogeography. *New Zealand Journal of Zoology* 16:473–83.

Panchen, A. L. 1992. *Classification, Evolution and the Nature of Biology.* Cambridge: Cambridge University Press.

Paradis, J., and G. C. Williams. 1989. *Evolution and Ethics: T. H. Huxley's "Evolution and Ethics" with New Essays on Its Victorian and Sociobiological Context.* Princeton, NJ: Princeton University Press.

Paterson, A. M., G. P. Wallis, and R. D. Gray. 1995. Penguins, petrels, and parsimony: Does cladistic analysis of behaviour reflect seabird phylogeny? *Evolution* 49:190–205.

Penny, D., and Hasegawa. 1997. The platypus put in its place. *Nature* 387:549–50.

Pierce, G. J., and J. G. Ollason. 1987. Eight reasons why optimal foraging theory is a complete waste of time. *Oikos* 49:111–25.

Pimm, S. 1991. *The Balance of Nature.* Chicago: University of Chicago Press.

Pinker, S. 1994. *The Language Instinct: How the Mind Creates Language.* New York: William Morrow and Co.

Pittendrigh, C. S. 1958. Adaptation, natural selection and behavior. In *Behavior and Evolution,* edited by A. Roe and G. C. Simpson. New York: Academic Press.

Platnick, N. 1992. Patterns of diversity. In *Systematics, Ecology and the Biodiversity Crisis,* edited by N. Eldredge. Oxford: Oxford University Press.

Plutchik, R. 1962. *The Emotions: Facts, Theories and a New Model.* New York: Random House.

———. 1970. Emotions, evolution and adaptive processes. In *Feelings and Emotions,* edited by M. B. Arnold, 3–24. New York: Academic Press.

———. 1980a. *Emotion: A Psychoevolutionary Process.* New York: Harper and Row.

———. 1980b. A general psychoevolutionary theory of emotion. In *Emotion: Theory, Research and Experience,* vol. 1, *Theories of Emotion,* edited by R. Plutchik and H. Kellerman, 3–33. New York: Academic Press.

———. 1984. Emotions: A general psychoevolutionary theory. In *Approaches to Emotion,* edited by K. Scherer and P. Ekman, 3–34. Hillsdale, NJ: Lawrence Erlbaum Associates.

Portin, P. 1993. The concept of the gene: Short history and present status. *Quarterly Review of Biology* 68(2):173–223.

Price, P. W. 1984. Communities of specialists: Vacant niches in ecological and evolutionary time. In *Ecological Communities: Conceptual Issues and the Evidence,* edited by D. R. Strong, D. Simberloff, L. Abele, and A. Thistle, 510–23. Princeton, NJ: Princeton University Press.

Profet, M. 1992. Pregnancy sickness as adaptation: A deterrent to maternal ingestion of teratogens. In *The Adapted Mind: Evolutionary Psychology and the Generation of Culture,* edited by J. H. Barkow, L. Cosmides, and J. Tooby, 327–65. New York: Oxford University Press.

Putnam, H. 1975. *Mind, Language and Reality: Philosophical Papers.* Vol. 2. Cambridge: Cambridge University Press.

———. 1978. *Meaning and the Moral Sciences.* London: Routledge and Kegan Paul.

Quammen, D. 1996. *The Song of the Dodo: Island Biogeography in an Age of Extinctions.* New York: Scribner.

Quiring, R., U. Walldorf, U. Gehring, and W. J. Kloter. 1994. Homology of the *Eyeless* gene of *Drosophilia* to the *Small Eye* gene in mice and *Aniridia* in humans. *Science* 265: 785–89.

Raff, R. 1996. *The Shape of Life: Genes, Development and the Evolution of Animal Form.* Chicago: University of Chicago Press.

Raup, D. H. 1991. *Extinction: Bad Genes or Bad Luck?* Oxford: Oxford University Press.

Ray, T. S. 1996. An approach to the synthesis of life. In *The Philosophy of Artificial Life,* edited by M. Boden, 111–45. Oxford: Oxford University Press.

Reader's Digest Complete Book of Australian Birds. 1977. Sydney: Reader's Digest Services Pty. Ltd.

Real, L. A., and J. H. Brown, eds. 1991. *Foundations of Ecology: Classic Papers with Commentaries.* Chicago: University of Chicago Press.

Reeve, H. K., and P. W. Sherman. 1993. Adaptation and the goals of evolutionary research. *Quarterly Review of Biology* 68:1–32.

Reice, S. R. 1994. Nonequilibrium determinants of biological community structure. *American Scientist* 82:424–35.

Richards, R. 1987. *Darwin and the Emergence of Evolutionary Theories of Mind and Behaviour.* Chicago: University of Chicago Press.

Ridley, Mark. 1985. *The Problems of Evolution.* Oxford: Oxford University Press.

———. 1986. *Evolution and Classification: The Reformulation of Cladism.* London: Longman.

———. 1989. The cladistic solution to the species problem. *Biology and Philosophy* 4: 1–16.

———. 1990. Dreadful beasts: A review of S. J. Gould's *Wonderful Life. London Review of Books* (28 June):11–12.

———. 1993a. Analysis of the Burgess Shale. *Paleobiology* 19:519–21.

———. 1993b. *Evolution.* Cambridge, MA: Blackwell Scientific.

Ridley, Mark, and A. Grafen. 1981. Are green beard genes outlaws? *Animal Behaviour* 29(3).

Ridley, Matt. 1993. *The Red Queen: Sex and the Evolution of Human Nature.* London: MacMillan.

Rose, M. R., and G. V. Lauder, eds. 1996. *Adaptation.* San Diego: Academic Press.

Rosenberg, A. 1985. *The Structure of Biological Science.* Cambridge: Cambridge University Press.

———. 1994. *Instrumental Biology or the Disunity of Science.* Chicago: University of Chicago Press.

———. 1997. Reductionism redux: Computing the embryo. *Biology and Philosophy* 12(4):445–70.

Rosenzweig, M. L. 1995. *Species Diversity in Space and Time.* Cambridge: Cambridge University Press.

Rossiter, A. 1995. The cichlid fish assemblages of Lake Tanganyika: Ecology, behaviour and evolution of its species flocks. *Advances in Ecological Research* 26:187–252.

Rozin, P. 1976. The evolution of intelligence and access to the cognitive unconscious. In

Progress in Psychobiology and Physiological Psychology, edited by J. M. Sprague and A. N. Epstein, vol. 6, 245–81. New York: Academic Press.

Rudwick, M. J. S. 1985. *The Great Devonian Controversy: The Shaping of Scientific Knowledge among Gentlemanly Specialists.* Chicago: University of Chicago Press.

Ruse, M. 1986. *Taking Darwin Seriously.* Oxford: Blackwell.

———. 1996. *Monad to Man: The Concept of Progress in Evolutionary Biology.* Cambridge, MA: Harvard University Press.

Ruse, M., and E. O. Wilson. 1986. Moral philosophy as applied science. *Philosophy* 61: 173–92.

Sarkar, S. 1996. Biological information: A sceptical look at some central dogmas of molecular biology. In *The Philosophy and History of Molecular Biology: New Perspectives,* edited by S. Sarkar, 187–232. Dordrecht: Kluwer.

———. 1997. Decoding "coding": Information and DNA. *European Journal for Semiotic Studies* 9(2): 227–98.

———. 1998. *Genetics and Reductionism.* Cambridge: Cambridge University Press.

Schaffner, K. 1967. Approaches to reduction. *Philosophy of Science* 34: 137–47.

———. 1969. The Watson-Crick model and reductionism. *British Journal for the Philosophy of Science* 20: 325–48.

———. 1993. *Discovery and Explanation in Biology and Medicine.* Chicago and London: University of Chicago Press.

———. 1996. Theory structure and knowledge representation in molecular biology. In *The Philosophy and History of Molecular Biology: New Perspectives,* edited by S. Sarkar, 27–43. Dordrecht: Kluwer.

———. In press. Genes, behavior and developmental emergentism: One process, indivisible? *Philosophy of Science.*

Schlosser, G. In press. Self re-production and functionality: A systems-theoretical approach to teleological explanation. *Synthese.*

Schneider, S. H., and P. J. Boston, eds. 1991. *Scientists on Gaia.* Boston: MIT Press.

Schoener, T. W. 1989. The ecological niche. In *Ecological Concepts: 29th Symposium of the British Ecological Society,* edited by J. M. Cherrett. Oxford: Blackwell Scientific Publications.

Schull, J. 1990. Are species intelligent? *Behavioral and Brain Sciences* 13: 63–108.

Seeley, T. D. 1989. The honey bee colony as a superorganism. *American Scientist* 77: 546–53.

———. 1996. *The Wisdom of the Hive.* Cambridge, MA: Harvard University Press.

Seger, J., and J. W. Stubblefield. 1996. Optimization and adaptation. In *Adaptation,* edited by M. R. Rose and G. V. Lauder, 93–124. San Diego: Academic Press.

Seligman, M. E., and J. L. Hager, eds. 1972. *Biological Boundaries of Learning.* New York: Appleton, Century, Crofts.

Sepkoski, J. J. 1984. A kinetic model of Phanerozoic taxonomic diversity: III. Post-Paleozoic families and mass extinctions. *Paleobiology* 10: 246–67.

———. 1989. Periodicity in extinction and the problem of catastrophism in the history of life. *Journal of the Geological Society of London* 146: 7–19.

Shannon, C. E., and W. Weaver. 1949. *The Mathematical Theory of Communication*. Urbana: University of Illinois Press.

Shapiro, J. A., and M. Dworkin, eds. 1997. *Bacteria as Multicellular Organisms*. Oxford: Oxford University Press.

Shepard, R. N. 1992. The perception of colours: An adaptation to regularities of the terrestrial world? In *The Adapted Mind: Evolutionary Psychology and the Generation of Culture*, edited by J. H. Barkow, L. Cosmides, and J. Tooby, 495–532. New York: Oxford University Press.

Sherman, P. W. 1988. The levels of analysis. *Animal Behavior* 36:616–19.

———. 1989. The clitoris debate and the levels of analysis. *Animal Behavior* 37:697–98.

Shrader-Frechette, K. S., and E. D. McCoy. 1993. *Method in Ecology*. Cambridge: Cambridge University Press.

Shykoff, J., and A. Widmer. 1998. Eggs first. *Trends in Ecology and Evolution* 13(4):158.

Sigmund, K. 1993. *Games of Life: Explorations in Ecology, Evolution and Behaviour*. London: Penguin.

Signor, P. W. 1994. Biodiversity in geological time. *American Zoologist* 34:23–32.

Signor, P. W., and J. H. Lipps. 1992. Origin and early radiation of the metazoa. In *Origin and Early Radiation of the Metazoa*, edited by P. W. Signor and J. H. Lipps, 3–23. New York: Plenum Press.

Simberloff, D. 1980. A succession of paradigms in ecology: Essentialism to materialism to probabilism. *Synthese* 43:3–39.

Simon, H. A. 1969. *The Sciences of the Artificial*. Cambridge, MA: MIT Press.

Sinervo, B., and C. M. Lively. 1996. The rock-paper-scissors game and the evolution of alternate male strategies. *Nature* 380:240–43.

Skuse, D. H., R. S. James, D. V. M. Bishop, B. Coppins, P. Dalton, G. Aamodt-Leeper, M. Bacarese-Hamilton, C. Creswell, R. McGurk, and P. A. Jacobs. 1997. Evidence from Turner's syndrome of an imprinted X-linked locus affecting cognitive function. *Nature* 387 (12 June):705–8.

Skyrms, B. 1996. *The Evolution of the Social Contract*. Cambridge: Cambridge University Press.

Small, M. 1993. *Female Choices: Sexual Behavior of Female Primates*. Ithaca, NY: Cornell University Press.

Smart, J. C. C. 1963. *Philosophy and Scientific Realism*. London: Routledge & Kegan Paul.

Smith, E. A. 1987. Optimization theory in anthropology: Applications and critiques. In *The Latest on the Best: Essays on Evolution and Optimality*, edited by J. Dupré. Cambridge, MA: MIT Press.

Smith, E. A., and S. A. Smith. 1994. Inuit sex-ratio variation: Population control, ethnographic error, or parental manipulation. *Current Anthropology* 35:595–624.

Smith, K. C. 1992. Neo-rationalism versus neo-Darwinism: Integrating development and evolution. *Biology and Philosophy* 7:431–52.

Sober, E. 1980. Evolution, population thinking and essentialism. *Philosophy of Science* 47:350–83.

———. 1983. Equilibrium explanation. *Philosophical Studies* 43:201–10.

————, ed. 1984a. *Conceptual Issues in Evolutionary Biology.* Cambridge, MA: MIT Press.

————. 1984b. *The Nature of Selection: Evolutionary Theory in Philosophical Focus.* Cambridge, MA: MIT Press.

————. 1988a. Apportioning causal responsibility. *Journal of Philosophy* 85 : 303–18.

————. 1988b. The conceptual relationship of cladistic phylogenetics and vicariance biogeography. *Systematic Zoology* 37 (3) : 245–53.

————. 1988c. *Reconstructing the Past: Parsimony, Evolution and Inference.* Cambridge, MA: MIT Press.

————. 1992. Models of cultural evolution. In *Trees of Life: Essays in the Philosophy of Biology,* edited by P. E. Griffiths, 1–17. Dordrecht: Kluwer.

————. 1993. *Philosophy of Biology.* Boulder: Westview Press.

————, ed. 1994. *Conceptual Issues in Evolutionary Biology.* 2d ed. Cambridge, MA: MIT Press.

————. 1996. Learning from functionalism: Prospects for strong artificial life. In *The Philosophy of Artificial Life,* edited by M. Boden. Oxford: Oxford University Press.

Sober, E., and R. Lewontin. 1984. Artifact, cause and genic selection. In *Conceptual Issues in Evolutionary Biology,* edited by E. Sober. Cambridge, MA: MIT Press. First published in *Philosophy of Science* (1982) 49 : 157–80.

Sober, E., and D. S. Wilson. 1994. A critical review of philosophical work on the units of selection problem. *Philosophy of Science* 61 : 534–55.

————. 1998. *Unto Others: The Evolution of Altruism.* Cambridge, MA: Harvard University Press.

Somit, A., and S. Peterson, eds. 1989. *The Dynamics of Evolution.* Ithaca, NY: Cornell University Press.

Spencer, F. 1992. Some notes on attempts to apply photography to anthropology during the second half of the nineteenth century. In *Anthropology and Photography,* edited by E. Edwards, 99–107. New Haven and London: Yale University Press, in association with The Royal Anthropological Institute, London.

Sperber, D. 1996. *Explaining Culture: A Naturalistic Approach.* Oxford: Blackwell.

Sperber, D., D. Premack, and A. J. Premack, eds. 1995. *Causal Cognition.* Oxford: Oxford University Press.

Stent, G. 1981. Strength and weakness of the genetic approach to the development of the nervous system. In *Studies in Developmental Neurobiology,* edited by W. M. Cowan, 288–321. Oxford: Oxford University Press.

————. 1994. Promiscuous realism. *Biology and Philosophy* 9 : 497–506.

Sterelny, K. 1990. *The Representational Theory of Mind: An Introduction.* Oxford: Blackwell.

————. 1992a. Evolutionary explanations of human behaviour. *Australasian Journal of Philosophy* 70 : 156–73.

————. 1992b. Punctuated equilibrium and macroevolution. In *Trees of Life: Essays in the Philosophy of Biology,* edited by P. E. Griffiths, 41–64. Dordrecht: Kluwer.

————. 1994. Science and selection. *Biology and Philosophy* 9 : 45–62.

————. 1995. The adapted mind. *Biology and Philosophy* 10 : 365–80.

————. 1996a. Explanatory pluralism in evolutionary biology. *Biology and Philosophy* 11 : 193–214.

————. 1996b. The return of the group. *Philosophy of Science* 63:562–84.

————. 1997a. Universal biology. *British Journal for the Philosophy of Science* 48(4):587–601.

————. 1997b. Where does thinking come from? A commentary on Peter Godfrey Smith's *Complexity and the Function of Mind In Nature. Biology and Philosophy* 12:551–66.

Sterelny, K., and P. Kitcher. 1988. The return of the gene. *Journal of Philosophy* 85:339–60.

Sterelny, K., K. Smith, and M. Dickison. 1996. The extended replicator. *Biology and Philosophy* 11:377–403.

Strahan, R. 1983. *The Australian Museum Complete Book of Australian Mammals: The National Photographic Index of Australian Wildlife.* London, Sydney, Melbourne: Angus and Robertson.

Sultan, S. 1987. Evolutionary implications of phenotypic plasticity in plants. *Evolutionary Biology* 21:127–78.

Symons, D. 1979. *The Evolution of Human Sexuality.* Oxford: Oxford University Press.

————. 1992. On the use and the misuse of Darwinism in the study of human behavior. In *The Adapted Mind: Evolutionary Psychology and the Generation of Culture,* edited by J. H. Barkow, L. Cosmides, and J. Tooby, 137–62. New York: Oxford University Press.

Tanner, N. M. 1981. *On Becoming Human.* Cambridge: Cambridge University Press.

Taylor, P. 1987. Historical versus selectionist explanations in evolutionary biology. *Cladistics* 3:1–13.

Templeton, A. 1989. The meaning of species and speciation: A genetic perspective. In *Speciation and Its Consequences,* edited by D. Otte and J. Endler, 3–27. Sunderland, MA: Sinauer Associates.

Thompson, J. N. 1994. *The Coevolutionary Process.* Chicago: University of Chicago Press.

Thornhill, R., and N. Thornhill. 1987. Human rape: The strengths of the evolutionary perspective. In *Sociobiology and Psychology,* edited by C. Crawford, M. Smith, and D. Krebs, 269–92. Hillsdale, NJ: Lawrence Erlbaum Associates.

————. 1992. The evolutionary psychology of men's coercive sexuality. *Behavioral and Brain Sciences* 15:363–421.

Tinbergen, N. 1952. Derived activities: Their causation, biological significance, origin and emancipation. *Quarterly Review of Biology* 27:1–32.

————. 1963. On the aims and methods of ethology. *Zeitschrift für Tierpsychologie* 20:410–33.

Tooby, J., and L. Cosmides. 1990. The past explains the present: Emotional adaptations and the structure of ancestral environments. *Ethology and Sociobiology* 11:375–424.

————. 1992. The psychological foundations of culture. In *The Adapted Mind: Evolutionary Psychology and the Generation of Culture,* edited by J. H. Barkow, L. Cosmides, and J. Tooby, 19–136. Oxford: Oxford University Press.

Trevarthen, C. 1984. Emotions in infancy: Regulators of contact and relationship with persons. In *Approaches to Emotion,* edited by K. Scherer and P. Ekman, 129–62. Hillsdale, NJ: Lawrence Erlbaum Associates.

Trewick, S. A. 1997. Flightlessness and phylogeny amongst endemic rails (Aves: Rallidae) of the New Zealand region. *Philosophical Transactions of the Royal Society of London, B* 352:429–46.

Trivers, R. L. 1971. The evolution of reciprocal altruism. *Quarterly Review of Biology* 46: 35–57.

———. 1985. *Social Evolution*. Menlo Park, CA: Benjamin/Cummings.

Turke, P. 1990. Which humans behave adaptively, and why does it matter? *Ethology and Sociobiology* 11:305–39.

Underwood, A. J. 1986. What is a community? In *Patterns and Processes in the History of Life,* edited by D. M. Raup and D. Jablonski, 351–67. Berlin: Springer-Verlag.

Valentine, J. W. 1990. The macroevolution of clade shape. In *Causes of Evolution: A Paleontological Perspective,* edited by R. Ross and W. Allmon. Chicago: University of Chicago Press.

Van Valen, L. 1976. Ecological species, multispecies and oaks. *Taxon* 25:233–39.

Vermeij, G. 1987. *Evolution and Escalation: An Ecological History of Life.* Princeton, NJ: Princeton University Press.

Vogel, S. 1988. *Life's Devices: The Physical World of Animals and Plants.* Princeton, NJ: Princeton University Press.

Vrba, E. S. 1984a. Evolutionary pattern and process in the sister-group Alcelaphini-Aepycerotini (Mammalia: Bovidae). In *Living Fossils,* edited by N. Eldredge and S. Stanley, 62–79. New York: Springer Verlag.

———. 1984b. Patterns in the fossil record and evolutionary processes. In *Beyond Neo-Darwinism: An Introduction to the New Evolutionary Paradigm,* edited by M.-W. Ho and P. Saunders, 115–42. London: Academic Press.

———. 1984c. What is species selection? *Systematic Zoology* 33:318–28.

———. 1989. Levels of selection and sorting with special reference to the species problem. *Oxford Surveys in Evolutionary Biology,* vol. 6, 111–68. Oxford: Oxford University Press.

———. 1993. Turnover-pulses, the red queen and related topics. *American Journal of Science* 293-A:418–52.

———. 1995. Species as habitat-specific complex systems. In *Speciation and the Recognition Concept: Theory and Applications,* edited by D. M. Lambert and H. G. Spencer, 3–44. Baltimore: Johns Hopkins University Press.

Waddington, C. H. 1957. *The Strategy of the Genes: A Discussion of Some Aspects of Theoretical Biology.* London: Ruskin House/George Allen and Unwin Ltd.

———. 1959. Canalisation of development and the inheritance of acquired characters. *Nature* 183:1654–55.

Wade, M. J. 1976. Group selection among laboratory populations of *Tribolium. Proceeedings of the National Academy of Sciences USA* 73:4604–7.

———. 1978. A critical review of the models of group selection. *Quarterly Review of Biology* 53:101–14.

Wagner, G. P. 1996. Homologues, natural kinds and the evolution of modularity. *American Zoologist* 36:36–43.

Wagner, G. P., G. Booth, and B. C. Homayoun. 1997. A population genetic theory of canalization. *Evolution* 51(2):329–47.

Wagner, P. 1995a. Diversity patterns amongst early gastropods: Contrasting taxonomic and phylogenetic descriptions. *Paleobiology* 21:410–39.

———. 1995b. Testing evolutionary constraint hypotheses with early Paleozoic gastropods. *Paleobiology* 21:248–72.

Walsh, D. M. 1996. Fitness and function. *British Journal for the Philosophy of Science* 47(4): 553–74.

Walsh, D. M., and A. Ariew. 1996. A taxonomy of functions. *Canadian Journal of Philosophy* 26:493–514.

Ward, P. 1992. *On Methuselah's Trail: Living Fossils and the Great Extinctions.* W. H. Freeman, New York.

———. 1994. *The End of Evolution.* New York: Bantam.

———. 1995. The K/T trial. *Paleobiology* 21:245–47.

Wason, P. 1968. Reasoning about a rule. *Quarterly Journal of Experimental Psychology* 20: 273–81.

Waters, C. K. 1994a. Genes made molecular. *Philosophy of Science* 61:163–85.

———. 1994b. Tempered realism about the forces of selection. *Philosophy of Science* 58: 553–73.

———. 1994c. Why the antireductionist consensus won't survive the case of classical genetics. In *Conceptual Issues in Evolutionary Biology,* 2d ed., edited by E. Sober, 401–18. Cambridge, MA: MIT Press. First published in *Proceedings of the Philosophy of Science Association* (1990), vol. 1, 125–39.

Watson, J. B. 1930. *Behaviourism.* New York: W. W. Norton.

Weber, B., and D. Depew. 1996. Natural selection and self-organization. *Biology and Philosophy* 11:33–65.

Weiner, J. 1994. *The Beak of the Finch: Evolution in Real Time.* London: Vintage.

Weinert, F., ed. 1995. *Laws of Nature: Essays on the Philosophical, Scientific and Historical Dimensions.* Berlin: De Gruyter.

Weinrich, J. D. 1980. Towards a sociobiological theory of emotions. In *Emotion: Theory, Research and Experience,* vol. 1, *Theories of Emotion,* edited by R. Plutchik and H. Kellerman, 113–40. New York: Academic Press.

Werren, J. 1994. Genetic invasion of the insect body snatchers. *Natural History* 103(6): 36–38.

West-Eberhard, M. J. 1992. Adaptation: Current usages. In *Keywords in Evolutionary Biology,* edited by E. F. Keller and E. Lloyd, 13–18. Cambridge, MA: Harvard University Press.

Wheeler, W. M. 1923. *Social Life Amongst The Insects.* New York: Harcourt Brace.

Whiten, A., and R. Byrne, eds. 1997. *Machiavellian Intelligence II: Extensions and Evaluations.* Cambridge: Cambridge University Press.

Whittington, H. S. 1974. *Yohoia* Walcott and *Plenocaris* n. gen, arthropods from the Burgess Shale, Middle Cambrian, British Columbia. *Geological Survey of Canada* 231:1–21.

———. 1985. *The Burgess Shale*. New Haven, CT: Yale University Press.

Wickler, W. 1968. *Mimicry in Plants and Animals*. London: Weidenfeld and Nicolson.

Wiley, E. O. 1978. The evolutionary species concept reconsidered. *Systematic Zoology* 27 : 17–26.

———. 1988. Vicariance biogeography. *Annual Review of Ecology and Systematics* 19 : 513–42.

Wilkinson, G. 1990. Food sharing in vampire bats. *Scientific American* 262(2) : 64–70.

Williams, G. C. 1966. *Adaptation and Natural Selection*. Princeton, NJ: Princeton University Press.

———. 1975. *Sex and Evolution*. Princeton, NJ: Princeton University Press.

———. 1992. *Natural Selection: Domains, Levels and Challenges*. Oxford: Oxford University Press.

Wilson, D. S. 1983. The group selection controversy: History and current status. *Annual Review of Ecology and Systematics* 14 : 159–87.

———. 1989. Levels of selection: An alternative to individualism in biology and the social sciences. *Social Networks* 11 : 257–72.

———. 1992. Group selection. In *Keywords in Evolutionary Biology,* edited by E. Keller and E. Lloyd, 145–68. Cambridge, MA: Harvard University Press.

———. 1997. Incorporating group selection into the adaptationist program: A case study involving human decision making. In *Evolutionary Social Psychology,* edited by J. Simpson and D. Kendrick, 345–86. Mahwah, NJ: Lawrence Erlbaum Associates.

———. 1998a. A critique of R. D. Alexander's views on group selection. *Biology and Philosopy.* In press.

———. 1998b. Hunting, sharing and multilevel selection: The tolerated theft model revisited. *Current Anthropology* 39 : 73–97.

Wilson, D. S., and E. Sober. 1989. Reviving the superorganism. *Journal of Theoretical Biology* 136 : 332–56.

———. 1994. Reintroducing group selection to the human behavioral sciences. *Behavioral and Brain Sciences* 17 : 585–654.

Wilson, E. O. 1975. *Sociobiology: The New Synthesis*. Cambridge, MA: Harvard University Press.

———. 1978. *On Human Nature*. Toronto, Bantam Books.

———. 1992. *The Diversity of Life*. New York: W. W. Norton.

———. 1994. *Naturalist*. Washington, DC: Island Press.

Wilson, J. B. 1990. Mechanisms of species coexistence: Twelve explanations for Hutchinson's "paradox of the plankton": evidence from New Zealand plant communities. *New Zealand Journal of Ecology* 13 : 17–42.

———. 1994. The "intermediate disturbance hypothesis" of species coexistence is based on patch dynamics. *New Zealand Journal of Ecology* 18 : 176–81.

Wilson, R. A., ed. 1999. *Species: New Interdisciplinary Essays*. Cambridge, MA: MIT Press.

Wimsatt, W. C. 1974. Reductive explanation: A functional account. In *Proceedings of the Philosophy of Science Association,* vol. 2, 671–710. East Lansing, MI: Philosophy of Science Association.

————. 1976. Reduction, levels of organisation and the mind/body problem. In *Consciousness and the Brain,* edited by G. Globus, G. Maxwell, and I. Savodnik, 199–267. New York: Plenum Press.

————. 1980a. Reductionistic research strategies and their biases in the units of selection controversy. In *Scientific Discovery: Case Studies,* edited by T. Nickles, 213–59. Dordrecht: Reidel.

————. 1980b. Units of selection and the structure of the multi-level genome. In *Proceedings of the Philosophy of Science Association,* edited by P. Asquith and T. Nickles, vol. 2, 122–83. East Lansing, MI: Philosophy of Science Association.

————. 1994. The ontology of complex systems: Levels of organization, perspectives and causal thickets. In *Biology and Society: Reflections on Methodology,* edited by M. Matthen and R. X. Ware, 207–74. Calgary: University of Calgary Press.

————. In press. *(Piecewise) Approximations to Reality: A Realist Philosophy of Science for Limited Beings.* Chicago: University of Chicago Press.

Wimsatt, W. C., and J. C. Schank. 1988. Two constraints on the evolution of complex adaptations and the means of their avoidance. In *Evolutionary Progress,* edited by M. H. Nitecki, 231–75. Chicago: University of Chicago Press.

Wolpert, L. 1995. Development: Is the egg computable or could we generate an angel or a dinosaur. In *What is Life? The Next Fifty Years,* edited by M. P. Murphy and L. A. J. O'Neill, 57–66. Cambridge: Cambridge University Press.

Worster, D. 1994. *Nature's Economy: A History of Ecological Ideas.* Cambridge: Cambridge University Press.

Wright, L. 1994. Functions. In *Conceptual Issues in Evolutionary Biology,* 2d ed., edited by E. Sober, 27–47. Cambridge, MA: MIT Press. First published in *Philosophical Review* (1973) 82:139–68.

Wright, R. 1994. *The Moral Animal: Why We Are the Way We Are.* New York: Pantheon Books.

Wynne-Edwards, V. C. 1962. *Animal Dispersion in Relation to Social Behaviour.* Edinburgh: Oliver and Boyd.

————. 1986. *Evolution through Group Selection.* Oxford: Blackwell Scientific.

Zahavi, A., and A. Zahavi. 1997. *The Handicap Principle: A Missing Piece of Darwin's Puzzle.* Oxford: Oxford University Press.

Index